Geology and Sedimentology of the Korean Peninsula

Geology and Sedimentology of the Korean Peninsula

Sung Kwun Chough
Earth & Environmental Sciences
Seoul National University
Seoul
South Korea

ELSEVIER

AMSTERDAM • BOSTON • HEIDELBERG • LONDON • NEW YORK • OXFORD
PARIS • SAN DIEGO • SAN FRANCISCO • SINGAPORE • SYDNEY • TOKYO

Elsevier
32 Jamestown Road, London NW1 7BY
225 Wyman Street, Waltham, MA 02451, USA

First edition 2013

British Library Cataloguing-in-Publication Data
A catalogue record for this book is available from the British Library

Library of Congress Cataloging-in-Publication Data
A catalog record for this book is available from the Library of Congress

ISBN: 978-0-12- 405518-6

For information on all Elsevier publications
visit our website at store.elsevier.com

This book has been manufactured using Print On Demand technology. Each copy is produced to order and is limited to black ink. The online version of this book will show color figures where appropriate.

Contents

Preface

Geology is the study of dynamic processes of the Earth. What causes uplift of mountains and subsidence of ocean basins? How are the rocks in the high mountains weathered and transported and eventually deposited in sedimentary basins? How have the surface environments changed and what forces have driven the changes? Geology focuses on the constituent materials and their formative processes as well as the life forms. A geological study starts with a field observation of outcrops or the constituent materials, including sediments, rocks, strata, and fossils. An accurate description of the constituent materials and their spatial and temporal relationships is most important for accurate interpretation of dynamic processes. The study commonly employs laboratory analyses whose results are commonly displayed as conceptual drawings or models, which explain how a natural system is built or behaves.

An initial geological survey of the Korean Peninsula was made in 1884 by a German geologist, C. Gottsche, followed by an extensive Japanese investigation of rocks, minerals, and fossils for the exploitation of economic mineral deposits, which helped establish stratigraphy and the outline of geological structures (Kobayashi, 1927, 1933, 1967; Tateiwa, 1976). A syllabus entitled *Geology of Korea* was first published by the Geological and Mineral Institute of Korea (Reedman and Um, 1975). It contained an outline of geological history based on the description of major rock units in the southern part of the peninsula, an account of geological mapping (1:50,000 scale) about 70% of the country. Another version was edited at the occasion of the Pacific Science Congress in Seoul (Lee, 1987a), primarily based on the notion of geosyncline evolution and crustal deformation.

Since the first publication, tremendous progress has been made in the geological understanding of the Korean Peninsula on land and under the sea with the advances in sophisticated study technique and reinforcement of research personnel. The number of research scientists in geosciences has been more than tripled in academic and research institutions. An updated review of tectonic and sedimentary evolution of the Korean Peninsula was made at the turn of the new millennium in view of the unifying theory of plate tectonics (Chough et al., 2000). The plate tectonics theory explains the whole range of geological processes, such as basin subsidence, sedimentation, mountain building, volcanism, and other processes that construct features on the Earth's surface. Plate tectonic processes are understood as controls of the overall architecture and evolution of crust and sedimentary basins.

However, the emergence of the concept of sequence stratigraphy since the late 1970s has brought a fundamental difference in thinking about the evolution of sedimentary basins. The principal tenet of sequence stratigraphy was that a change in relative sea level results in a change in stratal patterns. Sequence stratigraphic interpretations helped reveal the effect of sea-level controls on nearshore sedimentation and stratal architecture in a chronological framework. Recent development reveals, however, that there is significant variability in the expression of sequence stratigraphic units. This book summarizes the state of the art of the dynamic processes and environmental changes in the Korean Peninsula on land and under the sea during the Phanerozoic Era. The main focus is on the evolution of sedimentary basins, i.e., from sedimentary facies to plate tectonics.

Acknowledgments

I would like to thank the following publication and copyright holders for their cooperation: W.H. Freeman & Company, American Geophysical Union (Tectonics and Geophysical Research Letters), Geological Society of America (Geological Society of America Bulletin and Geology), SEPM (Journal of Sedimentary Research), Springer-Verlag (Geo-Marine Letters and Facies), Wiley-Blackwell (Sedimentology), Elsevier Science (Sedimentary Geology, Earth-Science Reviews, Marine Geology, Marine and Petroleum Geology, Cretaceous Research, Journal of Asian Earth Sciences, Gondwana Research, Tectonophysics, and Palaeogeography, Palaeoclimatology, Palaeoecology), *Geosciences Journal* and National Geographic Society.

I am indebted to the colleagues (C.E. Baag, D.K. Choi, M. Cho, S.M. Lee, and H.M. Jeong) of the Crust and Mantle Research Division, School of Earth and Environmental Sciences, Seoul National University, for their continuous dialogue and support. It has been a great pleasure to work with my former students and many colleagues for the past 30 years, all of whom have provided me with some scientific inspiration through discussion, especially S.S. Chun (Cheonnam National University), S.H. Yoon (Jeju National University), Y.K. Sohn (Gyeongsang National University), C.W. Rhee (Chungbuk National University), J.-H. Ree (Korea University), D.J. Lee and H.R. Jo (Andong National University), M.Y. Choe, J. Woo, and T.-Y. Park (Korea Polar Research Institute), H. Kim, I.G. Hwang, J.H. Jin, S.R. Lee, J.J. Bahk, Y.K. Kwon, Y.J. Shinn, and H.S. Lee (Korea Institute of Geoscience and Mineral Resources), H.J. Lee, K.S. Jeong, and S.H. Lee (Korea Ocean Research and Development Institute), K.S. Bahk (Centennial Technology Co.), K.C. Na (Chungbuk National University), S.B. Kim (STX Energy Co.), K.M. Yu and S.-T. Kwon (Yonsei University), S.J. Choh (Korea University), S.J. Park (Korea National Oil Corporation), W.H. Ryang (Chunbuk National University), K. Yi and C.S. Cheong (Korea Basic Science Institute Ochang Center), S.C. Park (Chungnam National University), D.C. Kim and G.H. Lee (Pukyong National University), K.S. Woo (Kangwon National University), S.-B. Lee (Korea National Science Museum), J.W. Kim (Daewoo International), H.J. Kang and K.E. Lee (Korea Maritime University), H.K. Ha (Korea Polar Research Institute), and Y. Huh (Seoul National University). Thanks are due to M. Cho, J.-H. Ree, H.J. Lee, S.H. Yoon, Y.K. Sohn, and H.R. Jo for helpful review of the manuscript.

I am grateful to A.D. Miall (University of Toronto), O.H. Pilkey, Jr. (Duke University), D.A.V. Stow (Heriot-Watt Univeristy), J.T. Wells (Virginia Institute of Marine Science), W.M. Moon (University of Manitoba), A. Taira (Japan Agency for Marine-Earth Science and Technology), Z. Han (Shandong University of Science and Technology), W.R. Fitches (Llandudno), and the late K. Tamaki (University of Tokyo) for dialogue over the years. I gratefully acknowledge the continuous discussions and editorial help by the graduate students of the Sedimentology Laboratory (sedlab@snu.ac.kr), J.-H. Lee, J. Chen, G.B. Kim, and J. Howell. Thanks are due to H.S. Kim for preparing figure artwork. I gratefully acknowledge the financial support of the Korea National Research Foundation during the past 30 years (Current Project, 2012054600). I thank J. Fedor, T. Miller, S. Viswanathan, and E. Hill-Parks (Elsevier) for successful publication of this work. Thanks to my family for continuous support of my work.

1 Introduction

1.1 Geological Questions

The Korean Peninsula consists of scenic mountains and valleys that run one after another (Figures 1.1–1.3). These features have descended from crustal deformation and the associated plutonic and volcanic activities. Rivers run through the valleys, carrying sediments mostly toward the west and the south, forming floodplains. Where the rivers meet the coastal plain, estuaries form. Along the western and southern coasts of the peninsula, tidal flats are extensive, rivaling those of the North Sea and the Bay of Fundy. The Yellow Sea is a shallow (55 m on average) epicontinental sea, surrounded by the landmass of China and Korea. Off the western and southern coasts of the peninsula, there are more than 3000 islands. However, the eastern continental shelf is narrow and transitional to a bowl-shaped deep basin, the Ulleung Basin, and the Korea Plateau, which are dotted by numerous volcanic seamounts and islands, including the Ulleung and Dok islands (Figure 1.4). The volcanic island of Jeju comprises a central crater, about 360 scoria cones, and tuff rings and cones.

The Korean Peninsula comprises denudation remnant of deformed basement rocks and sedimentary successions, concealing a long history of crustal deformation (Figure 1.5). Sedimentary rocks especially contain records of environmental change through Earth's history, including climate change, the rise and fall of sea level, and the movement of tectonic plates over millions of years. The peninsula presents a number of fundamental questions as to its origin and dynamic processes. What constitutes the scenic mountains and ridges? When and how did these mountain chains form? Was it due to the collision of the tectonic plates? How often was the continent submerged under the sea throughout Earth's history? How are the rocks in the peninsula linked to those of China and Japan? What are the origins of the deep basins and submarine plateaus in the East Sea? How did the continent under the Yellow Sea form and evolve? How would the coastal areas be affected by sea-level change over timescales of a few decades to centuries? These are commonly asked questions regarding dynamic processes and environmental changes of the Korean Peninsula on land and under the sea.

Geology and Sedimentology of the Korean Peninsula. DOI: http://dx.doi.org/10.1016/B978-0-12-405518-6.00001-2

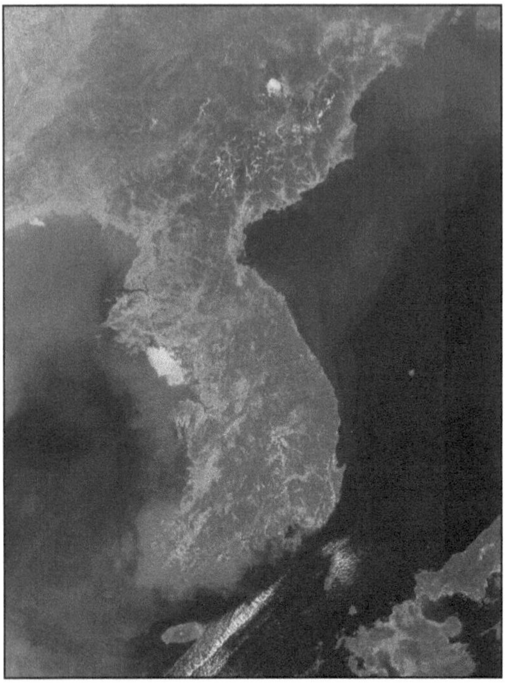

Figure 1.1 Satellite image of the Korean Peninsula. The peninsula consists largely of mountains (about 65%). There are about 3000 islands off the south and west coasts. (For interpretation of the references to color in this figure legend, the reader is referred to the web version of this book.)
Source: GOCI/COMS RGB Color Composite Image processed by Korea Ocean Satellite Center/Korea Ocean Research and Development Institute.

1.2 Sedimentary Facies Analysis

Sedimentary rocks are important constituents (more than 50%) of the crust of the Korean Peninsula. Sedimentary rocks (and sediments) are characterized by grain size and sedimentary structures, i.e., sedimentary facies, defined as a depositional unit characterized by a particular combination of grain size (lithology) and sedimentary (physical and biological) structures (Figure 1.6). Each unit of sedimentary facies represents distinct depositional processes that act on the sediments in particular environments (Figure 1.7). Genetically related facies can form a group, defined as a facies association or a sequence that has some environmental significance. An analysis of sedimentary facies thus leads the way to diagnosing depositional processes and environments (Dalrymple, 2010).

Sedimentary facies analysis is essential to the study of stratigraphy, which is primarily concerned with the recognition of distinct bodies of rocks in a chronological framework. A lithostratigraphic unit is defined by its lithologic characteristics in stratigraphic position relative to other bodies of sedimentary rock. A sedimentary rock unit can also be characterized by its fossil contents, biostratigraphy. Various other methods have also been used to define stratigraphic units, including magnetostratigraphy and chemostratigraphy.

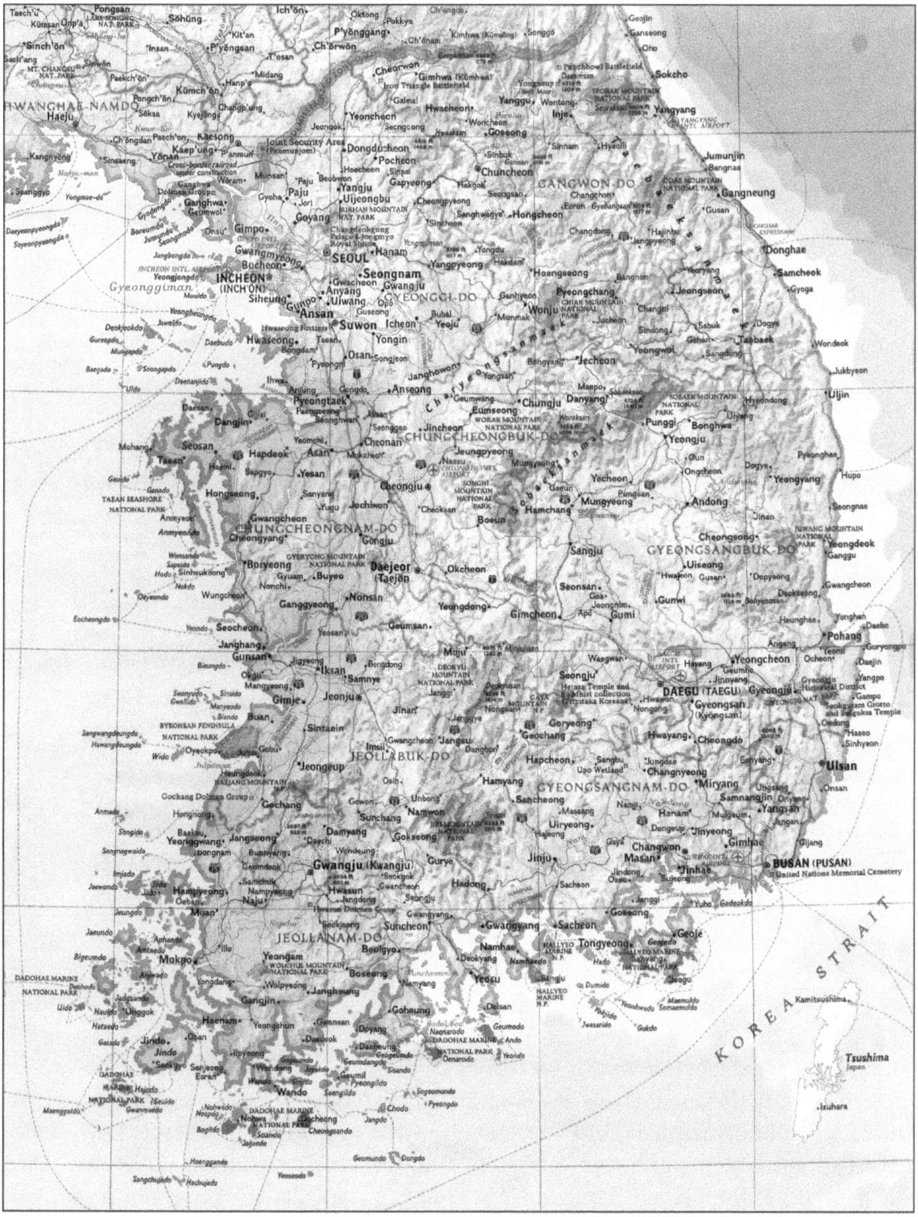

Figure 1.2 Geographic map of South Korea.
Source: National Geographic (2003) by permission of the National Geographic Society.

Figure 1.3 (A) View of a mountain chain in the southeast of Taebaek city, Gangwon Province. The ridges comprise sedimentary rocks (limestone and sandstone) that formed in shallow water environments during the Paleozoic (about 520–250 Ma) and deformed in the Mesozoic (ca. 250–150 Ma). The mountains represent remnant of continuous denudation for the last 150 million years. (B) Snow covered Wolak mountain in the background. It consists of granite intruded into metasedimentary rocks. (C) The meandering East River showing mid-channel and sidebars of gravelly sands.

Sedimentary facies analysis leads to an understanding of controls on basin evolution, including tectonics, climate changes, and sea-level changes (Miall, 2000, 2010). Plate movement causes subsidence of the crust, which provides the space for the accumulation of sediments. Sedimentary basins are commonly formed according to the plate tectonic regime: divergent, convergent, transcurrent, intraplate, and hybrid settings. Subsidence is induced by a thinning of the crust due to stretching, erosion, and magmatic withdrawal as well as tectonic and sedimentary loading and others (Busby and Azor, 2011). Uplift/subsidence also affects climate changes and the amounts of sediment supply. Along with these factors, sea-level changes control depositional processes and environments, especially in the shoreline and shallow waters.

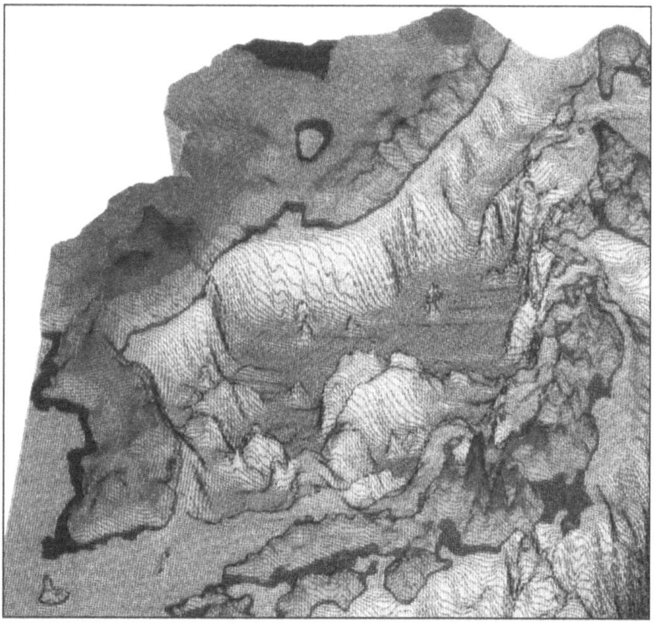

Figure 1.4 Topographic relief of the Korean Peninsula and the northeast Asian margin.
Source: Courtesy of K. Tamaki.

Sedimentary facies analysis eventually leads to sequence stratigraphy, which is the study of rock relationships within a chronostratigraphic framework of repetitive, genetically related strata bounded by surfaces of erosion or nondeposition, or their correlative conformities (Posamentier et al., 1988; Van Wagoner et al., 1990). Sequence stratigraphic analysis focuses on the geometric characters of stratal patterns and identifications of key surfaces to determine the chronological order of basin fills (Catuneanu, 2006; Emery and Myers, 1996). The underlying tenet of sequence stratigraphy was that a change in sea level would result in a change in stratal patterns, which could be compared among basins worldwide. This deductive generic view of sequence stratigraphy has hampered, however, to further process-based sedimentological approach to an integrated basin analysis. The high variability of bounding surfaces and stratigraphic units requires inductive analysis for individual rock records (Catuneanu et al., 2009, 2010). Modern sequence stratigraphy focuses on the changes in stratal stacking patterns in response to varying accommodation and sediment supply through time (Catuneanu et al., 2010; Miall, 2010). Attention is now given to the specifics of how stratal architecture can throw light on depositional processes and allogenic controls.

Over the past 30 years, extensive sedimentary facies analyses have been made in major sedimentary basins of the Korean Peninsula, including the siliciclastics, carbonates, and mixed siliciclastic–carbonate successions as well as the nonmarine

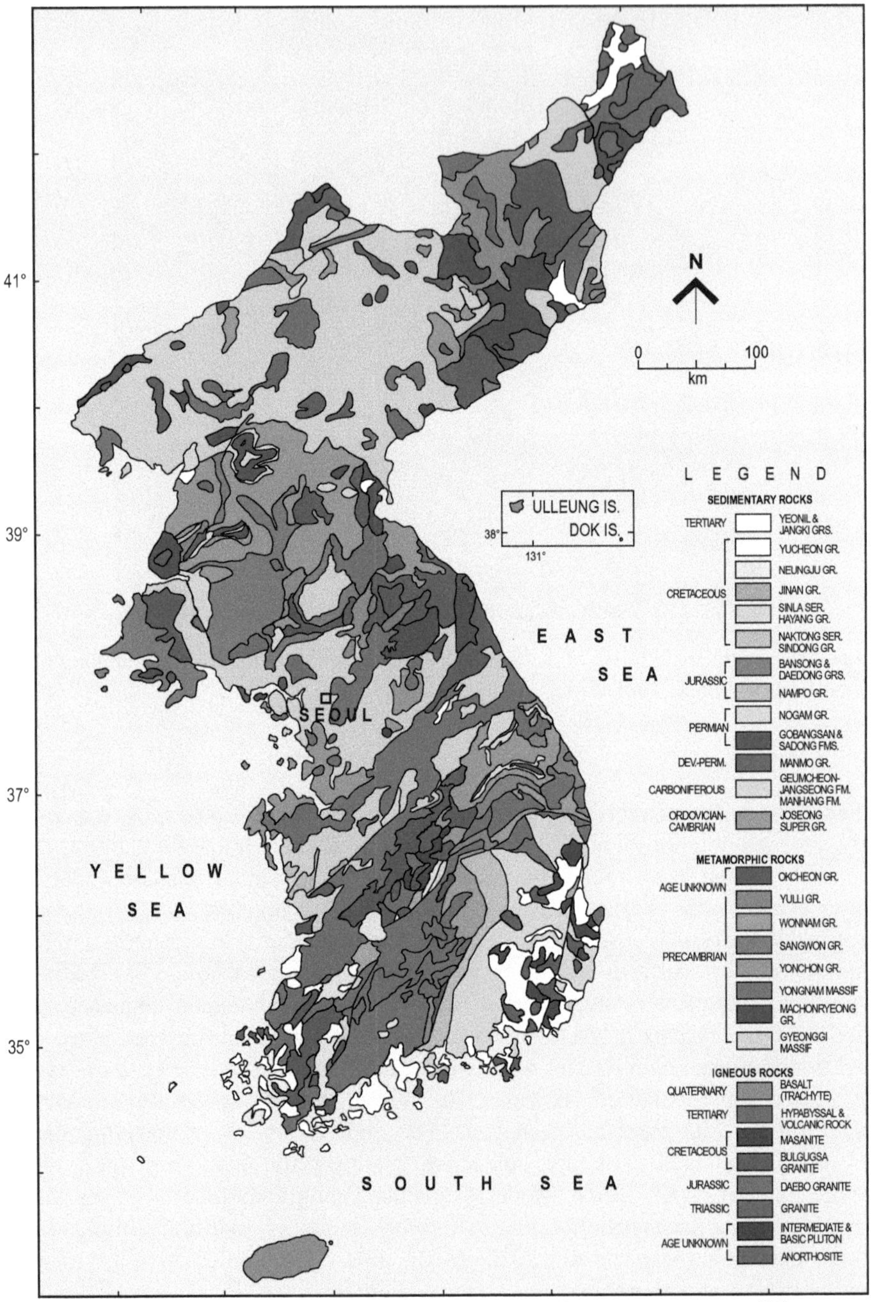

Figure 1.5 Geologic map of the Korean Peninsula.
Source: Korea Institute of Energy and Resources (1981) by permission of the Korea Institute of Geoscience and Mineral Resources.

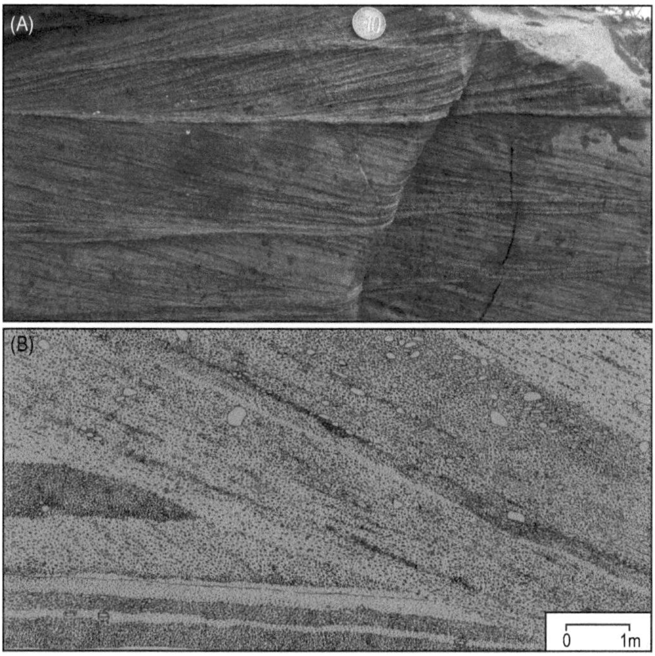

Figure 1.6 Sedimentary facies. (A) Bidirectional cross-bedded sandstone, Mantou Formation, Shandong Province, China. (B) Line drawing of large-scale scour at the base of foreset. The lower left part is characterized by cross stratified, disorganized, and openwork amalgamated beds of conglomerates similar to Gilbert-type topset facies. The foreset is characterized by steeply inclined beds of stratified and inversely graded conglomerates, Doumsan Fan Delta, Pohang Basin.

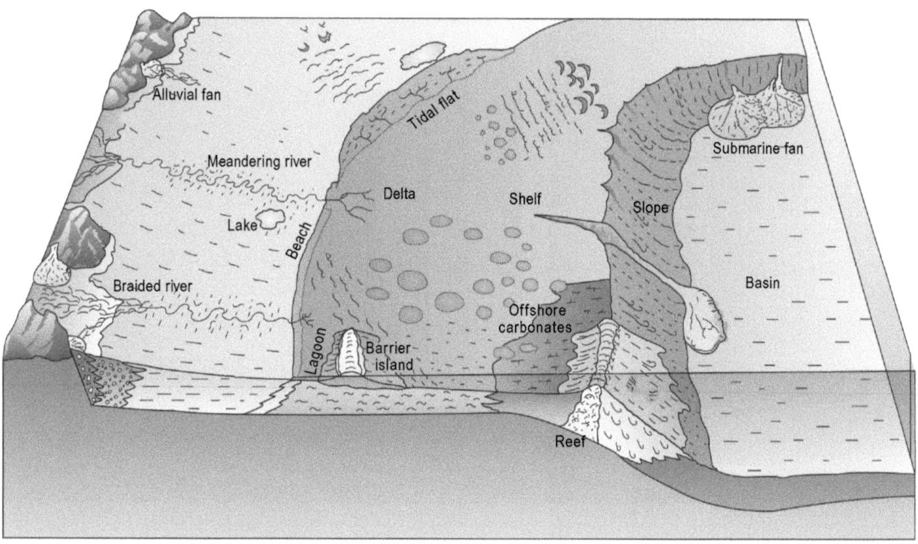

Figure 1.7 A large-scale view of depositional environments on Earth's surface.

deposits. The results have led to significant advancement in the understanding of depositional processes and environments as well as of dynamic crustal evolution of the peninsula on land and under the sea. In this book, I summarize the results of these works, focusing on the tectonic and sedimentary evolution and environmental changes in view of the global events.

2 Tectonic Setting

2.1 Basement Rocks

2.1.1 Introduction

The basement rocks of the Korean Peninsula consist of high-grade gneiss and schist, Precambrian massifs, which formed in the early stage of Earth's history (Figure 2.1). These rocks are unconformably overlain by metasedimentary rocks (schist, quartzite, marble, calcsilicate, and amphibolite) of the Middle to Late Proterozoic (Cho and Kim, 2005; Lee, 1987a). Subsidence of a large sedimentary basin, subduction of a cooling plate, and obduction of colliding plates can bury unconsolidated sediments and crystalline rocks to depths of several tens of kilometers where metamorphic minerals form, including garnet, kyanite, sillimanite, staurolite, and andalusite. Andalusite recrystallizes to form sillimanite at low pressure or kyanite at high pressure. The presence of one of these index minerals is indicative of a specific range of temperature and pressure of mineral formation in the Earth's interior. There are three massif bodies in the Korean Peninsula, namely the Nangrim Massif in the northern part and the Gyeonggi and Yeongnam massifs in the southern part (Figures 2.1 and 2.2).

2.1.2 Nangrim Massif

The Nangrim Massif underlies the Pyeongnam Basin and bounds the microblocks of Macheonryeong and Kwanmobong massifs (Figure 2.2). These blocks were fused together in the Paleoproterozoic (Li et al., 2011; Lyang et al., 2009). The entire massif consists of orthogneiss and granite gneiss and the overlying metasedimentary units of the lower Nangrim and Musan groups and the upper Jeungsan Group (Lower Paleoproterozoic) as well as the Hwanghae, Euiju, and Macheonryeong groups (Neoproterozoic) (Lee and Cho, 2012). The Nangrim Group consists of quartzite, marble, amphibolite, and biotite gneiss, and the interlayered hypersthene–biotite gneiss, which formed at 2.7–2.5 Ga (Lee and Cho, 2012). The Jeungsan Group consists largely of crystalline schists of low-grade metamorphism and weak granitization. The Euiju Group along the Aprok River also consists of crystalline schist, intruded by rapakivi granite. In the Myohyangsan area, it yields an age of 1861 Ma (Zhai et al., 2007). The Hwanghae Group in the southwestern part comprises metasedimentary rocks and volcanics, intruded by

Geology and Sedimentology of the Korean Peninsula. DOI: http://dx.doi.org/10.1016/B978-0-12-405518-6.00002-4

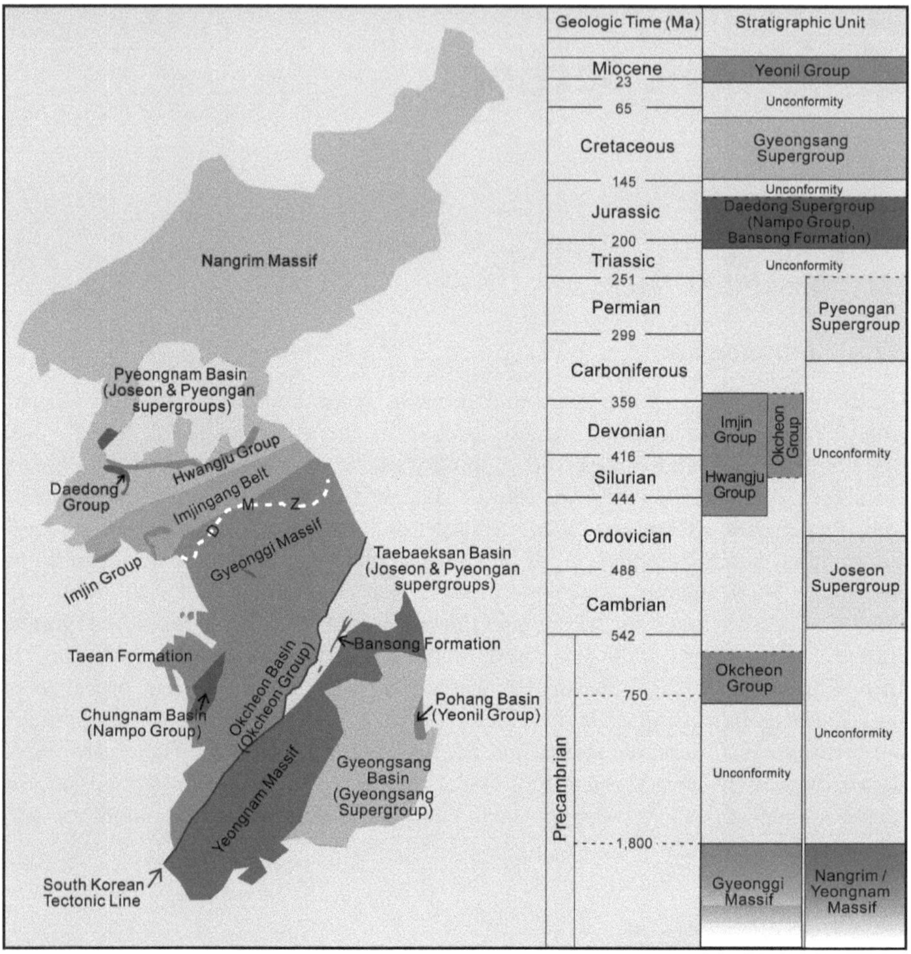

Figure 2.1 Major tectonic features and stratigraphy of the Korean Peninsula.

granite (Lee and Cho, 2012). The Macheonryeong Group, formed along the boundary between the Nangrim Massif and the Kwanmobong micromassif, consists of thick dolostone, metavolcanics, and schist, which were intruded by Paleoproterozoic granite (1847 Ma) (Paek et al., 1996). The uppermost part of the massif comprises the Sangwon Group in the Pyeongnam Basin and the Guhyun Group (upper Mesoproterozoic–Neoproterozoic). Age dates of granitic rocks and gneisses of the massif using a sensitive high-resolution ion microprobe (SHRIMP) and other methods at various localities (Myohyangsan, Gaesong, and others) yield two age groups, namely 2.1 Ga and 1.9–1.8 Ga (Lee and Cho, 2012; Song et al., 2009; Wu et al., 2007; Zhao et al., 2006) (Figures 2.3 and 2.4).

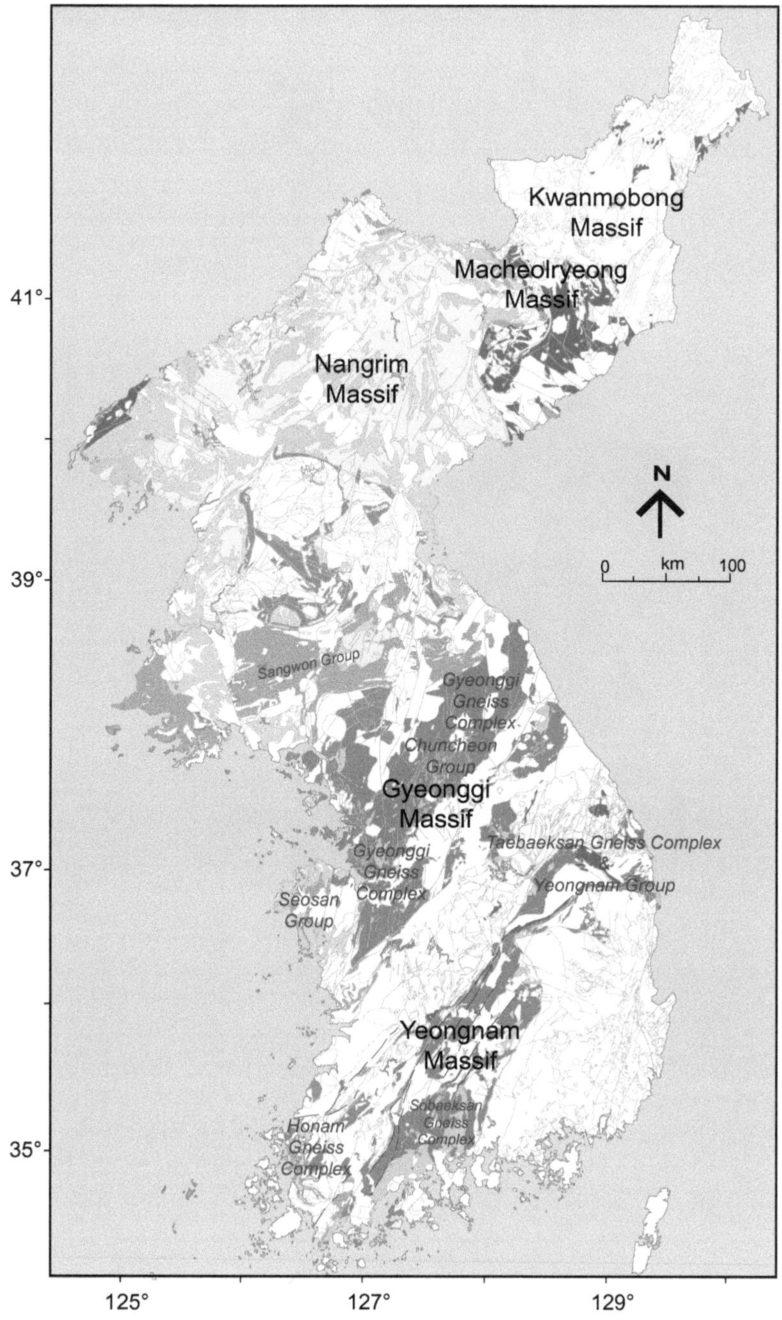

Figure 2.2 Distribution of Precambrian basement rocks (massifs) in the Korean Peninsula. *Source*: Lee and Cho (2012).

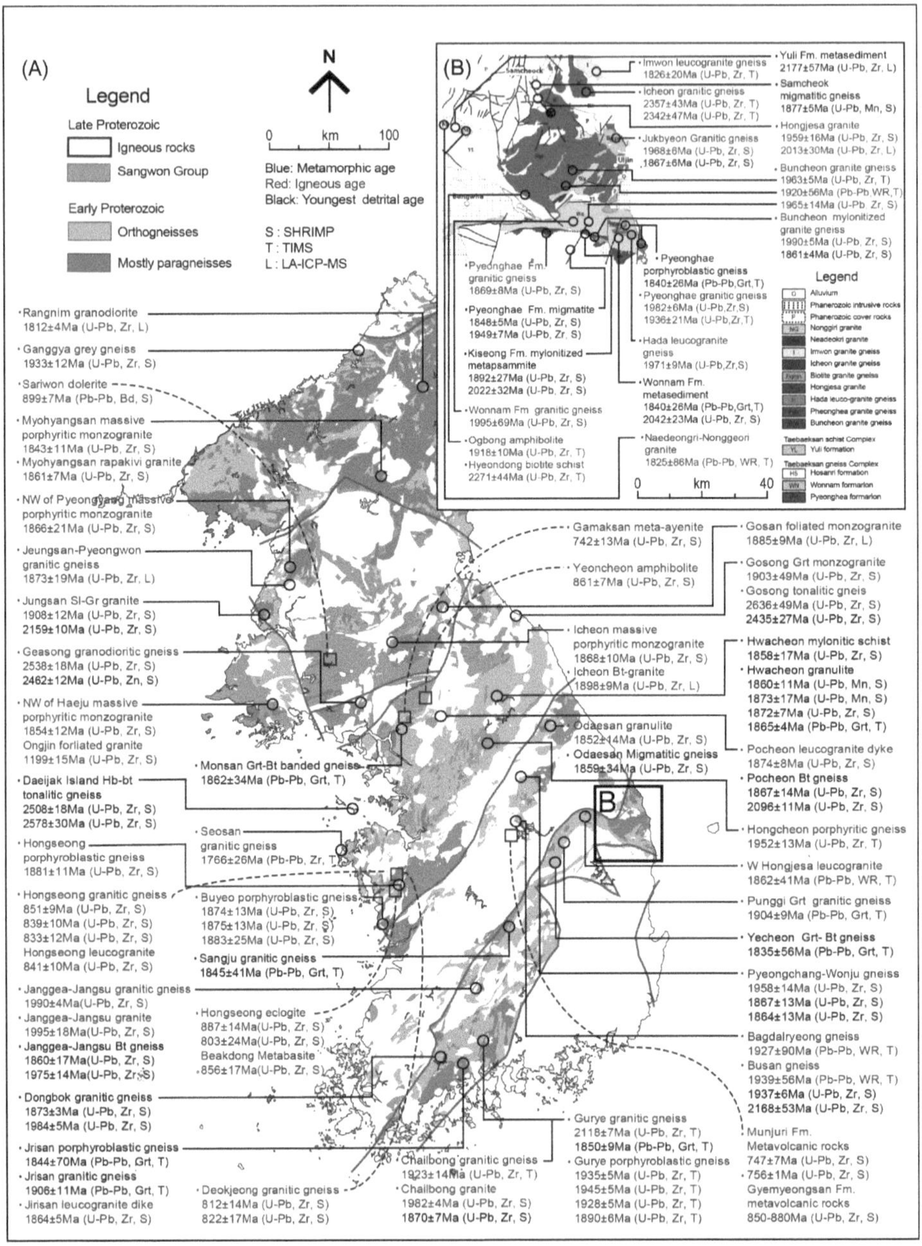

Figure 2.3 Age dates of the Precambrian basement rocks (massifs) in the Korean Peninsula. *Source*: Lee and Cho (2012).

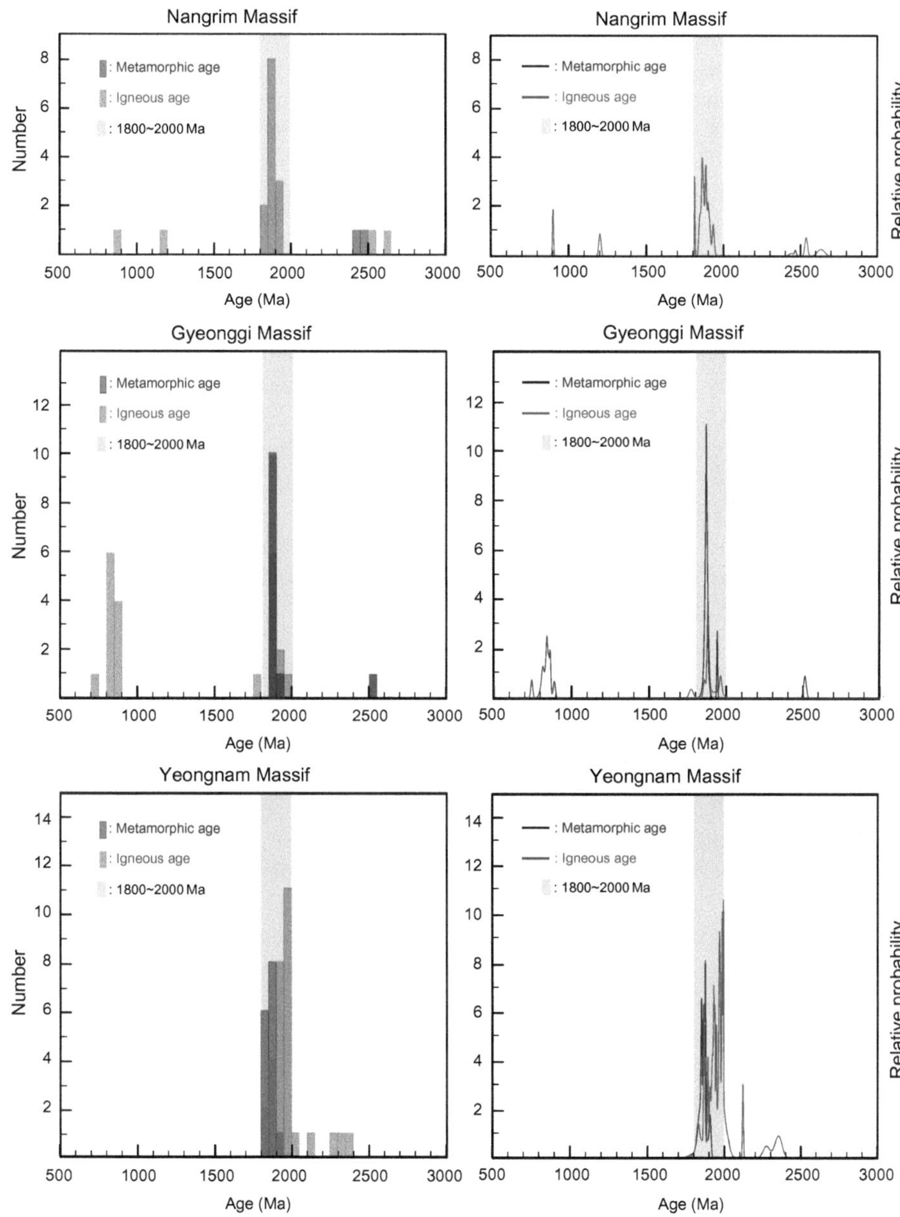

Figure 2.4 Age histograms and their relative probabilities in the massifs of the Korean Peninsula.
Source: Lee and Cho (2012).

2.1.3 Gyeonggi Massif

The Gyeonggi Massif consists of banded biotite gneiss, porphyroblastic garnet-bearing gneiss, migmatitic gneiss, granitic gneiss, biotite schist, crystalline limestone, and quartzite (Na, 1987). It comprises a number of metasedimentary rock units, including the Seosan Group, the Gyeonggi gneiss complex, and the Chuncheon Group (Figure 2.2). The Seosan Group consists largely of banded biotite gneiss and quartzite as well as quartz–biotite schist and marble. The granitic gneiss yielded a U–Pb age of 1870 Ma (Lee et al., 2000a); SHRIMP U–Pb age of detrital zircon in the quartzite ranges from 1781 to 1898 Ma (Paleoproterozoic) (Figure 2.3) (Cho et al., 2006a; Lee and Cho, 2012). The Gyeonggi gneiss complex consists of metasedimentary gneiss, quartzite, and marble (Lee and Cho, 2012). These rocks mostly formed at 2.3–1.9 Ga (Paleoproterozoic), but tonalite migmatite is as old as 2.51 Ga (Cho et al., 2008) (Figures 2.3 and 2.4). The Chuncheon Group consists largely of quartzite, biotite schist and gneiss, banded gneiss, marble, and amphibolite, whose lithostratigraphic relationships are unclear at present (Lee and Cho, 2012). Detrital zircons in schist and quartzite yielded an age of upper limit of 1866–1876 Ma (Lee and Cho, 2012). The massif experienced cratonization in the Paleoproterozic and magmatism in the Neoproterozoic (0.86–0.74 Ga) (Lee and Cho, 2012).

2.1.4 Yeongnam Massif

The Yeongnam Massif in the southeastern part of the peninsula comprises various gneisses (granitic, porphyroblastic, migmatitic, and banded), intruded and interlayered by the Paleoproterozoic granite such as the Hongjesa granite body (Figure 2.3). According to the regional distribution, these rocks are grouped into the Honam gneiss complex, the Sobaeksan gneiss complex, and the Taebaeksan gneiss complex, whose stratigraphic relations are not well defined (Lee and Cho, 2012). In the Jirisan region (the Honam complex), the granites intruded at 2.12 Ga and 1.98–1.89 Ga and were metamorphosed at 1.87–1.86 Ga (Lee and Cho, 2012).

2.1.5 Summary

The oldest basement rocks in the Korean Peninsula are dated at 2.51 Ga in the Gyeonggi gneiss complex (Cho et al., 2008), although inherited zircon cores yield an age as old as 3.6 Ga (Eoarchean) (Figure 2.5) (Jeon et al., 2007; Lee and Cho, 2012; Wu et al., 2007). Most of the crust-forming materials were extracted from the mantle about 2.7 Ga (Lee and Cho, 2012). In the Nangrim Massif, major event of cratonization occurred at about 2.3 Ga; in the Gyeonggi and Yeongnam massifs at about 2.5 Ga (Figure 2.5). These Paleoproterozoic metasedimentary and granitic rocks underwent repeated intracrustal differentiation, followed by the events of cratonization, i.e., regional metamorphism and igneous activity, at 1.9–1.8 Ga (Lee and Cho, 2012). In the Nangrim and Gyeonggi massifs, sediments deposited

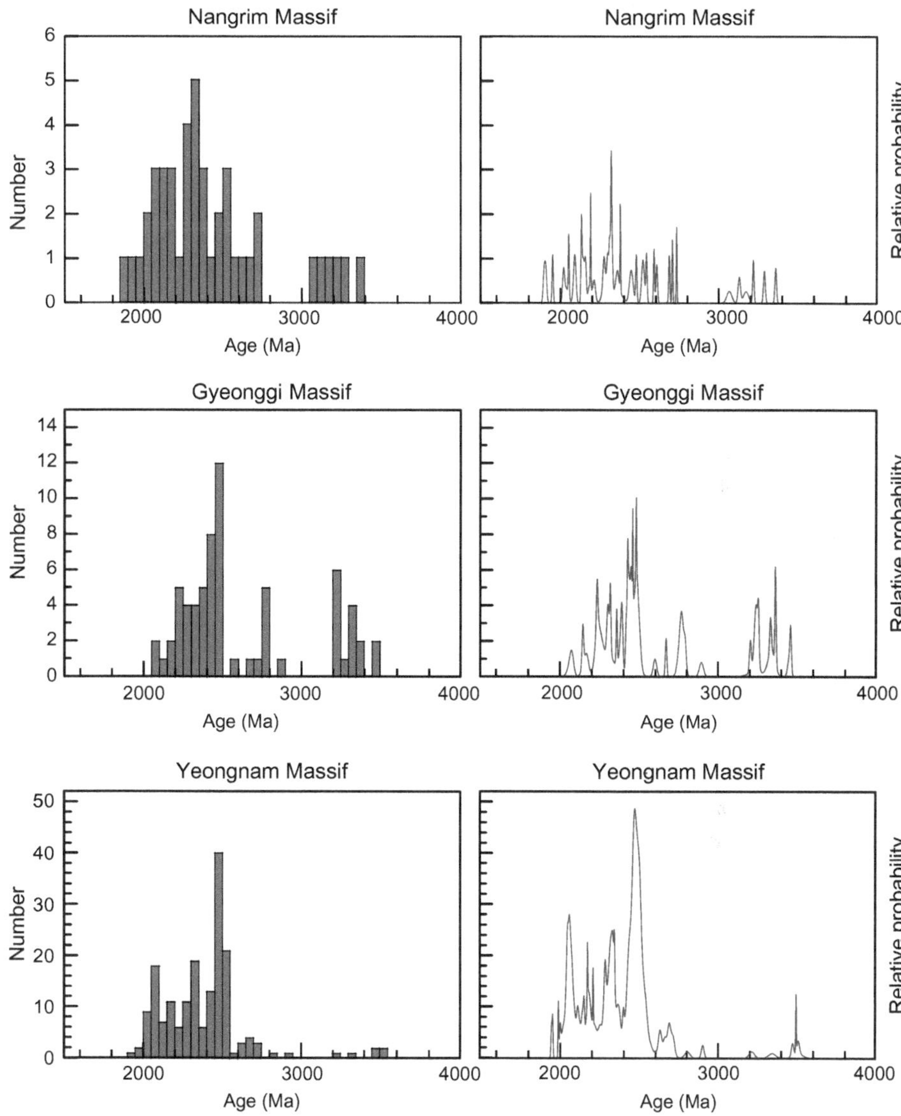

Figure 2.5 Age histograms and their relative probabilities derived from inherited zircons in the massifs of the Korean Peninsula.
Source: Lee and Cho (2012).

in the periphery basins during the Mesoproterozoic and Neoproterozoic leading to stabilization as basement of the peninsula.

2.2 Constituent Lithospheric Plates

The Earth's lithosphere is divided into 13 major rigid plates (Figure 2.6). The Korean Peninsula is part of the Eurasian Plate. During the Earth's evolution, however, the rigid plates have moved apart, creating new lithospheres, or collided to form a single amalgamated lithosphere (Figure 2.7). In the Phanerozoic Era, the Eurasian Plate comprised a few blocks, including the Sino-Korean (North China) Block and the South China Block within Gondwana (Choi et al., 2001; McKenzie et al., 2011). The Nangrim and Yeongnam massifs were part of the Sino-Korean Block, whereas the Gyeonggi Massif was part of the South China Block. The Nangrim and Gyeonggi massifs are separated by the Imjingang Belt, a collisional suture zone (Figure 2.1) (Ree et al., 1996). The Gyeonggi Massif bounds the Yeongnam Massif along the South Korean Tectonic Line. The Yeongnam Massif was part of the Nangrim Massif prior to the collision of the two blocks in the Mesozoic. The Yeongnam Massif is unconformably overlain by a thick (about 3 km) Paleozoic marine sedimentary succession, the Joseon Supergroup of the Taebaeksan Basin. The collision event during the Mesozoic is most important in the geological history of the peninsula.

2.3 Major Geologic Events

Major geologic events in the Korean Peninsula can be metaphorically compared to the Seoul tower (Figure 2.8). Gneiss and schist of the Gyeonggi and Yeongnam massifs formed in the middle Archean between approximately 2.7 and 1.8 Ga. In the Gyeonggi Massif, about 1 billion years had passed without record prior to the deposition of sediments in the Okcheon Basin, which was followed by sedimentation posterior to about 430 Ma (M. Cho, personal communication; Park et al., 2011). In the Yeongnam Massif, deposition of sediments was initiated in the Taebaeksan Basin in the Early Cambrian (about 520 Ma). This unconformity also represents a significant time gap (about 1.3 billion years) (Figure 2.9). Sedimentation in the Taebaeksan Basin terminated in the Upper Ordovician. The basin was then uplifted during the Silurian and Devonian periods, forming another unconformity. There is virtually no geological record of these periods in either the Taebaeksan Basin or the North China Platform, Shandong Province. The continent was once again submerged under the sea in the Carboniferous and the early Triassic periods. The fact that the Gyeonggi Massif, including the Okcheon Basin, and the Yeongnam Massif, including the Taebaeksan Basin, had different histories in the Paleozoic Era suggests that the

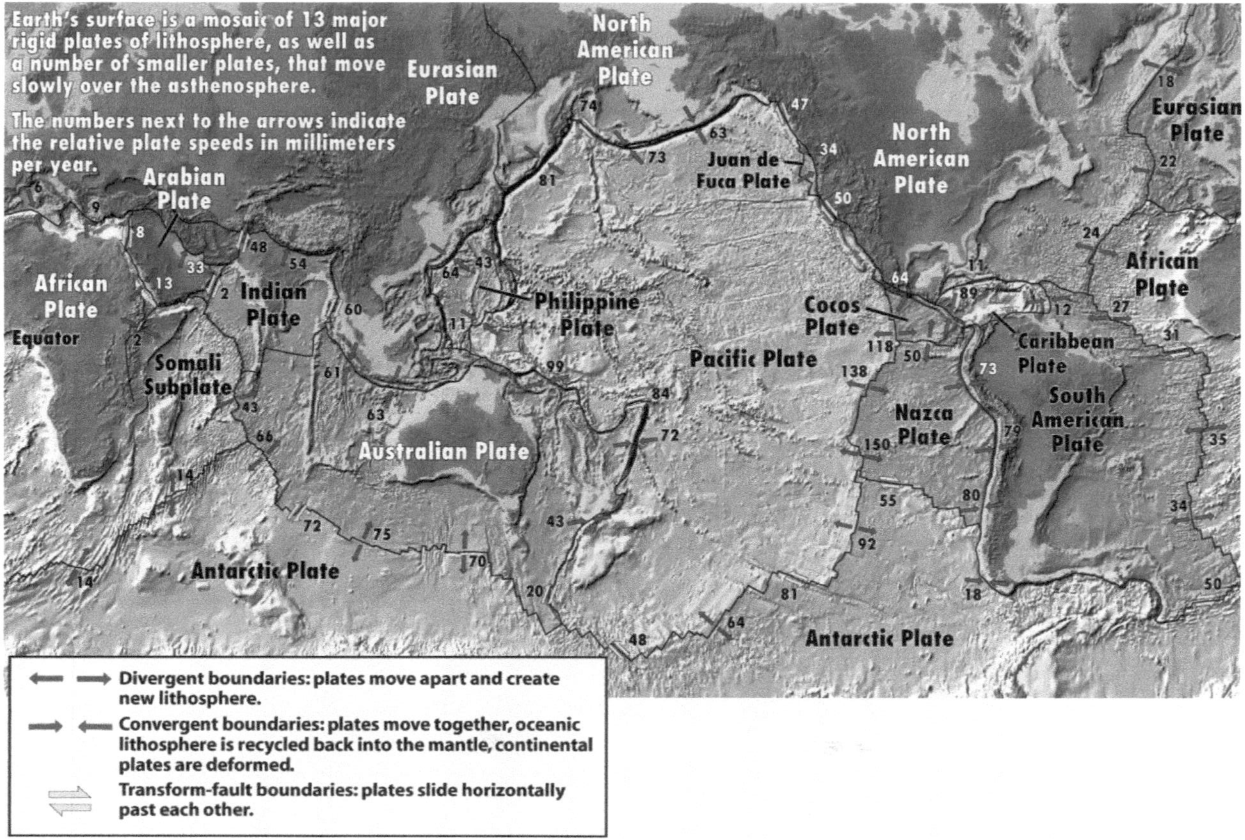

Figure 2.6 Lithospheres (plates) and their boundaries.
Source: Grotzinger et al. (2007) by permission of the W.H. Freeman & Company.

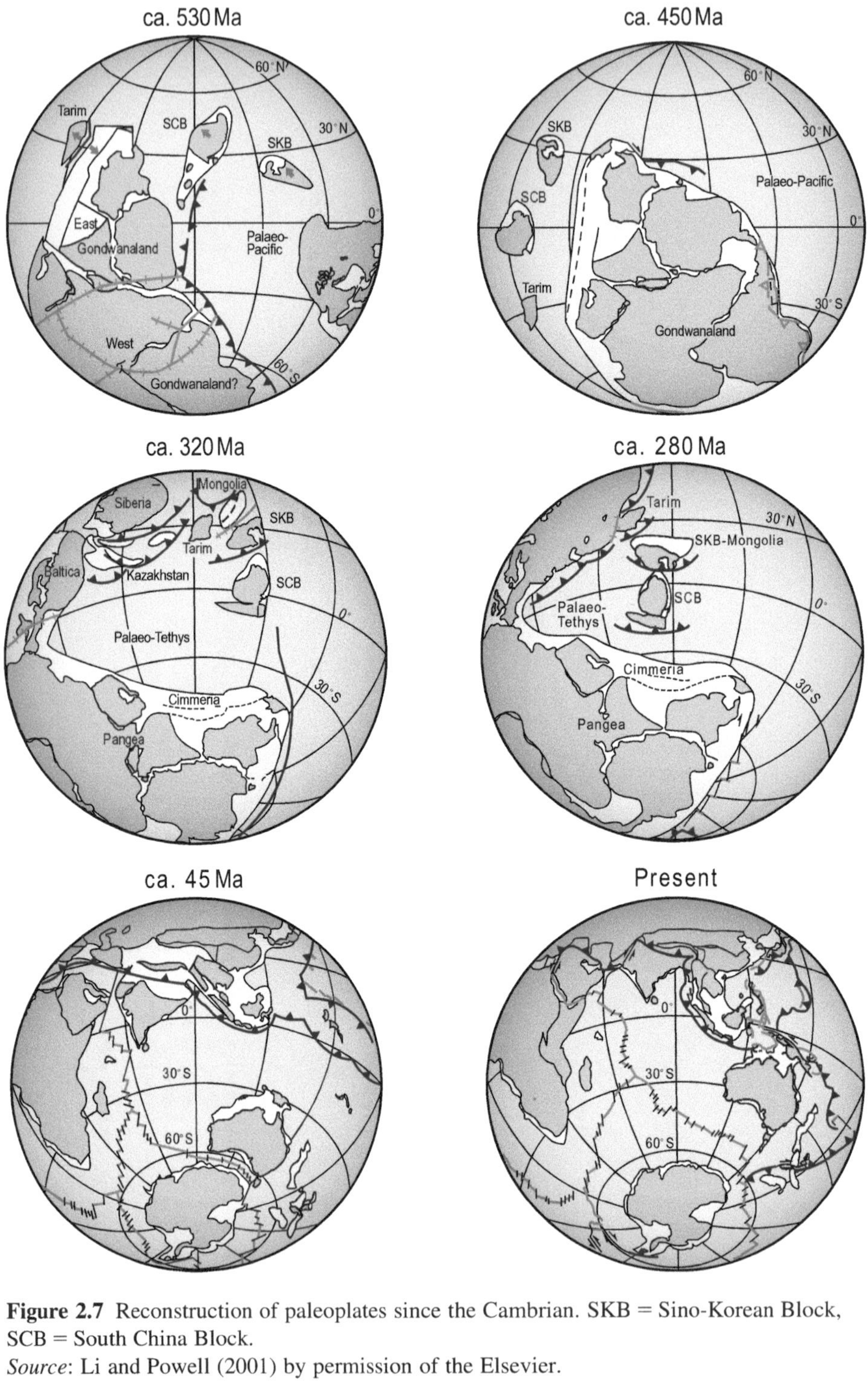

Figure 2.7 Reconstruction of paleoplates since the Cambrian. SKB = Sino-Korean Block, SCB = South China Block.
Source: Li and Powell (2001) by permission of the Elsevier.

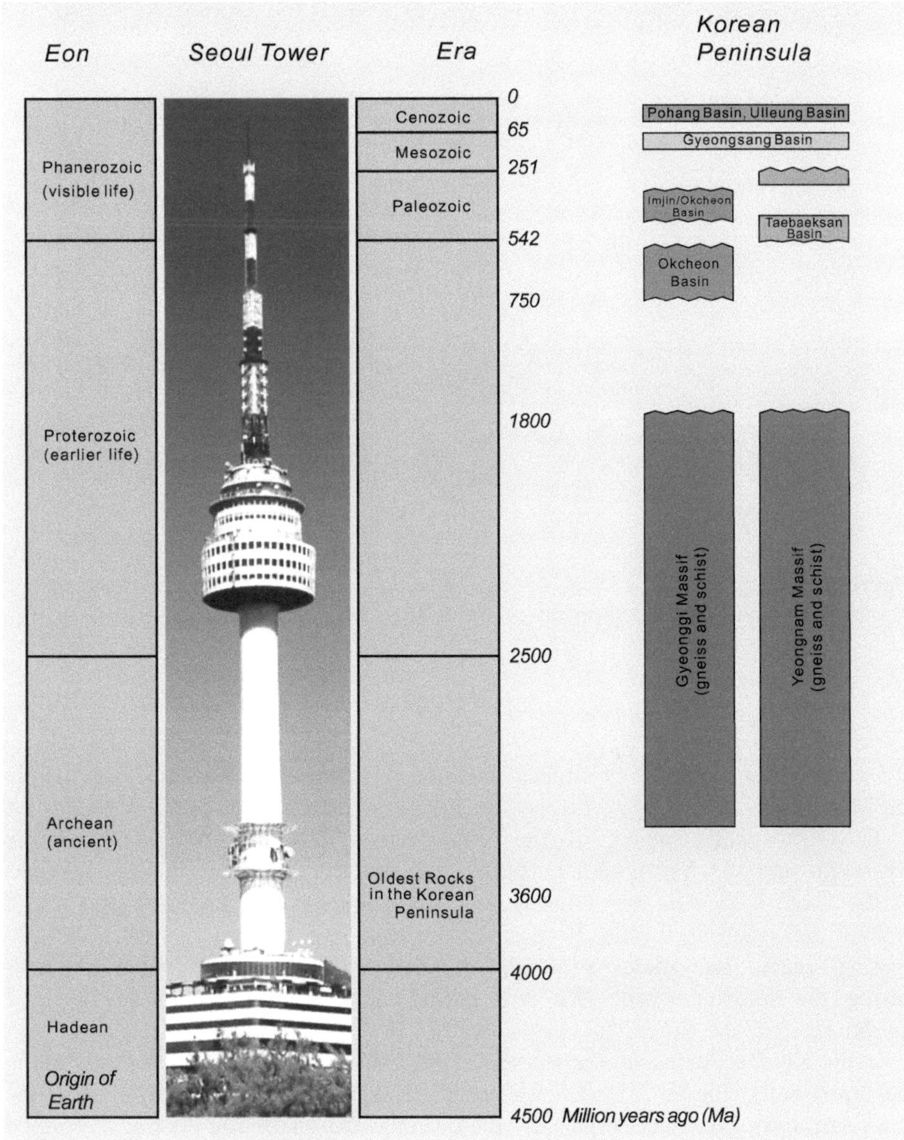

Figure 2.8 Geological column of the Korean Peninsula compared with the Seoul Tower. Ma = millions of years before present. A simplified geological timescale in millions years. The Gyeonggi Massif is unconformably overlain by the succession of the Okcheon Basin, whereas the Yeongnam Massif is unconformably overlain by the succession of the Joseon Supergroup, Taebaeksan Basin.

Figure 2.9 Photograph of the Dongjeom stream section, showing Precambrian basement (granite gneiss) and the unconformably overlying Myeonsan Formation (Cambrian).

two continental blocks were situated at different locations until fused together in the Mesozoic.

The lithospheric plates of the Earth have constantly moved. Both the Gyeonggi and the Yeongnam massifs were sutured against each other as part of the large plates in the Triassic and Jurassic periods, recording the major crustal deformation in the Korean Peninsula. The event spanned about 70 million years. The plates probably traveled for a few hundred kilometers during the suturing event. We will come back to these events in *Tectonic Evolution*.

In the Korean Peninsula, the records of Jurassic sedimentation only occur along the thrust belts (Figure 2.1) (Chough et al., 2000). In the Cretaceous Period, the large part of the peninsula, especially the Gyeongsang and Jeolla Provinces, formed continental backarc basins into which large amounts of siliciclastic and volcanic sediments were delivered from the surrounding continent and volcanic arc. This was accompanied by massive intrusion of granitic magma. Large and small lakes formed in these basins but eventually filled with volcanic materials from the arc. Dinosaurs dominated the continent until their extinction at the end of the Cretaceous. Magmatism continued in the Cenozoic under the arc setting. In the Miocene Epoch, sediments were deposited in marine backarc basins, including the Pohang and Ulleung basins.

Eon	Era	Period	Millions of years	Event
Phanerozoic	Cenozoic (modern life)	Quaternary	2.6	Glaciation First humans
		Tertiary: Neogene	23.0	
		Tertiary: Paleogene	65.5	Evolutionary expansion of mammals Grass dominant
	Mesozoic (middle life)	Cretaceous	145.5	Extinctinction of dinosaurs First flowering plants
		Jurassic	201.6	First birds;dinosaurs prominent Break-up of Pangea
		Triassic	251	First mammals and dinosaurs Mammal-like reptiles prominent
	Paleozoic (ancient life)	Permian	299	End-Permian extinction Extinction of trilobites Reptiles dominant
		Carboniferous	359	Major coal-forming swamps First reptiles and ferns Formation of Pangea
		Devonian	416	Late-Devonian Extinction First trees; fish dominant First amphibians
		Silurian	444	Trilobites decline First bony fishs (with jaws)
		Ordovician	488	Gondwana glaciation First land plants
		Cambrian	542	Cambrian Explosion Diversification of invertebrates First vertebrates
Precambrian	Proterozoic	Ediacaran	630	First multicellular organisms
				Break-up of Rodinia
			1000	Formation of Rodinia
			2500	Great oxidization event
	Archean		3400	First organisms

Figure 2.10 Major geologic events.

Although multicellular organisms first appeared at about 610 Ma (Ediacaran), abundant invertebrates thrived in the Cambrian (Figure 2.10). Trilobites were the most prominent animals on the seafloor of an extensive marine realm in the Cambrian Period, along with crinoids and corals (Figure 2.10). In the late Paleozoic, other life forms became dominant, such as fish and ammonoids. Trilobites perished at the end of the Paleozoic, whereas crinoids and corals were

Figure 2.11 Trilobites of the Joseon Supergroup: 1–18 and 28 from Gonggiri section (Yeongwol) and 19–27 from Seokgaejae section (Taebaek). Scale bars are 1 mm. 1, 2 *Glyptagnostus stolidotus*, 3, 4 *Glyptagnostus reticulatus*, 5 *Olenus asiaticus*, 6 *Pseudoyuepingia asaphoides*, 7, 8 *Pseudagnostus securiger*, 9, 10 *Ivshinagnostus hunanensis*, 11, 12 *Ivshinagnostus alatus*, 13 *Kormagnostus inventus*, 14 *Peratagnostus orientalis*, 15, 16 *Micragnostus hisakoshi*, 17, 18 *Agnostotes orientalis*, 19 *Mictosaukia* sp., 20 *Pagodia* sp., 21 *Haniwa* sp., 22 *Pseudokoldiniodia perpetis*, 23 *Onychopyge* sp., 24 *Richardsonella* sp., 25 *Litagnostus* sp., 26 *Sinosaukia* sp., 27 *Quadraticephalus elongates*, 28 *Irvingella megalops*.
Source: Courtesy of D.K. Choi.

almost wiped out in the Permian, about 250 million years ago (Figure 2.10). Abundant trilobite fossils occur in the shale and limestone beds of the Joseon Supergroup in Gangwon Province, including more than 60 genera and 110 species (Figure 2.11) (Choi, 1998; Park and Choi, 2010). Crinoids are also abundant. These fossils can be correlated with those of the North China Platform and other continents which were close together in Gondwana during the Paleozoic Era (Choi et al., 2001; McKenzie et al., 2011).

3 Phanerozoic Granites

3.1 Occurrence

The Seoul area comprises overlapping mountain ridges and peaks with a winding river (Han River) running through it (Figure 3.1). The ridges consist of Precambrian gneiss of the Gyeonggi Massif, as described in the previous chapter. However, the mountain peaks such as the Bukhansan consist of Jurassic granites (Figure 3.2). The granites cover a large area. The Bukhansan Mountain, for example, spans an area of 2000 km^2, comprising a number of scenic pinnacles and peaks (Figure 3.1). On a large scale, these rock bodies are largely oriented northeast to southwest in the entire peninsula (Figure 3.3).

The granites in the peninsula largely range in age from the Permian to the Jurassic and from the Cretaceous to the Tertiary (Figure 3.3) (Kwon and Sagong, 1998). The Permian–Triassic granites largely occur in scattered areas, whereas the Jurassic granites prevail in the middle part of the peninsula (Figure 3.3). The latter comprise a batholith, several tens of kilometers wide belts, and consist of medium-to-coarse crystalline granite, granodiorite, and tonalite (Kwon et al., 1995b; Sagong et al., 1997). They are foliated in places, especially along the shear zone (Kwon and Ree, 1997). However, the Cretaceous–Tertiary granites are generally fine- to medium-grained and form stocks, which are associated with volcanic rocks (Figure 3.3).

The emplacement of granites was regulated by the changes in subduction mode of the Pacific Plate (Cheong and Kim, 2012). In the Early Permian, magmatic intrusion occurred in the northern part of the Gyeongsang Basin. In the Early Triassic, tonalities, trondhjemites, granodiorites, and monzonites were emplaced in the middle part of the Yeongnam Massif under an arc-related setting (Cheong and Kim, 2012). In the Late Triassic, postcollisional shoshonite and high-K series and A-type granitic rocks were emplaced in the southwestern Gyeonggi Massif and the central Okcheon Basin. The Early Jurassic calcalkaline granites occur in large areas of the northeastern Gyeonggi Massif and the middle part of the Yeongnam Massif, whereas those of the Middle Jurassic prevail in the central part of the Gyeonggi Massif and along the South Korean Tectonic Line (Figure 3.3). These Jurassic granites formed during the orthogonal subduction of the Pacific Plate under the Asian continent. The Cretaceous–Paleogene granites occur in the Gyeongsang Basin, which were associated with the arc magmatism and volcanism during the oblique subduction of the Pacific Plate (Chough and Sohn, 2010).

Geology and Sedimentology of the Korean Peninsula. DOI: http://dx.doi.org/10.1016/B978-0-12-405518-6.00003-6

Figure 3.1 Bukhansan Mountain looking north across downtown Seoul. The mountain consists of granite that was intruded in the Jurassic.

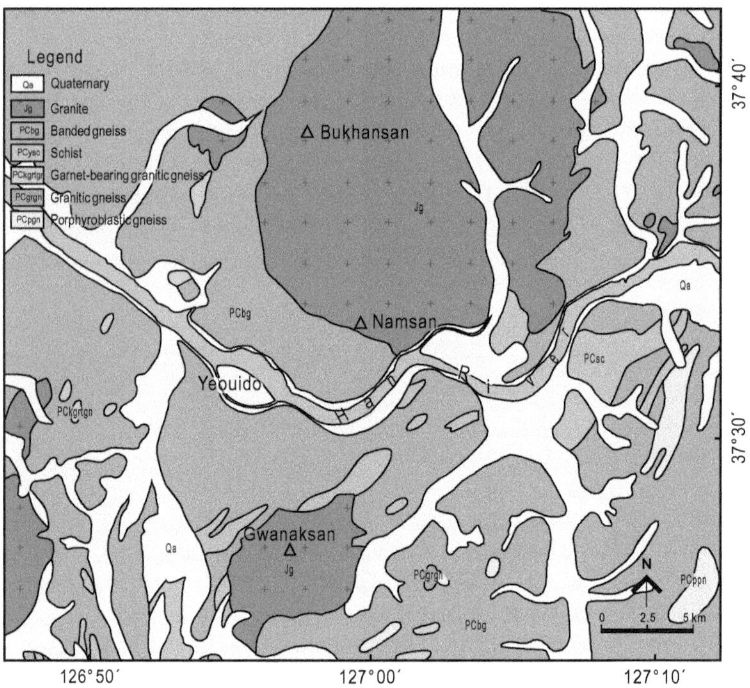

Figure 3.2 Geologic map of metropolitan Seoul. Precambrian gneiss, part of the Gyeonggi Massif, was intruded by granite in the Jurassic.
Source: Lee et al. (1999a).

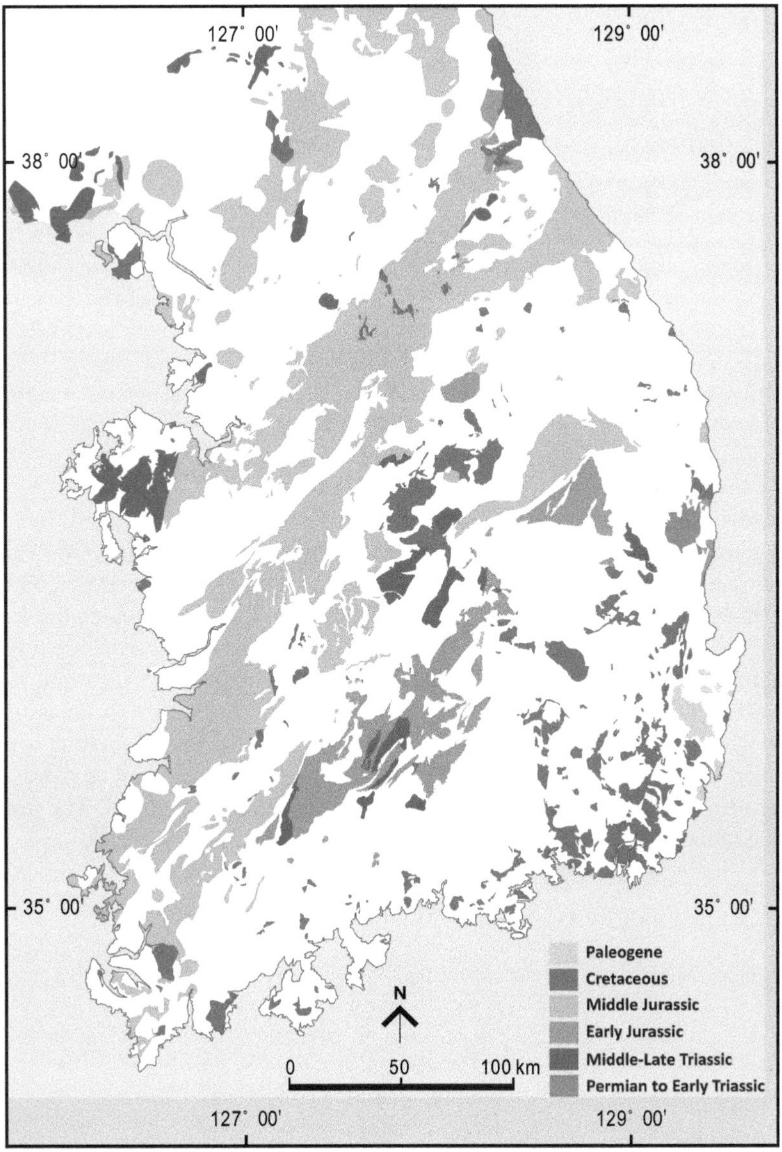

Figure 3.3 Distribution of Phanerozoic granites.
Source: Cheong and Kim (2012).

3.2 Origin of Granites

The granite batholiths were emplaced more than 10 km subsurface into the overlying gneiss and sedimentary rocks. The batholiths comprise biotite, hornblende–biotite, and biotite–muscovite granite and granodiorite (Kwon et al., 1995a; Sagong et al., 1997).

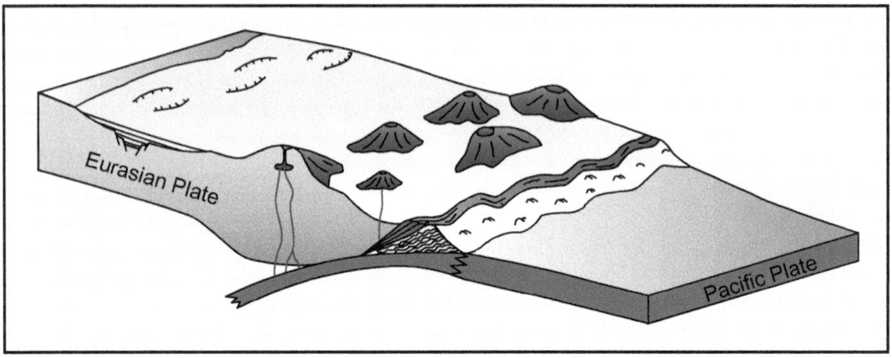

Figure 3.4 Granite forms by slow cooling and crystallization of large crystals (quartz and feldspar) in the deep Earth's crust above a subduction zone.

They also contain garnet and megacrysts of K-feldspar (Sagong and Jwa, 1997) and are generally subalkaline to calcalkaline. The granites are peraluminous to metaaluminous and belong to I-type (Cho, 1992). The intrusion of magma occurred in a convergent margin where the Pacific Plate subducted at low angle under the Eurasian Plate (Figure 3.4) (Kee et al., 2010). The magma originated from partial melting above the subduction zone. Granite generally contains more than 70% silica and forms at depths where continental crust partially melts. Magma rises up into areas of lower pressure where it cools and solidifies slowly to form granite. Partial melting is of great importance in producing magmas of varying compositions. Removal of the overlying rocks by erosion and weathering has exhumed since the emplacement. The integrated uplift rate of the batholiths is about 0.1 mm/year (Cho and Kwon, 1994).

4 Okcheon Basin

4.1 Okcheon Group

4.1.1 Introduction

The oldest sedimentary succession in the peninsula, Okcheon Group, comprises metasedimentary rocks (metapelites, quartzites, clast-bearing phyllites, marbles, and calcsilicates) and metavolcanic rocks, which unconformably overlies the Gyeonggi Massif (South China Block). The succession in the mid-central part of the peninsula is represented by a number of lithologic units (Figure 4.1) (Chough and Bahk, 1992; Kim, 1970; Lee, 1987a; Lee and Park, 1965; Lee et al., 1972, 1980; Reedman et al., 1973). The sedimentary facies distribution is indicative of deposition in a failed rift basin (Chough, 1981). Since there are no reliable body fossils in the succession, definitive geologic age and stratigraphy are still uncertain. Choi et al. (2012) have assigned a Neoproterozoic age for the succession, based on the stratigraphic correlation with that of the Nanhua Basin, South China. Isotopic ages of the metavolcanics (Munjuri Formation) suggest that the succession formed in the Late Proterozoic (about 750 Ma). Bimodal distribution of metavolcanics is suggestive of a rift basin origin (Cluzel, 1992; Cluzel et al., 1990, 1991; Kang et al., 2012; Kwon and Lan, 1991; Kwon and Lee, 1992). Recent age dates of detrital zircons using an SHRIMP in quartzites (Midongsan Formation) indicate, however, that sedimentation resumed posterior to 430 Ma (M. Cho, personal communication; Park et al., 2011). The entire succession experienced medium-pressure (5–9 kb) metamorphism associated with the deformation between 300 and 270 Ma as well as 250 and 220 Ma (Cho and Kim, 2005).

4.1.2 Lithologic Units

4.1.2.1 Metasandstones and Metavolcanics

In the northwestern margin of the basin, the succession consists of magnetite-bearing metavolcanics and metapelites with minor quartzite, dolomitic marble, and metavolcanic breccias, which are strongly foliated and lineated (Gyemyeongsan Formation) (Figure 4.1). It extends to the southwest and is bounded by metasandstones, phyllites, and volcanics (Munjuri Formation) and carbonates (Hwajeonri Formation) whose stratigraphic relationships are unclear. Further east (near Suanbo), the pebble-bearing Hwanggangri Formation contains a number of

Geology and Sedimentology of the Korean Peninsula. DOI: http://dx.doi.org/10.1016/B978-0-12-405518-6.00004-8

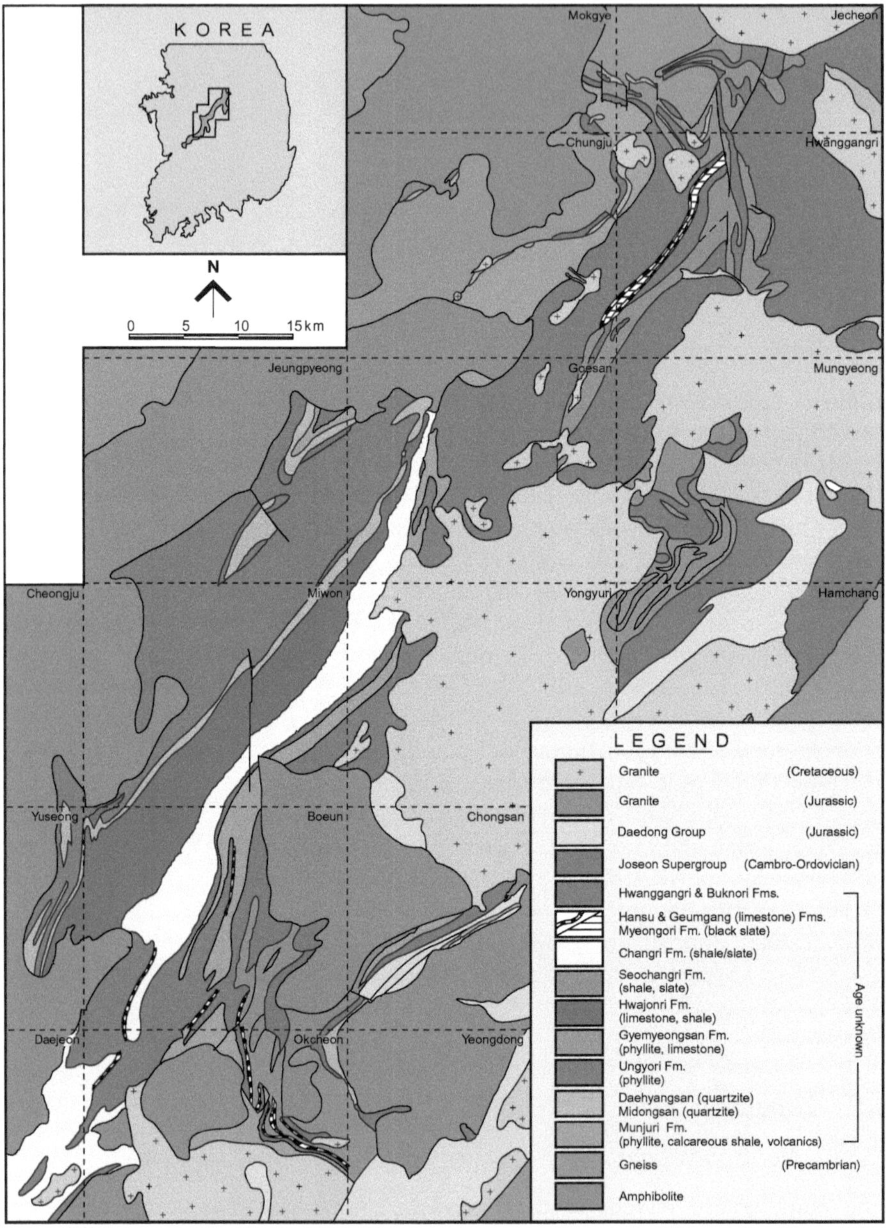

Figure 4.1 Simplified geologic map of the Okcheon Basin.
Source: Chough and Bahk (1992).

depositional features such as inverse grading, debris plugs, load casts, and pseudo-nodules (Chough and Bahk, 1992). These features are suggestive of a stratigraphic order of Seochangri, Buknori, Geumgang limestone, and Hwanggangri formations (Lee and Park, 1965). Cluzel et al. (1991) and Kim et al. (1995b) proposed that the distinct discontinuity between the older Munjuri Formation ($>$700 Ma) and the younger Hwanggangri Formation (age unknown) represents a thrust contact.

In the northwestern part, the basin margin is characterized by an extensive occurrence of quartzite, siltstone, and interlayered dolostone (Ungyori Formation) (Figure 4.2). According to an age dating of detrital zircons using an SHRIMP, the quartzite formed posterior to 430 Ma (M. Cho, personal communication). The interlayered sandstones and siltstones are well sorted, indicating deposition in subtidal environments, whereas the quartzites were most likely deposited in shoreface environments. Further east, the metasedimentary rock units are interlayered with black shale and slate, previously designated as the Changri Formation (Figure 4.1). Lim et al. (2007) recently identified plant fossils (*Calamites* sp., *Cordaites* sp., *Pecopteris* sp., and *Taeniopteris* sp.) in the shale and slate beds, which are suggestive of deposition in the Late Carboniferous to Early Permian. A refined map shows that these metasedimentary rock units are juxtaposed by thrust deformation.

4.1.2.2 Phyllites, Black Shales, and Slates

Dark brown phyllites (shale with tripartite foliations) intercalated with black slates (Seochangri Formation) largely occur in the northern part of the basin, interlayered with thin limestone and marble layers (Geumgang and Hansu formations) (Figure 4.1). These layers contain no primary structures but are laterally persistent along the boundary between the Hwanggangri and Myeongori formations. The shale beds most likely formed in a deep, carbonaceous euxinic basin.

4.1.2.3 Pebble-Bearing Phyllites

The pebble-bearing phyllites of the Hwanggangri Formation occur in an extensive area (Figure 4.1). Clasts constitute 3–30% of the rock by volume. Quartzite clasts are predominant, which are well rounded (Figure 4.2). Some clasts of quartzite are large (more than 2 m in diameter). Others comprise pebble-size quartzite, elongated shale, limestone, and dolomite (Figure 4.3) (Chough and Bahk, 1992). Shale clasts are of the same composition as the matrix, displaying a wide range of deformed shape (Figure 4.3). Limestone clasts are also abundant, including chips, pebbles, and boulders, and show various deformational structures such as flattening, foliation, and folding. Other clasts are minor, including granite, sandstone, and metamorphic rock fragments. Matrices are either siliciclastic or calcareous (calciclastic) and poorly sorted. They range from 70% to 97% by volume. Assemblages of these clasts differ from region to region (Chough and Bahk, 1992).

The pebble-bearing phyllites are interpreted as tillites (Reedman and Um, 1975; Reedman et al., 1973) or glaciogenic deposits of snowball Earth (Choi et al., 2012). Although the occurrence of isolated dropstones suggests melting of icebergs,

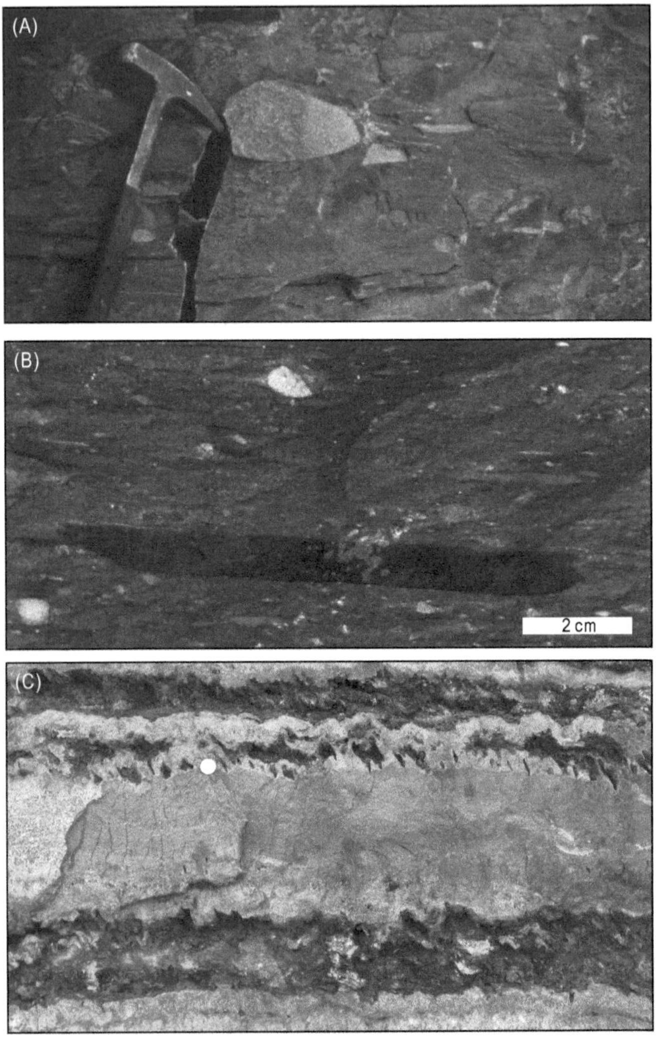

Figure 4.2 Pebble-bearing phyllite of the Hwanggangri Formation. Clasts comprise quartzite, granite, limestone, and shale (A) and quartzite (B). (C) Metasandstone with interlayered limestone, Ungyori Formation.

the associated sedimentary facies successions are not indicative of extensive glaciation. The existence of contemporaneous limestone and shale chips suggests that these deposits formed by resedimentation during slump/slide and subsequent debris flows in slope and base-of-slope environments (Chough, 1981; Chough and Bahk, 1992). Quartzites and granites originated from local denudation, most likely during the period of sea-level fall (Figure 4.4). The clasts of shale and limestone chips

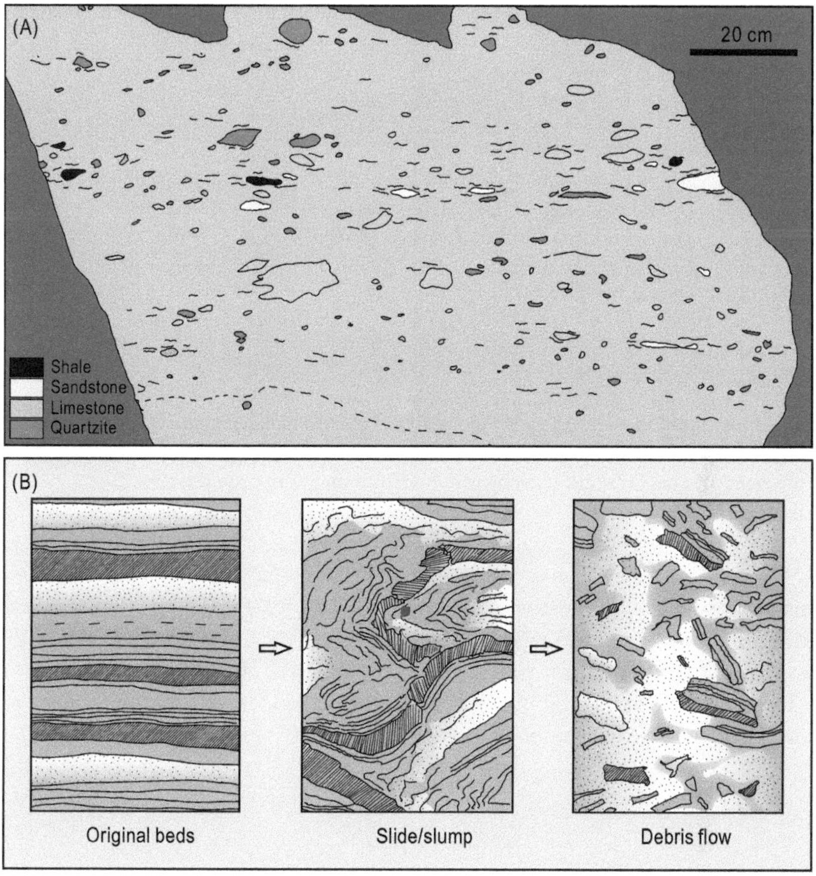

Figure 4.3 (A) Line drawing of pebble-bearing phyllite (Hwanggangri Formation) showing various clasts of quartzite, limestone, sandstone, and black shale in the Suanbo area. (B) A schematic diagram showing slope failure and progressive development of debris flow. *Source*: Chough and Bahk (1992).

originated from slope failure of the contemporaneous substrates. The breakage of contemporaneous mud commonly occurs in slope and base-of-slope environments (Chough and Bahk, 1984–1985; Yoon et al., 1991). The occurrence of isolated and dispersed clasts is suggestive of open marine environments where floating icebergs melt.

A hypothesis is put forward to argue for synchronous Neoproterozoic glaciation, namely snowball Earth, for periods of up to 10 million years (Hoffman, 2008). They argue that ice covered the oceans of both polar and tropical regions to depths of a kilometer and shut down all hydrological and biological activities. The

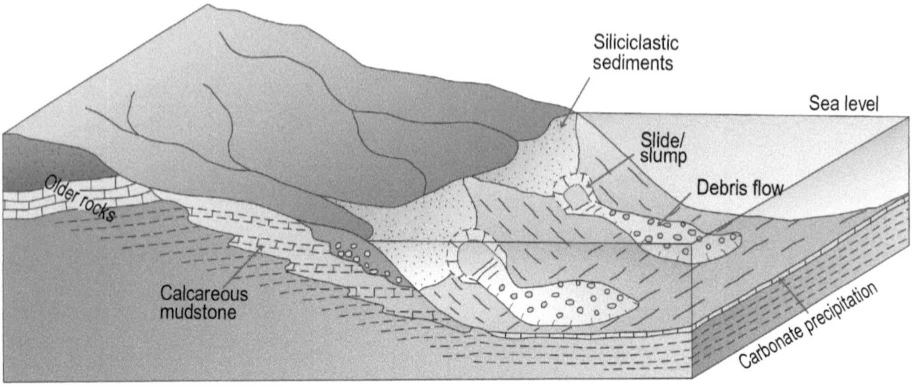

Figure 4.4 Depositional model of the Hwanggangri Formation. During lowstand of sea level, the Okcheon Basin was in euxinic condition. Coarse-grained terrigenous sediments shed into the head of shelf margin causing slope failure and subsequent debris flows. *Source*: Chough and Bahk (1992).

dropstones are, however, suggestive of formation in an ice-free deep-water environment, preferably during interglacial periods. Alternatively, the regional-scale diachronous glaciations could be tied to the progressive breakup and rifting of Rodinia; ice covers waxed and waned as tectonic processes elevated crust and formed marine glaciomarine facies (Eyles and Eyles, 2010).

5 Taebaeksan Basin

5.1 Geological Setting

The Taebaeksan Basin formed unconformably, overlying the granite gneiss of the Yeongnam Massif. The basin comprises mixed siliciclastics and carbonates of the Cambrian–Ordovician, Joseon Supergroup, and unconformably overlying siliciclastic rocks of the Pyeongan Supergroup (Carboniferous–early Triassic) (Figure 5.1). Choi (1998) redefined the Joseon Supergroup into a number of lithologic units according to the regional occurrence, including the Taebaek, Yeongwol, Yongtan, Pyeongchang, and Mungyeong groups. Among these units, the Taebaek and Yeongwol groups are well delineated in terms of litho- and biostratigraphy (Table 5.1) (Choi, 1998; Choi and Chough, 2005; Choi et al., 2004). On the other hand, the Yongtan, Pyeongchang, and Mungyeong groups were subject to strong deformation, which made it difficult to map the lithologic units. For these reasons, sedimentary facies analyses are only valid for the Taebaek Group (Kim and Lee, 1998; Kwon et al., 2006; Paik, 1987; Woo, 1999; Woo and Chough, 2007; Woo and Park, 1989) and part of the Yeongwol Group (Paik and Lee, 1989; Paik et al., 1991; Yoo et al., 1994).

The Yongtan Group in the Jeongseon area consists of metasedimentary rocks of shallow-water carbonate facies, including laminated and bedded lime-mudstone, limestone breccia, wackestone, and packstone/grainstone. Hisakoshi (1943) and Kim et al. (1973) proposed a lithostratigraphy (seven formations) similar to that of the Taebaek Group. The litho- and biostratigraphic scheme of Cheong et al. (1979b) and Lee (1980, 1982a) is invalid, because both the lithologic (formation) boundaries and the geologic ages are ill defined. A preliminary log of the metasedimentary facies in the Jeongseon area shows that the succession comprises deformed limestone facies, including thin and thick banded limestone, pseudoconglomerate, folded lime-mudstone, and foliated limestone (Figure 5.2). These metasedimentary facies are generally characterized by intermingled, fractured, brecciated, and folded beds, which are indicative of strong deformation.

For the Pyeongchang Group, Hukasawa (1943) and Kobayashi (1966) initially proposed a four-formation scheme, whereas Cheong et al. (1979a) suggested a scheme of six formations, similar to that of the Taebaek Group.

Geology and Sedimentology of the Korean Peninsula. DOI: http://dx.doi.org/10.1016/B978-0-12-405518-6.00005-X

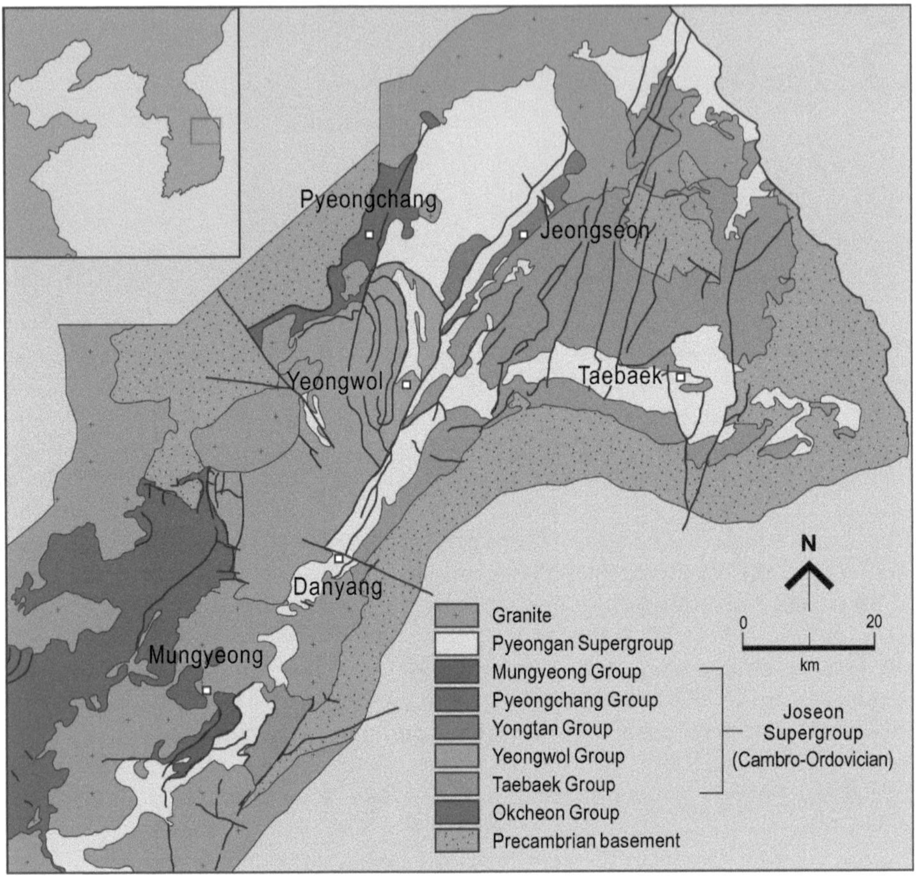

Figure 5.1 Geologic map of the Taebaeksan Basin.
Source: Chough et al. (2000) by permission of the Elsevier.

The stratigraphic scheme of the Mungyeong Group is also not clear at present, although efforts were made to establish various lithologic units (Aoti, 1942; Choi, 1998; Kobayashi et al., 1942). These metasedimentary successions experienced strong deformation and weak–moderate metamorphism. For these reasons, it warrants an extensive reexamination of the succession in synthetic view of sedimentary facies, fossil contents, metamorphic facies, and geologic structures.

Table 5.1 Lithostratigraphic and Biostratigraphic Summary of the Taebaek and Yeongwol Groups, Taebaeksan Basin

	Age	Taebaek Formation	Taebaek Biozone	Yeongwol Formation	Yeongwol Biozone
Ordovician	'Ashgillian'				
Ordovician	'Caradocian' Darriwilian 'Arenigian' Tremadocian	Duwibong		Yeongheung	*Kayserapsis*
		Jigunsan	*Dolerobasilicus*		
		Makgol			
		Dumugol	*Kayserapsis* *Protopliomerops* *Asaphellus*	Mungok	*Shumardia pellizzarii* *Kainella euryrachis* *Yosimurapsis vulgaris*
Cambrian	Furongian	Dongjeom	*Richardsonella* *Pseudokoldinioidia* *Eosaukia* Fauna	Wagok	*Fatocephalus hunjiangensis*
		Hwajeol	*Quadraticephalus* *Asioptychaspis*	Machari	*Pseudoyuepingia asaphoides* *Agnostotes orientalis* *Eochuangia hana* *Eugonocare longifrons* *Hancrania brevilimbata* *Proceratopyge tenuis* *Glyptagnostus reticulatus*
		Sesong	*Kaolishania* *Chuangia* *Prochuangia* *Neodrepanura* *Stephanocare*		*Glyptagnotus stolidotus* *Lejopyge armata* *Tonkinella*
	Middle	Daegi	*Cyclolorenzella* *Amphoton* *Crepicephalina*		
		Myobong	*Bailiella* *Mapania* *Elrathia* *Redlichia*	Sambangsan	*Megagraulos semicircularis* *Metagraulos smpoensis*
	Early	Jangsan/Myeonsan			

Source: Choi and Chough (2005), Kang and Choi (2007), Sohn and Choi (2007), Lee and Choi (2011).

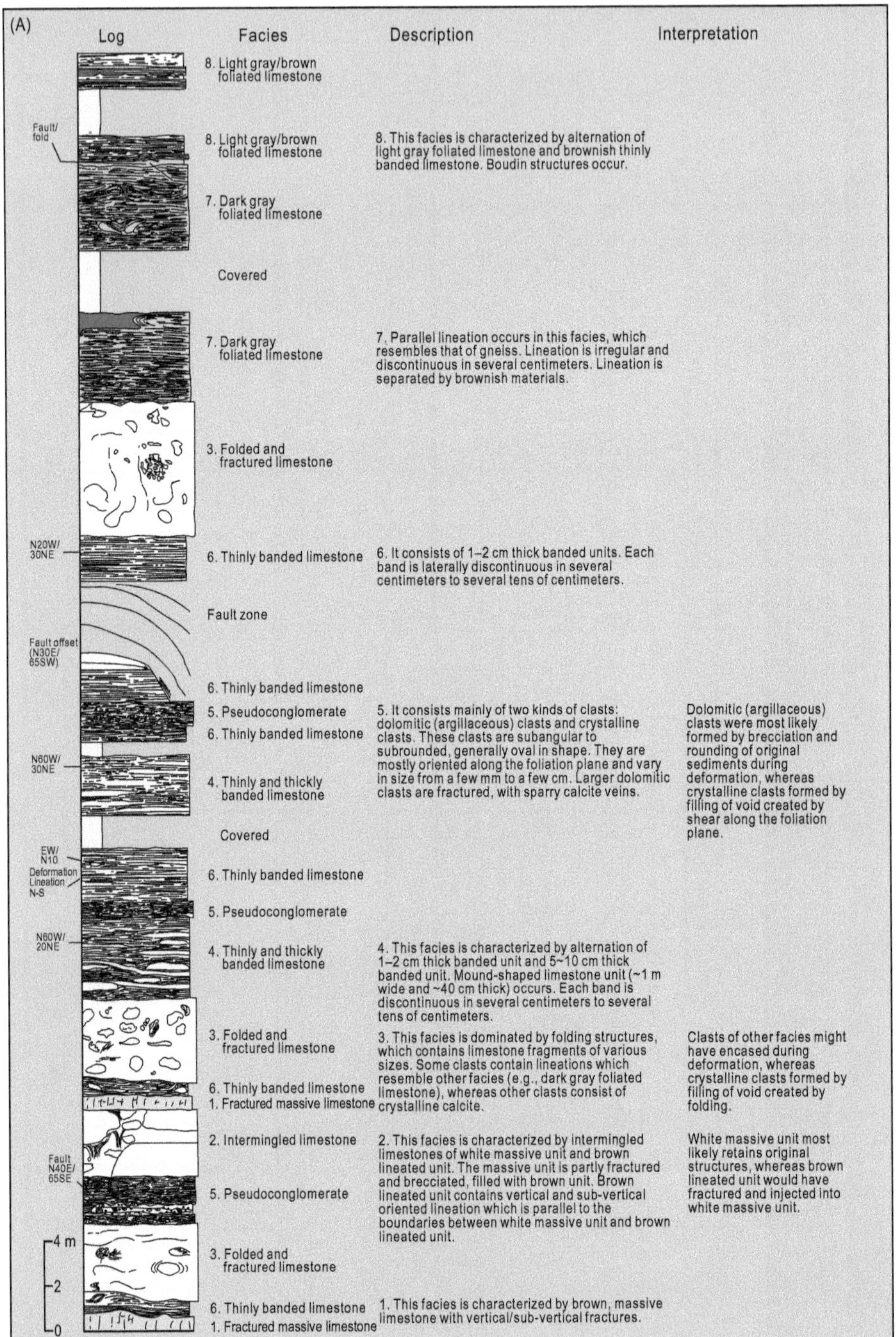

Figure 5.2 (A) Sedimentary log of metasedimentary rocks in the Jeongseon area. (B) 1. Fractured massive limestone. 2. Intermingled limestone. 3. Folded and fractured limestone. 4. Thinly and thickly banded limestone. 5. Pseudoconglomerate. 6. Thinly banded limestone.

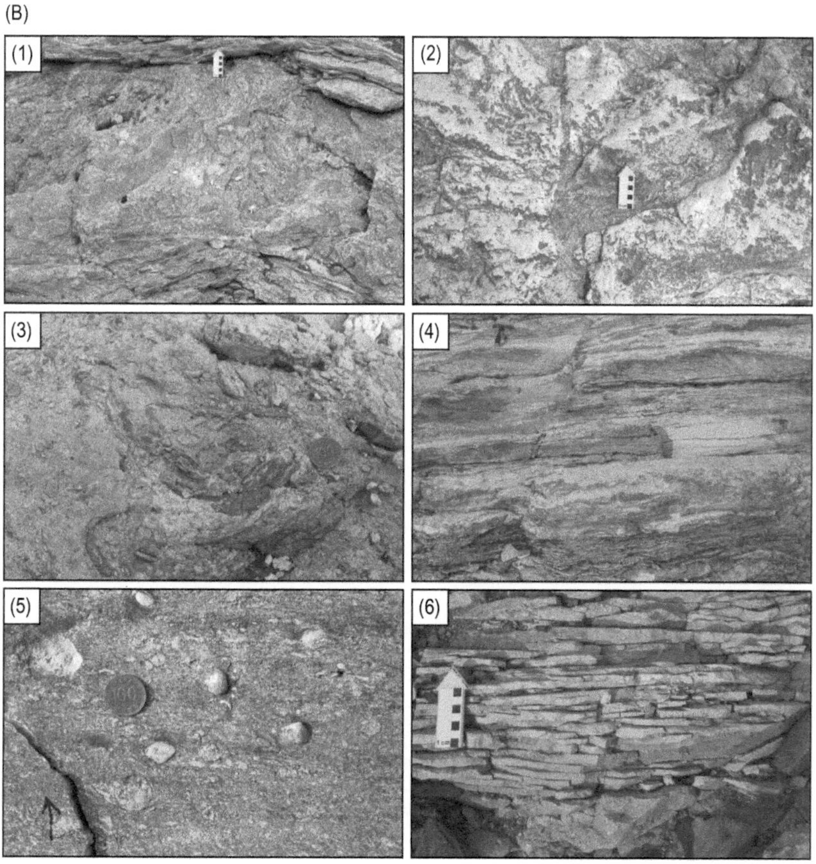

Figure 5.2 (Continued)

5.2 Taebaek Group

5.2.1 Introduction

The Taebaek Group consists of a thick succession of siliciclastic and shallow-marine carbonates (Kobayashi, 1966; Woo, 1999; Woo and Park, 1989), including a number of lithologic units (Table 5.2) (Choi et al., 2004). The succession unconformably overlies Precambrian granite gneiss and metasedimentary rocks of the Yeongnam Massif (Figure 2.9). The outcrops are well exposed along the road-cuts and mountain trails, including the Seokgaejae, Dongjeom, and Sangdong sections (Figures 5.3 and 5.4). The lower part of the succession consists of approximately 200-m-thick quartzite (Jangsan Formation), including a conglomerate and sandstone deposit (Myeonsan Formation). The Jangsan Formation is unfossiliferous, whereas the overlying Myobong Formation contains trilobites. The overlying Daegi Formation formed in the middle Middle Cambrian, whereas the Sesong Formation formed in the late Middle Cambrian to early Late Cambrian (Kobayashi, 1966). The Hwajeol Formation formed in the uppermost Cambrian (Choi et al., 2004). The Cambrian–Ordovician boundary most likely occurs within the *Richardsonella* fauna of the Dongjeom Formation (Choi et al., 2003). The Dumugol Formation formed in the Tremadocian

Table 5.2 Lithostratigraphy, Component Lithologies, and Biozones in the Taebaek Group

Age (Ma)			Formation (Thickness: meter)	Lithology	Biozone
Ordovician	445	Ashgillian			
	460	Caradocian	Duwibong (40-80)	Massive packstone to grainstone Bioturbated wackestone to grainstone Oncoid/ooid grainstone Limestone-shale couplet	*Acinoceroids*
	468	Darriwillian	Jigunsan (30-60)	Homogeneous mudstone Laminated lime-mudstone Massive or laminated grainstone Limestone pebble conglomerate	*Dolerobasilicus*
		Arenigian	Makgol (250-300)	Massive dolostone (breccia) Stromatolitic limestone Finely-laminated dolomitic limestone Bioturbated wackestone to grainstone	*Sigmorthoceras* *Polydesmia* *Manchuroceras* *Clarkella*
	478	Arenigian / Tremadocian	Dumugol (150-250)	Calcareous shale Limestone-shale couplet Bioturbated or flaser-bedded limestone Massive grainstone Limestone pebble conglomerate	*Kayseraspis* *Protopliomerops* *Asaphellus*
	488	Tremadocian	Dongjeom (80-120)	Massive sandstone Laminated or cross-stratified sandstone Calcareous shale Massive grainstone	*Richardsonella* *Pseudokoldinioidia* *Eosaukia Fauna*
Cambrian		Furongian	Hwajeol (50-150)	Calcareous shale Limestone-shale couplet Bioturbated or flaser-bedded limestone Massive grainstone Limestone pebble conglomerate	*Quadraticephalus* *Asioptychaspis*
	501	Furongian / Middle	Sesong (20-50)	Homogeneous or laminated mudstone Massive or laminated sandstone Limestone pebble conglomerate	*Kaolishania* *Chuangia* *Prochuangia* *Neodrepanura* *Staphanocare*
		Middle	Daegi (150-250)	Anastomosing wackestone to grainstone Massive grainstone Oncoid/ooid grainstone Homogeneous or laminated mudstone	*Cyclorenzella* *Solenoparia* *Crepicephalina*
	513	Middle	Myobong (70-200)	Homogeneous or laminated mudstone Laminated sandstone Massive sandstone Normally-graded sandstone	*Bailiella* *Mapania* *Elrathia* *Redlichia*
	542	Early	Jangsan/Myeonsan (50-200/40-60)	Massive sandstone Cross-or parallelly-stratified sandstone Disorganized conglomerate Foreset-bedded sandstone	

Source: Kwon et al. (2006).

and Early Arenigian (Kim et al., 1991a). The Makgol Formation is generally poorly fossiliferous, whereas the Jigunsan and Duwibong formations yield trilobites, graptolites, brachiopods, bivalves, gastropods, cephalopods, bryozoans, and ostracods. Conodont assemblages (Lee and Seo, 2004) suggest that the Makgol, Jigunsan, and Duwibong formations range in age from the Arenigian to the Caradocian (Table 5.2).

Sedimentation was initiated in the Early Cambrian with global rise in sea level on the stable Sino-Korean Block (Kwon et al., 2006). Erosion rates were high since there were no plants on the surrounding lands and islands. The siliciclastics formed mainly in shoreface environments under a strong tidal regime. During the long-standing transgression for tens of millions of years, the shallow platform was under mild climatic conditions with an explosive production of calcium carbonates by organisms such as algae, crinoids, brachiopods, mollusks, and trilobites. In subtidal and intertidal environments, cyanobacteria formed mounds and trapped carbonate mud. Shale and fine-grained carbonates formed

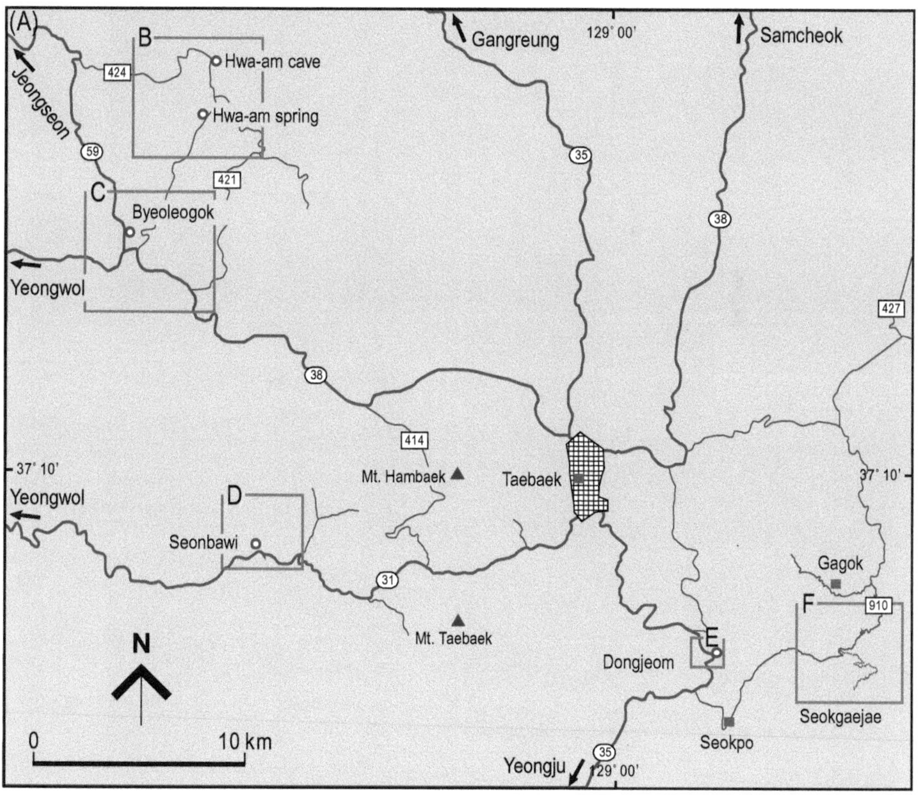

Figure 5.3 (A) Road map showing the locations of outcrop sections. (B) Hwa-am area. (C) Byeoleogok area. (D) Seonbawi section. (E) Dongjeom section. (F) Seokgaejae section. *Source*: Geologic maps modified from the GICTR (1962) and Chough et al. (2000)

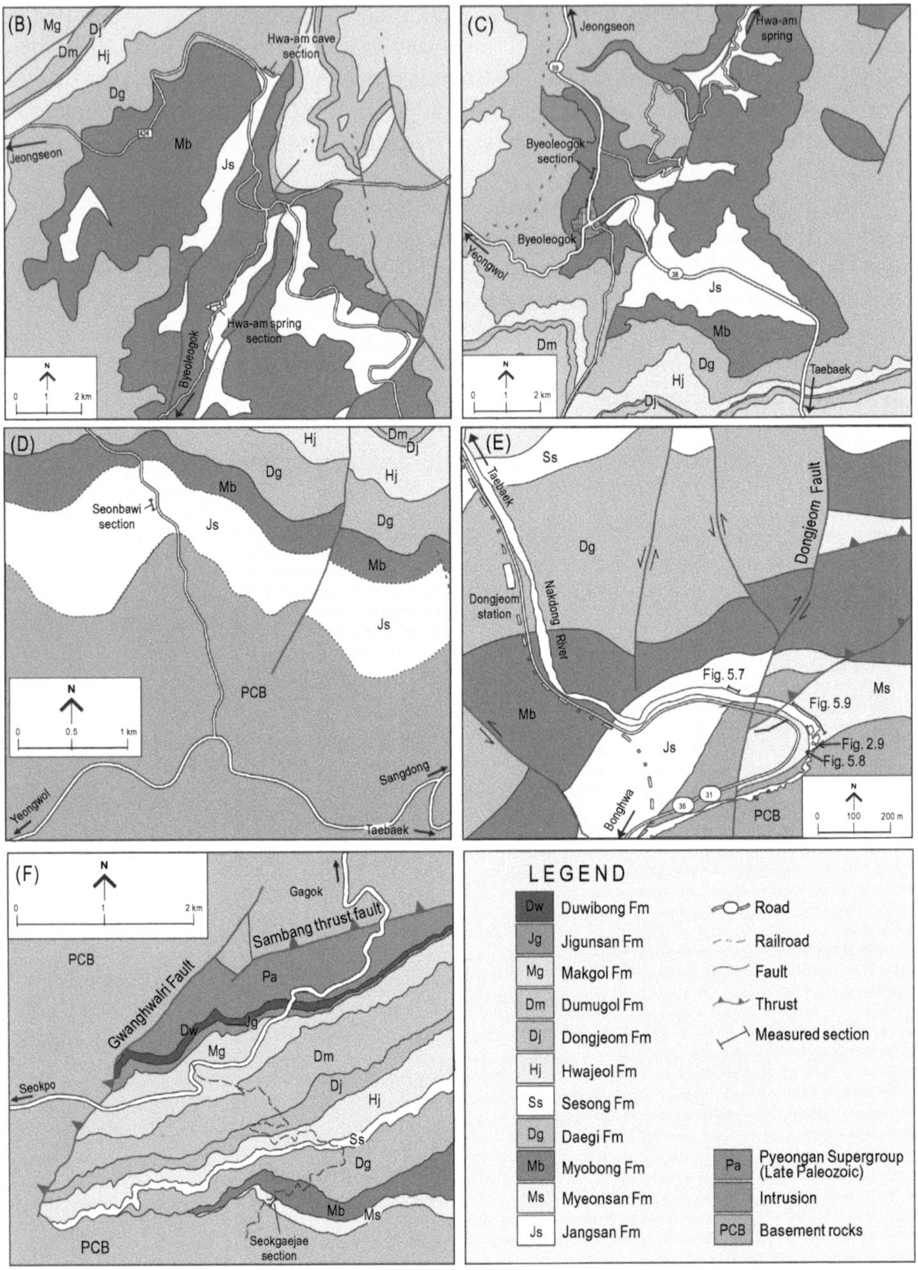

Figure 5.3 (Continued)

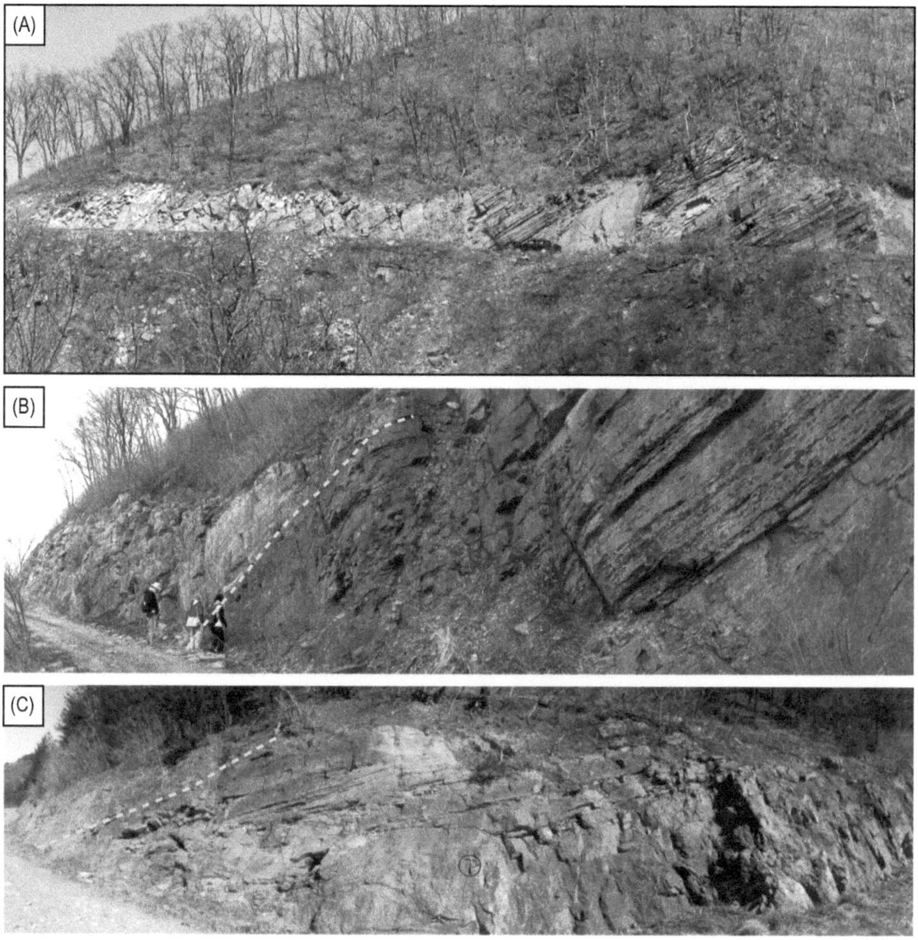

Figure 5.4 Panoramic views of the Seokgaejae section. See Figure 5.3 for location. (A) Inclined beds of limestone, oncoid, and oolite of the Daegi Formation, underlain by shale of the Myobong Formation. (B) The boundary between the lower and upper members of the Daegi Formation showing 10-m-thick mudstone unit overlain by carbonate facies. (C) Sponge bioherms and the associated facies in the uppermost part of the Dumugol Formation. Dashed line indicates the boundary between the Dumugol and Makgol formations. Hammer for scale (circle) is 27 cm long. (D) Road-cut exposure along the provincial road 911 showing the middle and upper members of the Makgol Formation. (E) The uppermost part of the Seokgaejae section showing a thick succession of carbonates of the Duwibong Formation. The yellow arrow denotes the boundary between the Jigunsan and Duwibong formations. (For interpretation of the references to color in this figure legend, the reader is referred to the web version of this book.)

Figure 5.4 (Continued)

in ramp and peritidal environments. These shallow ramp strata include numerous meter-scale shallowing-upward cycles.

5.2.2 Jangsan Formation

5.2.2.1 Sedimentary Facies

Sedimentary facies can be observed in the sections of the Hwa-am, Byeoleogok, Seonbawi, and Dongjeom areas (Figure 5.3). The sedimentary facies include massive and crudely stratified sandstone, cross-stratified sandstone, and foreset-bedded sandstone.

Massive and Crudely Stratified Sandstone

This facies comprises well-sorted medium coarse to very coarse quartzose massive and crudely stratified sandstone. Sand grains comprise spherical and angular quartz and heavy minerals (zircon and tourmaline); clasts of metasedimentary rocks occur in lesser amounts (Figure 5.5). Each facies unit is usually tabular and up to several meters thick. This facies most likely formed by rapid sedimentation of sands from suspension and bedload transport during strong currents and storms in shoreface environments.

Cross-Stratified Sandstone

It comprises well-sorted medium-to-coarse sandstone. The cross-strata are several decimeters in wavelength and about 10 cm in height. The dip directions of the trough cross-beds are mostly unidirectional to weakly bidirectional. This facies is characterized by tabular small-scale cross-sets with relatively high-angle (about 20°) laminae. The base shows a low-angle concave-up truncation surface. The tabular cross-stratified beds most likely formed by migration of two-dimensional dunes under unidirectional currents, whereas the trough ones formed by small three-dimensional dunes. The smaller tabular cross-sets resulted from two-dimensional dunes.

Foreset-Bedded Sandstone

This facies comprises foreset-bedded quartzose sandstone. The foreset is slightly tangential to the base and angular to the top (Figure 5.6A and B). Symmetrical ripples (wavelength of about 5 cm) occur on the bed surface (Figure 5.6C). The ripple crests are slightly rounded and straight. This facies probably formed by migration of subaqueous dunes. The tabular cross-beds are indicative of migration of two-dimensional dunes. The symmetric and round-crested ripples most likely formed under oscillatory waves.

5.2.2.2 Depositional Model

Cross- and tabular sets of meter-scale foreset beds, alternating with horizontally bedded massive sandstone, are indicative of deposition in the lower shoreface and inner shelf region where active migration of dunes prevails (Figure 5.7). Massive sandstone, characterized by abundant pebble-grade clasts and wedge-shaped geometry, and stratified sandstone show a consistent southwestward dip direction

Figure 5.5 Pebble-bearing very coarse sandstone (quartzite) of the Jangsan Formation, Hwa-am cave section. Pebbles from a thin layer. Lens cap is 5.2 cm in diameter. For location, see Figure 5.3.

which is suggestive of west- to southwestward currents, although the position of the paleoshoreline is uncertain. The extensive tabular sets of the foreset-bedded sandstone suggest, however, that currents lasted for considerable duration. The facies assemblages of low-angle trough cross- and horizontally stratified sandstones, as well as the frequent occurrence of medium-angle trough cross-beds, are also indicative of shoreface environments dominated by subaqueous dunes (Figure 5.7). The consistent unidirectional trough cross-stratifications with subordinate ripples are also indicative of wave-generated longshore currents. Sporadic offshore storm currents may have also caused migration of shoreface dunes. The well-sorted planar-stratified sandstone with very low-angle set boundary is probably suggestive of deposition in a swash zone of foreshore environments. The finer cross-bedded facies most likely formed by migration of various dunes in the inner shelf or lower shoreface, whereas the overlying coarser massive pebbly sandstone formed by storm activity.

5.2.3 Myeonsan Formation

5.2.3.1 Sedimentary Facies

The Myeonsan Formation is well exposed in the Dongjeom section (Figure 5.3E). It comprises disorganized conglomerates and cross-bedded and laminated sandstones (Woo et al., 2006).

Disorganized Conglomerate

This facies occurs in the lowermost part, unconformably overlying the Precambrian basement rocks (Figure 5.8). It is commonly disorganized and composed of pebble- to cobble-grade clasts of quartzite, slate, and gneiss. Quartzite and gneiss clasts are up to a meter in diameter. The matrix-rich conglomerate alternates with clast-rich

Figure 5.6 Photographs of the Jangsan Formation, Byeoleogok section. (A, B) Prograding foreset beds. The contact of the lower surface is tangential. Note thin fine sandstone layers between the foreset-bedded units (arrows). (C) Ripples on the surface of foreset beds. Coin is about 2.2 cm in diameter. Inset shows dip directions and orientation of ripples (broken line). For location, see Figure 5.3.

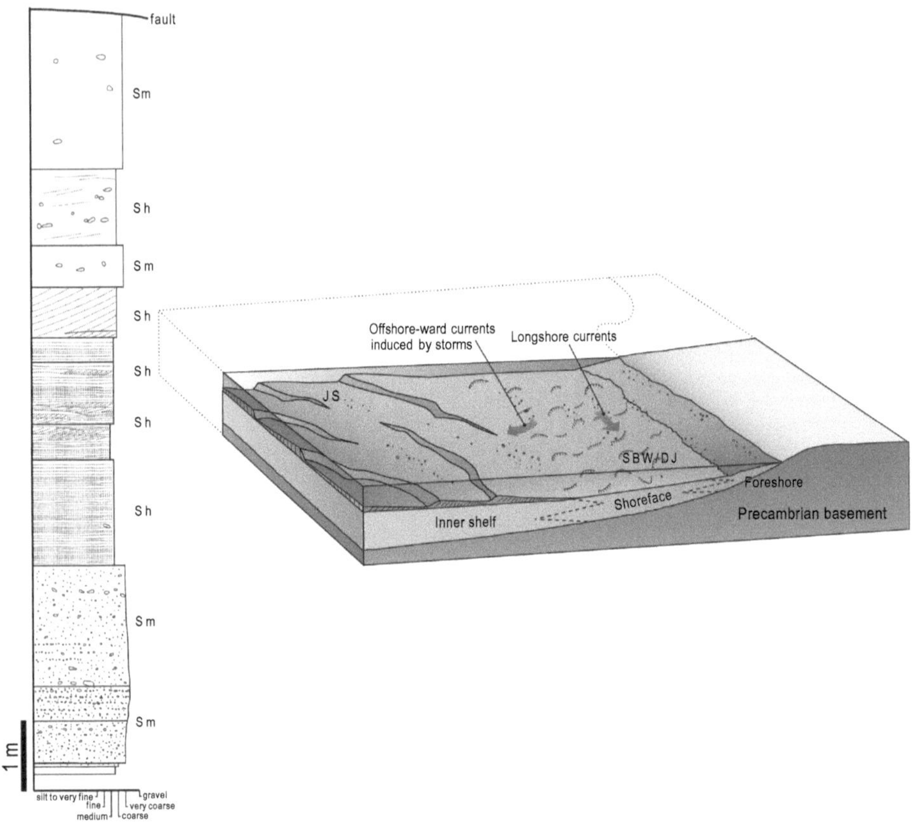

Figure 5.7 Sedimentary log of the Jangsan Formation, Dongjeom section, and depositional model of the Jangsan Formation (for abbreviation, see Table 5.3). JS, Jeongseon; SBW, Seonbawi; DJ, Dongjeom.
Source: Woo et al. (2006) by permission of *Geosciences Journal*.

conglomerate. Poor sorting and the dominant sand matrix are representatives of deposition from sandy debris flows.

Thin-Bedded Conglomerate

This facies consists mainly of granule- to pebble-grade mudstone clasts and quartz grains with sandy matrix (Figure 5.9). Clasts are subrounded to angular, poorly sorted, and slightly parallel to bedding plane. The scoured walls are steep and commonly exhibit a sharp erosional base. The mudstone clasts at the base are interpreted as channel lag deposits.

Trough Cross-Bedded Sandstone

This facies consists of relatively thick (0.1–0.5 m) sandstone beds, composed mainly of poorly sorted, medium to very coarse sandstone and pebbly sandstone.

Figure 5.8 The lowermost unit of the Myeonsan Formation overlying Precambrian basement rock. The matrix-rich unit with sparse clasts is overlain by a clast-rich unit with an indistinct boundary.

The sandstone is dominantly arkosic arenite and contains feldspar (less than 40%) and detrital grains of hematite and ilmenite. This facies displays small- to medium-scale sets of cross-beds (flaser, wavy, and lenticular) (Figure 5.9). The cross-beds are separated by a mud drape with a tangential contact at the base. The mud-draped cross-beds indicate suspension settling during periods of slack water. The unidirectional pattern of the cross-beds is indicative of the prevalence of either ebb or flood currents.

Ripple Cross-Laminated Sandstone

This facies is characterized by fine- to coarse-grained sandstone with thin intercalation of laminated mudstone. The lower boundary appears to be sharp and erosional with partial lag deposits, whereas the upper boundary is gradational. Granule-grade sand grains and mudstone chips occur in the ripple trough. The wavy to lenticular beds with ripple cross-laminae reflect alternating periods of ripple migration and suspension settling during tidal cycles. The wave- or current-generated ripple beds are common in sandy tidal flats or relatively shallow tidal channel margins.

Planar to Wavy-Laminated Sandstone

This facies consists of parallel-laminated fine sandstone. Individual laminae are slightly wavy to irregular and can be traced laterally for a few meters. The distinct alternation of laminated mudstone layers is suggestive of suspension settling of mud during intermittent pauses in wave- or wind-generated currents.

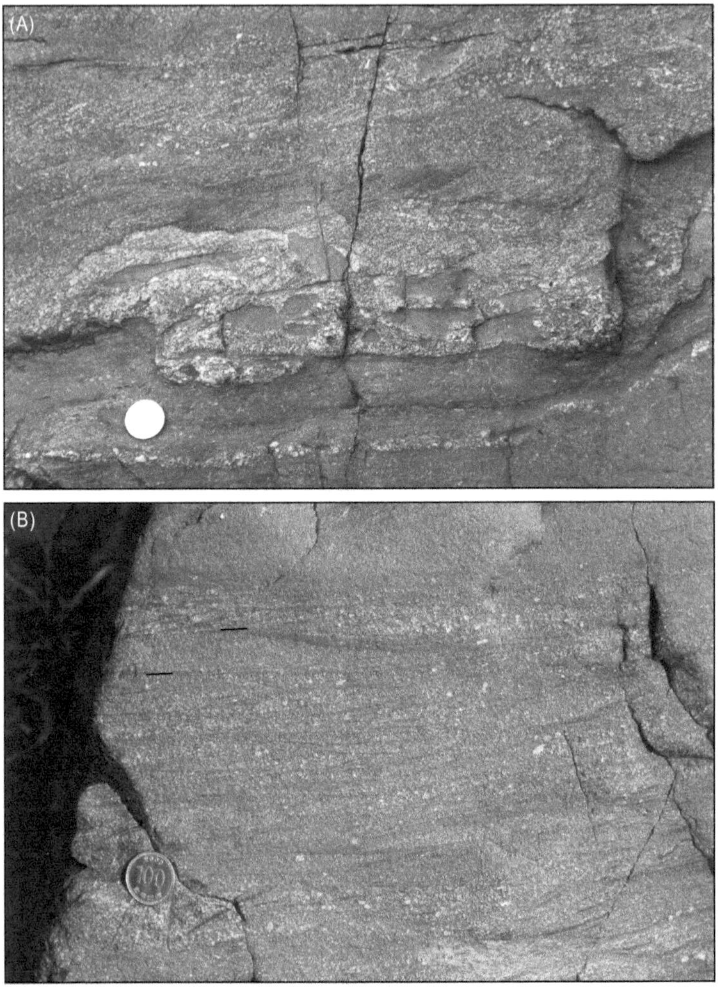

Figure 5.9 Outcrop photographs of the Myeonsan Formation. (A) Thin-bedded conglomerate with rip-up clasts. (B) Trough cross-bedded sandstone. Cross-beds are interlaminated with thin mud drapes. Coin for scale is 2.2 cm in diameter.

5.2.3.2 Depositional Environments

Ungraded clasts and sandy matrices of the disorganized conglomerate suggest that deposition occurred from cohesionless sandy debris flows. The poor lateral continuity of the conglomerate beds is suggestive of localized mass-flow events. The overlying trough cross-bedded, ripple cross-laminated, and mud-draped, cross-bedded sandstones are collectively suggestive of deposition in subtidal channels. Unidirectional dune cross-beds formed by ebb- or flood-dominant tidal currents in the channels (Figure 5.10). Ripple-topped sandstone beds most likely formed

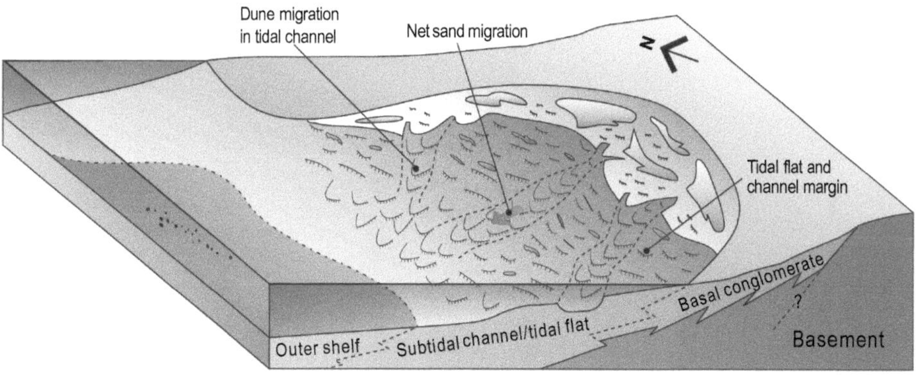

Figure 5.10 Depositional model of the Myeonsan Formation.
Source: Woo et al. (2006) by permission of *Geosciences Journal*.

during the upward growth of dune crests. Thin sandstone beds with a variety of small-scale cross-beds, which commonly alternate with thin beds of laminated or massive mudstone, represent sand–mud couplets of tidal origin. Trough cross-bedded gravelly sandstones with a large number of pebble-grade mudstone clasts in channel-shaped beds most likely represent tidal gullies (Figure 5.10).

5.2.4 Myobong Formation

The Myobong Formation (about 100–200 m thick) consists of greenish gray homogeneous mudstone and shale, intercalated with thin massive or normally graded sandstone layers (Table 5.3; Figure 5.11B). Layers of siltstone and fine sandstone commonly show a sharp erosional base. Mudstone facies are interpreted as deposits of hemipelagic settling, whereas massive or graded sandstone layers are interpreted as deposition by density and turbidity currents.

5.2.5 Daegi Formation

5.2.5.1 Lower Part

The lower part of the Daegi Formation comprises a grainstone-dominated succession, including oncoid/ooid grainstone, massive grainstone, and gravelly grainstone (Figure 5.11C). It is overlain by calcareous shale and limestone–shale couplet with intercalation of hummocky-stratified gravelly grainstone, wackestone to grainstone, and bioturbated wackestone to grainstone (Figure 5.11D). The grainstone is composed of spherical or oval oncoids, ranging from 2 to 10 cm in diameter, and ooids with concentric structures (1–3 mm in diameter) (Table 5.3) (Figure 5.11C).

This deposit most likely formed in a subtidal shoal environment under high-energy conditions. Ooids and oncoids are suggestive of subtidal shoal environments, less than 5 m in water depth. The overlying shale-dominated succession represents

Table 5.3 Sedimentary Facies of Taebaek Group

	Description	Interpretation
Siliciclastic facies		
Disorganized conglomerate facies (Cd)	Pebble- to cobble-grade, rounded to subrounded clasts of quartzite, granitic gneiss, and granite; disorganized, partly slightly imbricated and partly normally graded; tabular beds and irregular, sharp and erosional base	Basal lags
Foreset-bedded sandstone facies (Sf)	Medium to coarse sandstone; tabular and laterally extensive; about 1.5 m thick in average; slightly tangential to the base of the set; unidirectional; associated with pebbly sandstone/conglomerate, and horizontally	Migration of two-dimensional dune and superimposed oscillatory waves
Horizontally stratified sandstone facies (Sh)	Stratified coarse sandstone; ripples on top of some foreset beds moderately sorted, medium to coarse sandstone; alternated with dark gray granular bands or pebble trains; crudely laminated or stratified with various spacing between layers separated by discontinuous granule layers	Deposition from strong currents
Cross-stratified (bedded) sandstone facies (Sx)	Dark gray, medium to coarse sandstone with granule- to pebble-grade gravel lags; inclined, discontinuous granule bands (layers) or fine sand layers; bounded by flat base and upper boundary; commonly capped by relatively thin (about 10–20 cm thick) Sl layers	Deposition from migration of dunes
Massive sandstone facies (Sm)	Fine to coarse sands with thin laterally discontinuous granule- to pebble-grade clast trains; slight bioturbations; alternate with Sl with diffuse unit boundary	Rapid suspension settling and bioturbation
Laminated sandstone facies (Sl)	Well-sorted fine to medium sandstone; parallel laminations; distinct boundaries and some intercalated coarser-grained layers are laterally discontinuous; interlayered with Sm or Ss/Sx with diffuse lower and distinct upper boundaries	Shoreface deposition from waves and currents

Normally graded sandstone facies (Sg)	Normally graded units of moderately to poorly sorted fine to coarse sands; granule- or pebble-graded gravels in the lower part of the bed; rip-up mud clasts; some clasts as trains in the middle and upper parts of the bed; commonly overlies mudstone or fine sandstone with a sharp and erosional lower boundary	Deposition from storms or turbidity currents
Crudely stratified calcareous sandstone facies (Scs)	Light gray to brownish light gray coarse sandstone, massive to crudely stratified and more calcareous than Scx; floated quartz grains in calcite cement; small carbonate pebbles	Shoreface deposition from waves and currents
Cross-stratified calcareous sandstone facies (Scx)	White to light brown colored coarse to very coarse sandstone cemented with calcareous matrix; partly herring-bone cross-stratified	Migration of low amplitude bedforms
Intermingled fine to medium sandstone facies (Si)	Light colored fine- to medium-grained (calcareous) sandstone intermingled with dark gray siltstone to fine-grained sandstone; poorly to well stratified depending on the ratio between the dark colored/lighter colored sediments; lenticular, nodule-like, or patch-like and partly finely laminated lighter colored parts; irregular or chaotic (confluent/bifurcated) darker colored fine-grained part; diffuse boundary between darker and lighter colored sediments; bioturbated	Deposition by wave- and tide-induced currents; intermingled by bioturbation
Laminated fine sandstone facies (fSl)	Fine-grained sandstone-laminated by fine- to medium thickness dark gray mudstone laminae; generally massive but occasional parallel lamination or normal grading; generally planar, continuous, partly wavy, discontinuous or confluent/bifurcated mudstone laminae; characterizing the transition interval between the Dongjeom and Hwajeol formations	Shoreface deposition by storm-, wave-, and tide-induced currents
Dark gray homogeneous mudstone facies (Mhd)	Dark or blackish gray homogeneous mudstone; ranges in thickness from a few centimeters to a few decimeters; commonly interbedded with Sg and thin-bedded Sm; commonly intercalated thin, laterally discontinuous very fine to fine sand layers; pyrite and platy mica	Deposition by hemipelagic settling

(*Continued*)

Table 5.3 (Continued)

	Description	Interpretation
Greenish gray homogeneous mudstone facies (Mhg)	Light or dark, greenish gray homogeneous mudstone; partly intercalated thin, laterally discontinuous fine to very fine sand layers; some intensively bioturbated units; mainly associated with Sg or Sm	Deposition by hemipelagic settling
Parallel-laminated mudstone facies (Ml)	Very thinly laminated greenish gray mudstone; ranges in thickness from a few decimeters to several meters; thin, laterally continuous or discontinuous very fine- to silt-grade grain layers; interbedded with Mhg, Sg, and Sm	Alternation of background settling deposits and distal siltstone by storms or currents
Nodule-bearing shale facies (Sn)	Alternation of calcareous shale and discontinuous oval-shaped calcareous nodule layers; common parallel laminations; various nodules shapes, such as tabular, oval, and spherical; some nodules with parallel laminations and medium-sand-grade lags; abundant fossils of trilobites, brachiopods, and cephalopods	Deposition from turbidites or suspension settling
Calcareous shale facies (CS)	Brownish gray to reddish gray in color; either crudely laminated or homogeneous; occasional fine, parallel, nongraded laminations and thin-bedded; consisting of calcite, dolomite, and clay minerals, such as illite, chlorite, and smectite; highly variable in thickness, ranging from 10 cm to 2 m; slightly bioturbated; rare fragments of trilobites, echinoderms, and brachiopods; gravelly grainstone layers in some facies units; few tabular calcareous nodules	Deposition by hemipelagic settling
Limestone–shale couplet facies (L–S)	Alternation of planar or nodular, light gray limestone layer and dark gray dolomitic shale layer; limestone layers consisting of peloids and bioclasts of echinoderms and brachiopods; gradational transition between limestone and dolomific shale layers; dark gray or brownish dolomitic shale layers with rhombohedral dolomite crystals and argillaceous muds; anastomosing micro-stylolites; well-preserved fossil fragments in some limestone layers	Annual varves or decadal (centennial) bandings

Carbonate facies		
Crudely laminated lime-mudstone facies (LMcl)	Crude alternation of lime-mudstone and dolomitic argillaceous (or organic-rich) laminae; less than 5 mm in thickness and indistinct boundaries; few peloids and intraclasts (0.25–2.5 mm in diameter) in lime-mudstone; pyrites commonly associated with stylolitic structures	Suspension settling in low-energy condition such as lagoon or deep subtidal platform
Stromatolitic lime-mudstone facies (LMs)	Two distinctive forms of stromatolites (cryptalgal stromatolites and laterally linked hemispheroid (LLH) stromatolites); alternation of algal-rich layer and sediment-rich layer; parallel to bedding plane and 0.5- to 2-mm-thick individual laminae; dolomitization of algal-rich layer; common association with bird's eye structures, evaporite mineral casts, and desiccation cracks	Intertidal deposition
Finely laminated dolomitic lime-mudstone facies (LMdl)	Bluish gray in color and horizontally laminated; consisting of alternations of algal-rich silty dolomitic layer and carbonate-rich layer; parallel to bedding plane and 0.5–2-mm-thick individual laminae; containing dark brown to red stylolites; ranges in thickness from 10 to 50 cm; anhydrite and gypsum casts; partly stromatolitic	Supratidal dolomites, formed by strong currents
Flaser wackestone to packstone facies (W/Pf)	Alternation of crudely laminated, bluish gray limestone layer and dark gray argillaceous layer; 1 cm to a few centimeters thick limestone layer and thinner argillaceous layer; limestone layer consisting of lime-mudstone, wackestone and packstone; ripple cross-laminations and parallel laminations; normal grading with lags of peloids, intraclasts, and fossils; argillaceous layer composed of clay minerals, micrites, and dolomite rhombs; wispy "stylolite-like" seams and micro-stylolite swarms; bioturbated	Subtidal deposition

(*Continued*)

Table 5.3 (Continued)

	Description	Interpretation
Anastomosing wackestone to grainstone facies (W/Ga)	Wackestone to grainstone with interlayered shale; anastomosing structures by interlayered shale; very abundant fossils, such as algae, brachiopods, trilobites, and crinoids, and commonly interbedded with Gm and Go layers; intensively bioturbated; severely deformed by compaction and water escape; stylolites very common	Subtidal deposition and modification by compaction
Bioturbated wackestone to grainstone facies (W/Gb)	Dark gray wackestone with mottled texture caused by selective dolomitization of burrows; ichnofacies index-3 or index-4; more than 1 m in thickness; bioclastic fragments, including trilobites, echinoderms, and gastropods; horizontal or subhorizontal and subordinately vertical burrows replaced by fine to medium-sized euhedral dolomites	Subtidal deposits modified by bioturbation
Massive pack-grainstone facies (P/Gm)	Massive mixture of skeletal grains and lime mud/sparry calcite matrix; skeletal grains such as brachiopods, cephalopods, gastropods and trilobites; partly laminated	Deposition in moderately agitating shallow subtidal platform
Laminated calcisiltite facies (CZl)	Parallel, ripple cross-laminated or low-angle cross-laminated (hummocky or swaley) laminae of alternating calcareous and shaly materials; calcareous lamina consist of silt-sized particles, darker argillaceous lamina consist of unclear finer grains; distinct, sharp and erosional bases of each unit; faint, gradational, and occasionally wavy upper boundaries horizontal, oblique and v-shaped burrows each unit from several millimeters to 15 cm in thickness	Unidirectional current with oscillatory movement of water induced by storm on low-angle platform
Wavy and lenticular grainstone–mudstone couplet facies (G–M)	Wavy and lenticular beddings and ripple cross-laminations with bidirectional dip; alternations of grainstone layer and lime-mudstone layer; consisting mainly of peloids, fossil	Tidal couplets

	fragments, and intraclasts of peloidal grainstone and subordinately of ooids and quartz; continuous wavy thin beds to isolated lenses: a flat-bottom and a ripple-form top; continuous lime-mudstone interlayer; draping ripple-form tops of underlying bed by thinning over the ripple crests and thickening over the ripple troughs	
Massive grainstone facies (Gm)	Consisting of fossil fragments such as brachiopods, echinoderms, trilobites, ostracods, and gastropods and subordinately of peloids, ooids, and intraclasts; lime-micrites, marine fibrous calcite cements, and meteoric equant calcite cements; ranges in thickness from 10 cm to 1.4 m; common sharp base; diffuse and gradational upper boundary; tabular, but occasional channel geometry; including hardgrounds with sharp upper and diffuse lower contacts	Deposits of storm sedimentation
Reef mound facies (RFm)	Lithistid sponges, stromatolites, void-filling lime muds, and bioclastic debris; forming skeletal mounds and having a concave base and a planar (or irregular) top, averaging 1 m in height and 2 m in length; capped by limestone–shale couplet facies; transitional or abrupt laterally change into thin-bedded bioclastic grainstone, packstone, dolomitic grainstone, and pebble conglomerate; some grainstone facies showing hummocky cross-stratification with bioturbation; pebble conglomerates comprising granule- to pebble-grade clasts; mostly flat-lying and subordinately imbricated	Bioherms in abandoned or sheltered channels
Dark gray massive dolostone facies (Dmd)	Showing various dolomitization fabric from slight to complete dolomitization; slight dolomitization: generating fine-crystalline dolomites and preserving precursory texture such as lamination; common in finely laminated dolo-mudstone; individual laminae consisting of clay minerals, silt-sized quartz, and coarser	Dolomitization by a mixed water model

(*Continued*)

Table 5.3 (Continued)

	Description	Interpretation
	dolomite crystals; bird's eye structures and evaporite mineral casts complete dolomitization: obliterating precursory texture and producing medium to coarse crystalline dolomites; common reddish brown stylolites	
Light gray massive dolostone facies (Dml)	Massive in fabric and light bluish gray in color; dolomitization to preserve precursory texture such as lamination to complete dolomitization to obliterate precursory texture; slight dolomitization generating fine-crystalline dolomites mostly less than 20 μm in size; occurring in the upper part of the Makgol Formation, associated with Dmd	Dolomitization of peritidal carbonates
Laminated dolostone facies (Dl)	Crude lamination in dark gray or light gray massive dolostone; parallel to bedding plane; highly variable in intervals of laminae, 0.5–3-mm-thick individual laminae	Dolomitization of finely laminated carbonates
Dolostone breccia facies (Bd)	Mosaic dolostone breccia: autoclastic, white gray, massive dolostone breccia with many evaporite mineral casts and bird's eye structure; mosaic-fit, granule- to pebble-grade tabular clasts; muddy matrices; laterally discontinuous and variable in thickness, ranging from 20 to 60 cm; gradational or irregular flat-bottom and flat-top	Deposits of solution-related brecciation

Source: Kwon et al. (2006) by permission of the Elsevier.

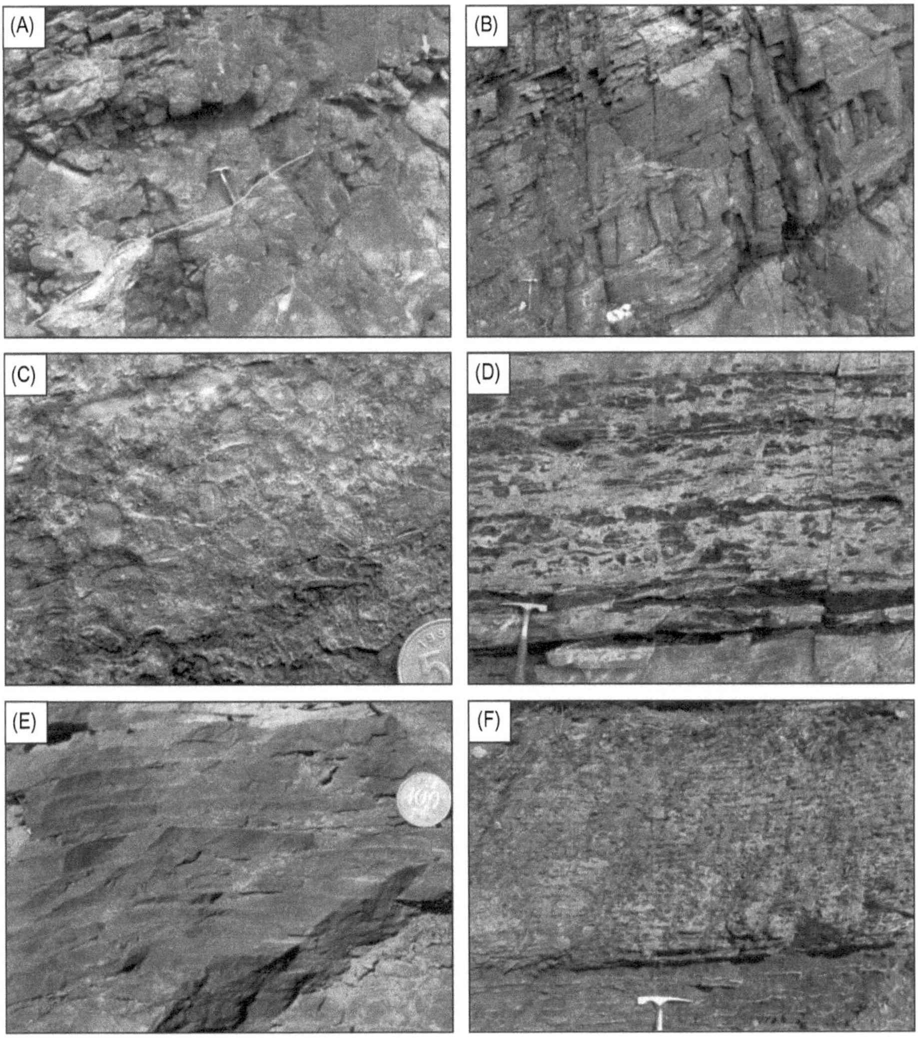

Figure 5.11 Representative outcrop photographs of the Taebaek Group in the Seokgaejae section. (Hammer for scale is 27 cm long; coin for scale in (E), (G), (H), (L), and (R) is 2.4 cm in diameter; coin for scale in (Q) is 2.3 cm in diameter; coin for scale in (C) is 2.1 cm in diameter; scale bar in (P) is in cm.) (A) The boundary (yellow line) between the Precambrian basement and the unconformably overlying disorganized conglomerate (Cd) of the Myeonsan Formation. (B) Typical lithologic feature of the lower part of the Myobong Formation comprising alternation of laminated mudstone (Ml) and sandstone facies (Sm, Sl, Sg, and Sx). (C) Oncoid grainstone (Go) in the lowermost unit of the Daegi Formation. (D) Nodule-bearing shale (Sn) facies common in the lower member of the Daegi Formation. (E) Laminated sandstone (Sl) facies of the Sesong Formation. (F) Nodule-bearing shale (Sn) facies of the Hwajeol Formation.

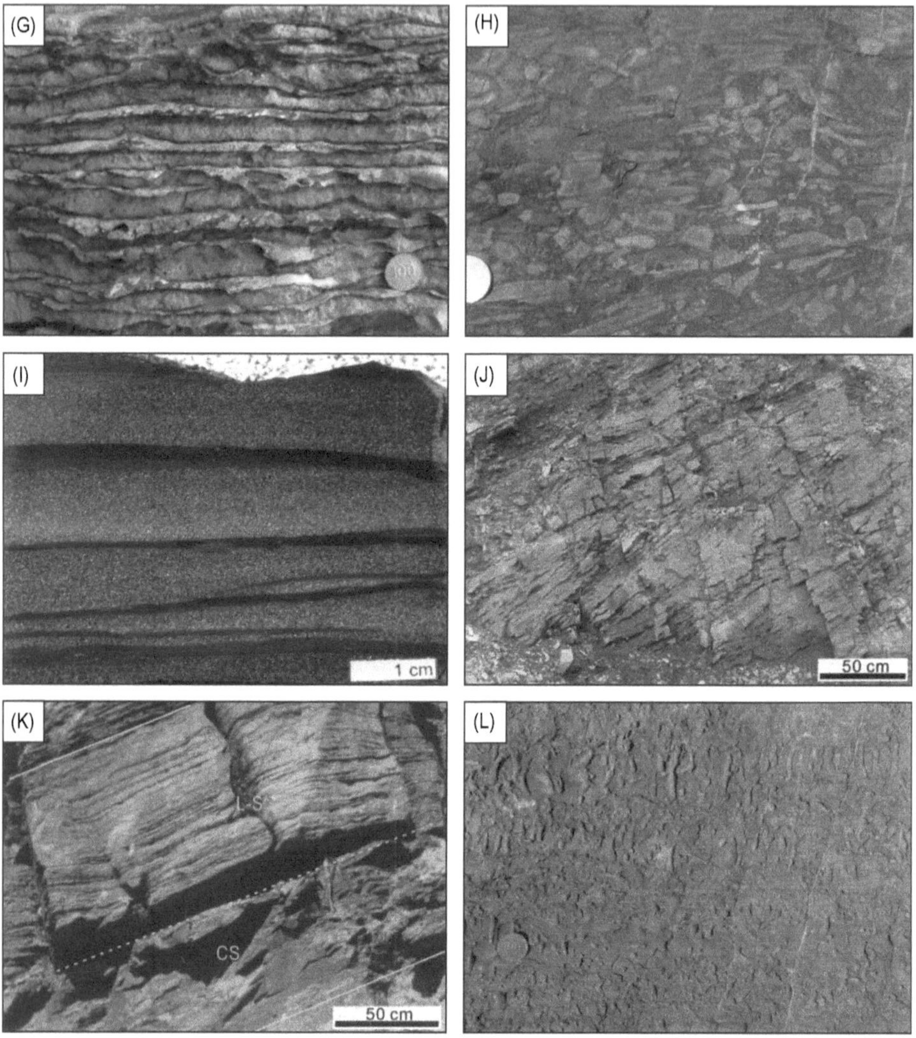

Figure 5.11 (Continued). (G) Limestone–shale couplet (L–S) facies of the Hwajeol Formation. (H) Limestone pebble conglomerate (Clp) facies of the Hwajeol Formation showing subparallel clast fabric and bioclast-bearing sparitic matrix. (I) Laminated fine sandstone (fSl) facies of the Dongjeom Formation. (J) Cross-stratified calcareous sandstone (Scx) in the Dongjeom Formation. (K) A meter-scale subtidal cycle in the Dumugol Formation consisting of calcareous shale (CS) and limestone–shale couplet (L–S) facies. (L) Bioturbated wackestone to grainstone (W/Gb) facies.

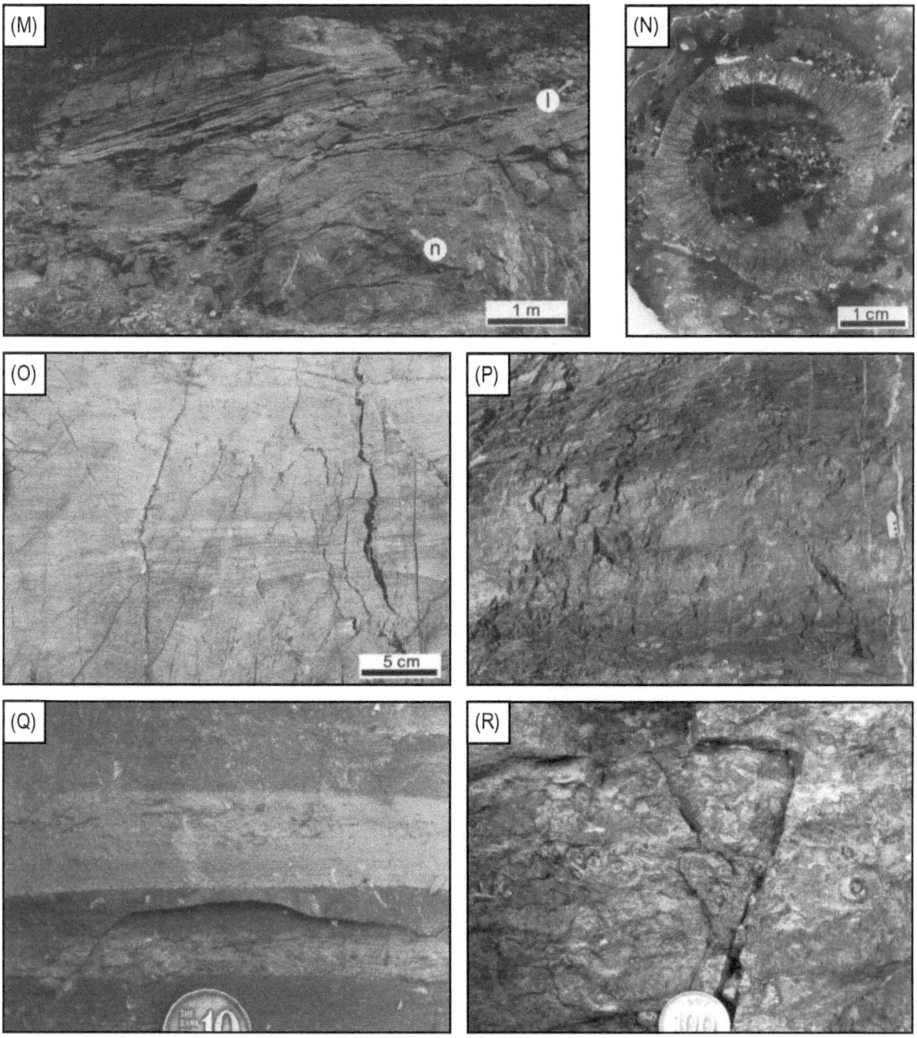

Figure 5.11 (Continued). (M) Sponge bioherm (RFm) and associated facies in the Dumugol Formation. Arrows with letter indicate positions of photographs (L) and (N). (N) Transverse thin section of reef building sponge Archaeoscyphia showing radially arranged trabs around spongocoel. (O) Stromatolitic lime-mudstone (LMs) facies of the Makgol Formation. (P) Alternating crudely laminated lime-mudstone (LMcl; light gray in the middle of photograph) and dark gray homogeneous mudstone (Mhd; dark gray in the upper and lower parts of photograph) facies in the lowermost part of the Jigunsan Formation. (Q) Laminated calcisiltite (CZl) facies alternating with dark gray homogeneous mudstone (Mhd) facies of the Jigunsan Formation. (R) Massive packstone to grainstone (P/Gm) characterized by abundant bioclasts (e.g., gastropods and cephalopods). (For interpretation of the references to color in this figure legend, the reader is referred to the web version of this book.)

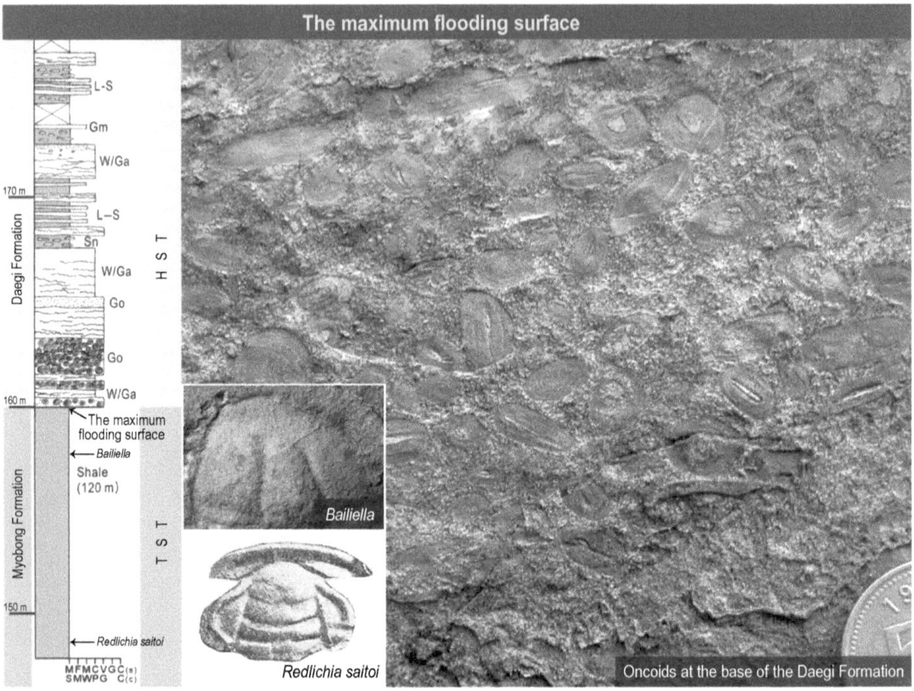

Figure 5.12 Maximum flooding surface between the Myobong and Daegi formations in the Seokgaejae Section. TST, transgressive systems tract; HST, highstand systems tract. For sedimentary facies code, see Table 5.3. For section location, see Figure 5.3.

progradation of fine-grained sediments in low-energy environments (backshoal/ lagoon). The oncoid and ooid layer, the first carbonate bed of the Taebaek Group, represents the maximum flooding surface (Figure 5.12).

5.2.5.2 *Upper Part*

The upper part of the Daegi Formation is characterized by peritidal facies of ooid grainstone, wackestone/grainstone, massive grainstone, microbial boundstone, and limestone–shale couplet (Table 5.3) (Figure 5.13). The grainstone is overlain by wackestone and grainstone and, in turn, overlain by a thick grainstone bed (more than 30 m thick). The uppermost part is represented by wackestone and grainstone with a few oolite beds.

This facies succession represents deposition in shallow subtidal and peritidal environments. Although limestone–shale couplet facies are commonly interpreted as deep and intermediate ramp deposits below wave base, wackestone and packstone facies are generally indicative of deposition in shallow subtidal environments. Wackestone and grainstone facies with stromatolites are also indicative of intertidal deposition.

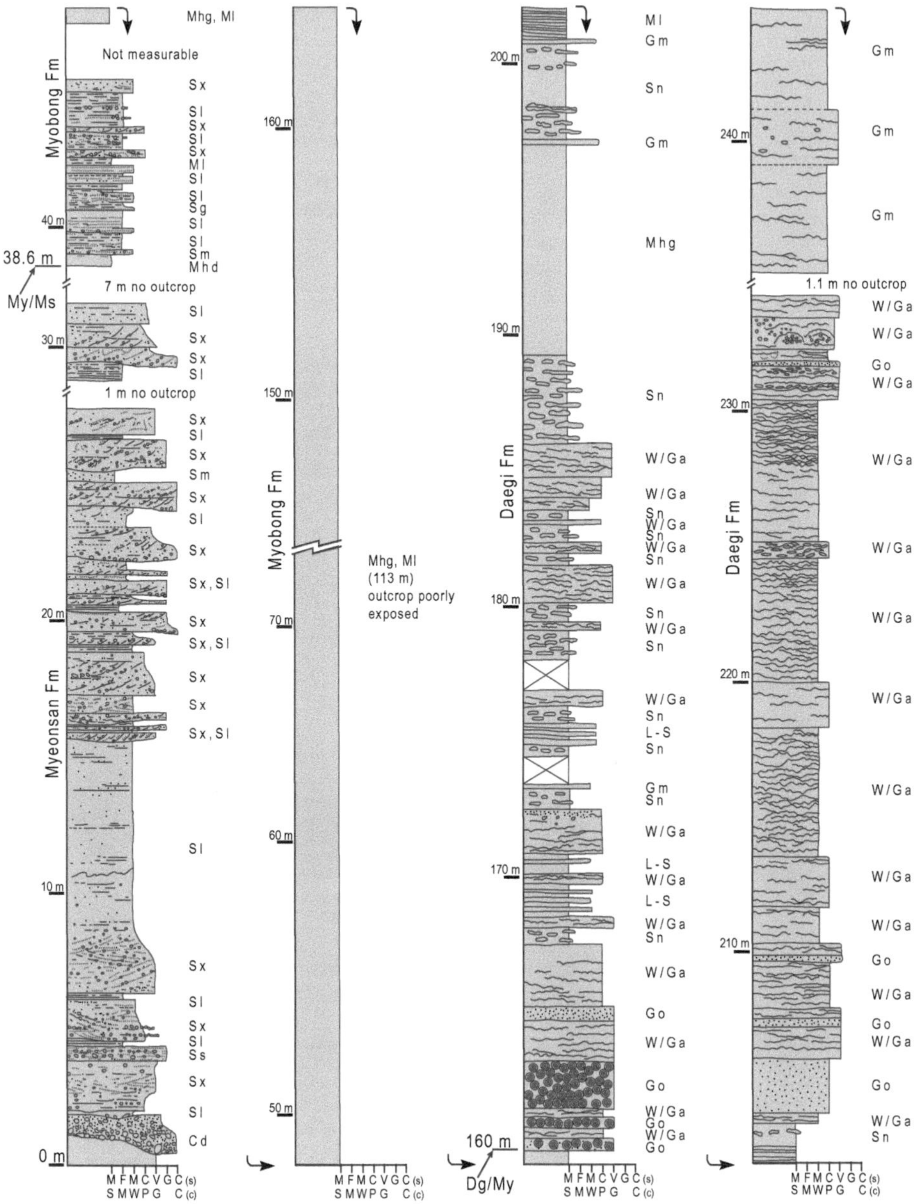

Figure 5.13 Sedimentary log of the Taebaek Group, Seokgaejae section (for sedimentary facies code, see Table 5.3). Ms, Myeonsan Formation; My, Myobong Formation; Dg, Daegi Formation; Ss, Sesong Formation; Hj, Hwajeol Formation; Dj, Dongjeom Formation; Dmg, Dumugol Formation; Mg, Makgol Formation; Js, Jigunsan Formation; Db, Duwibong Formation.

Source: Kwon et al. (2006) by permission of the Elsevier.

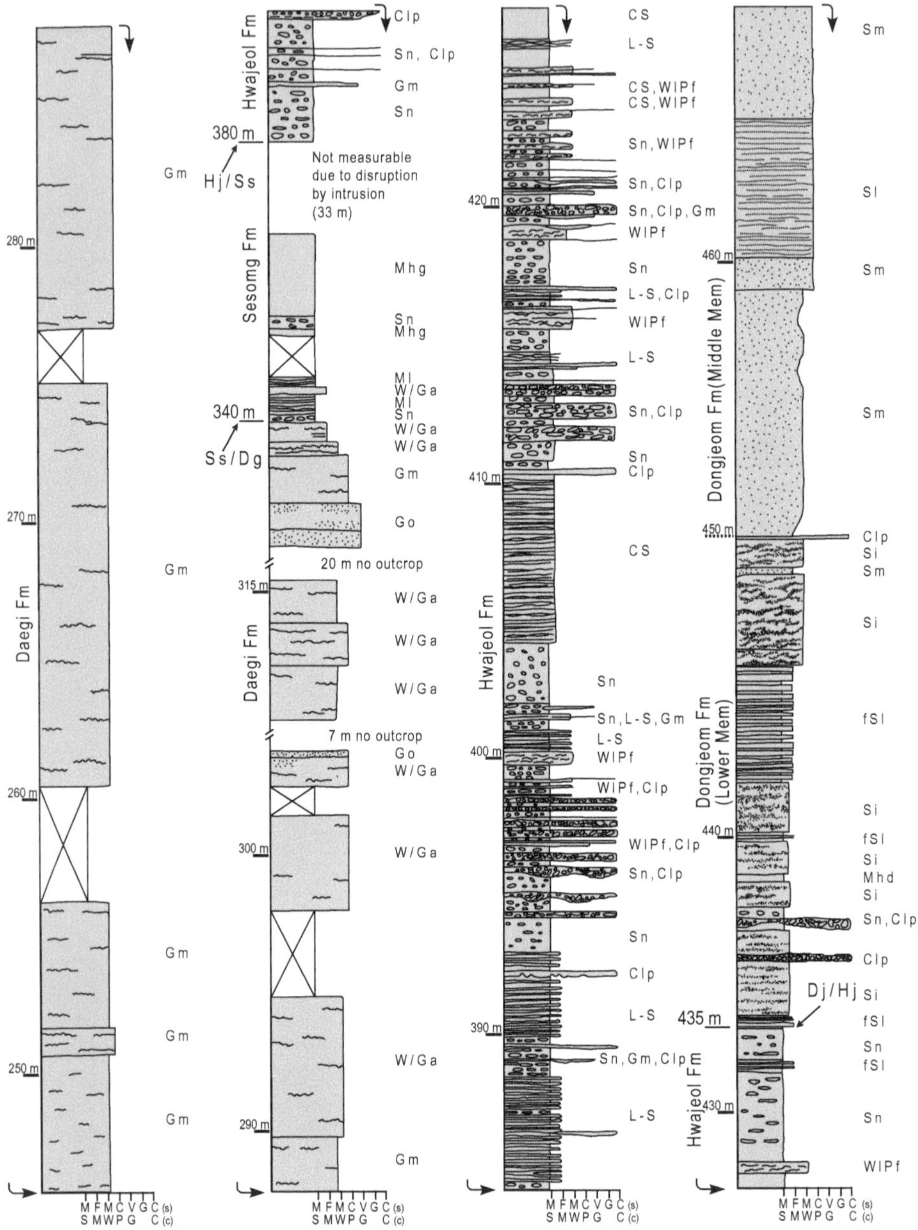

Figure 5.13 (Continued)

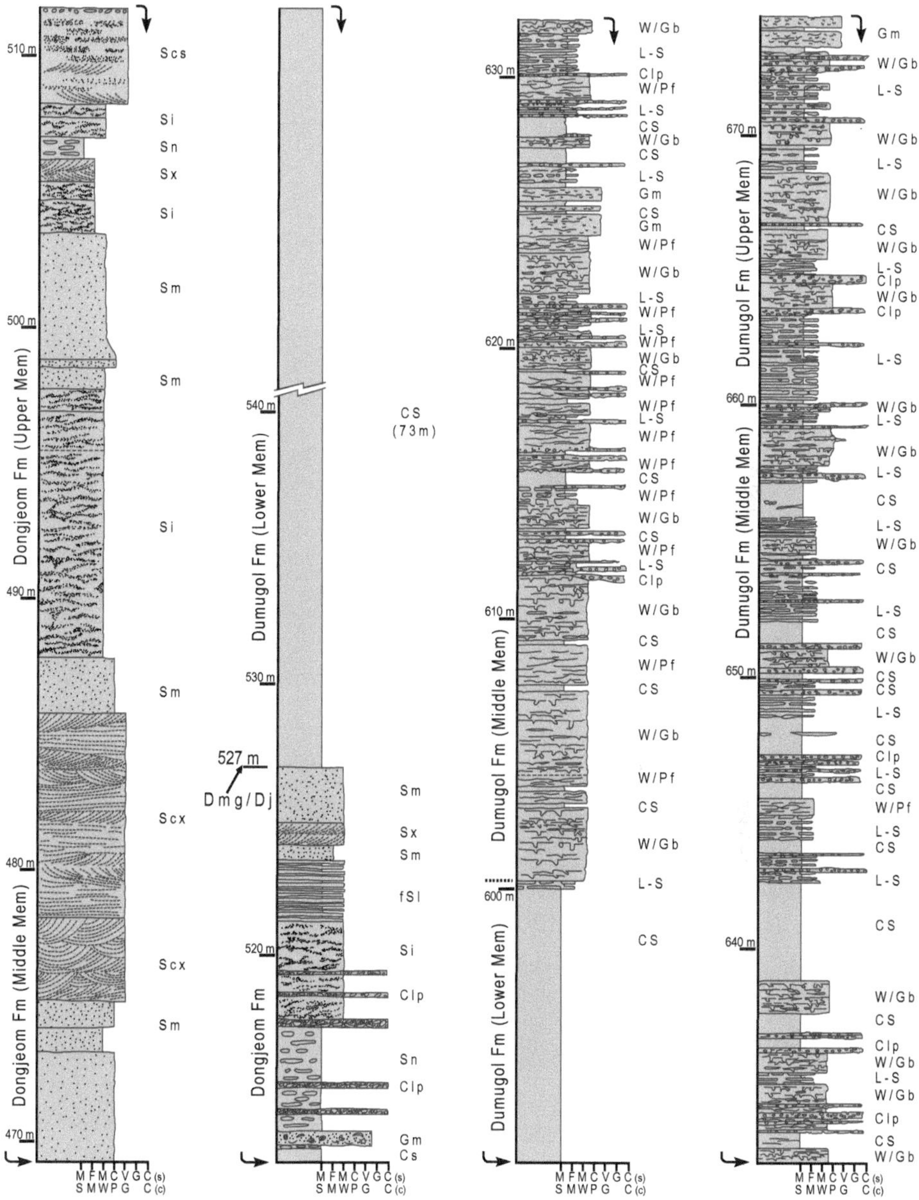

Figure 5.13 (Continued)

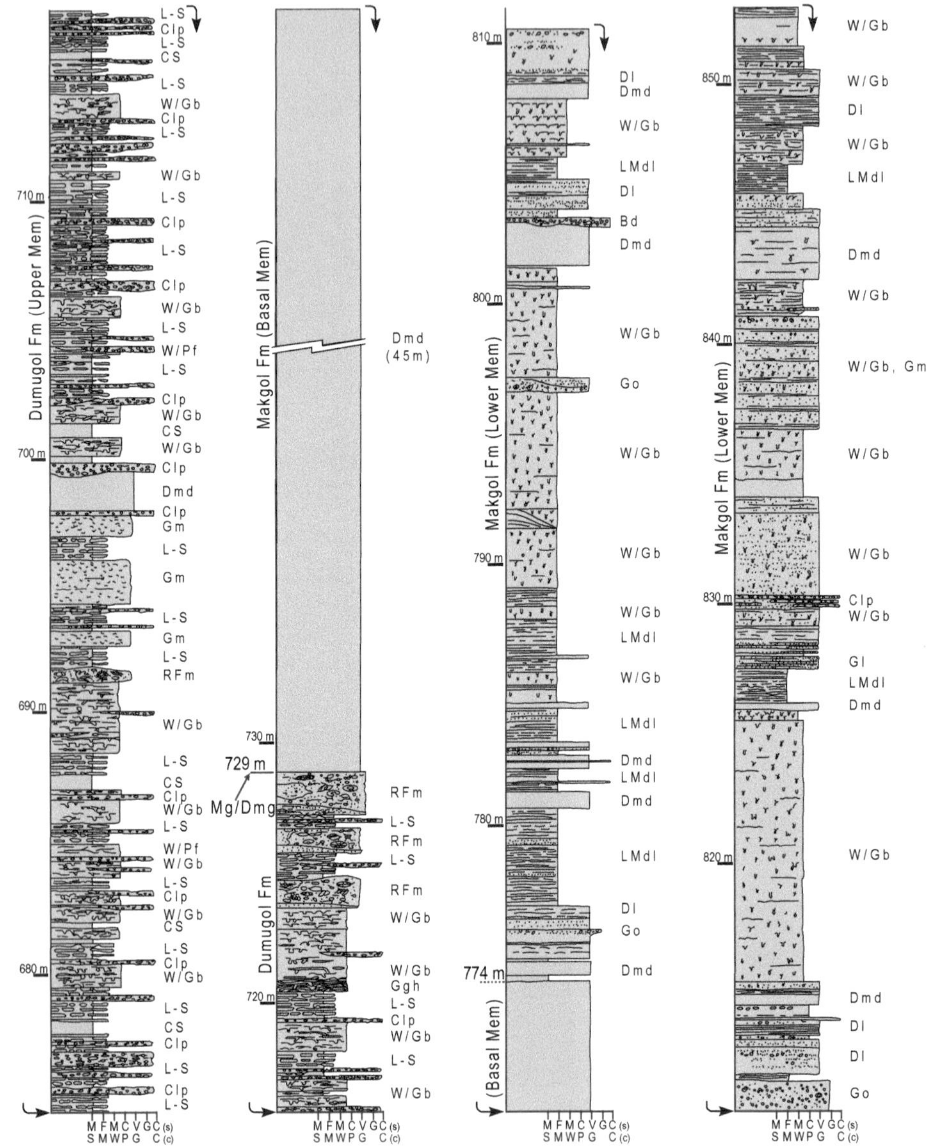

Figure 5.13 (Continued)

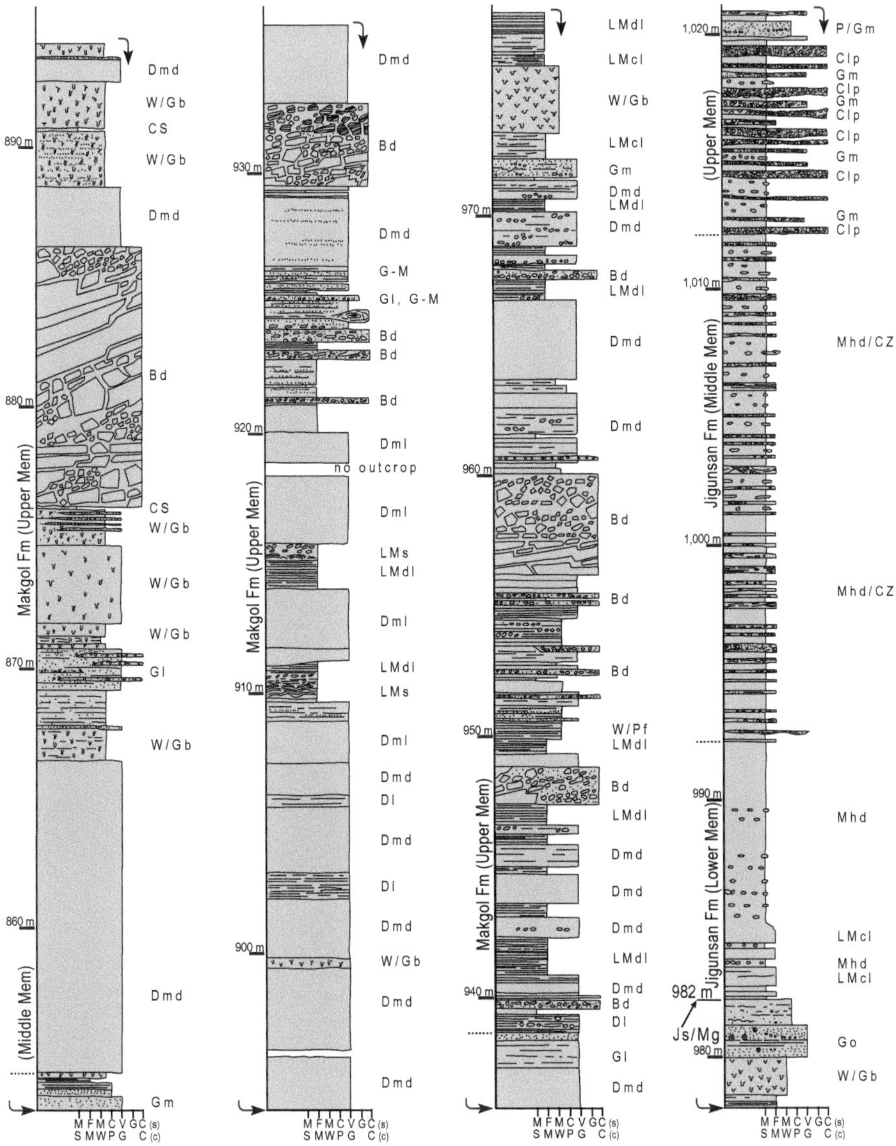

Figure 5.13 (Continued)

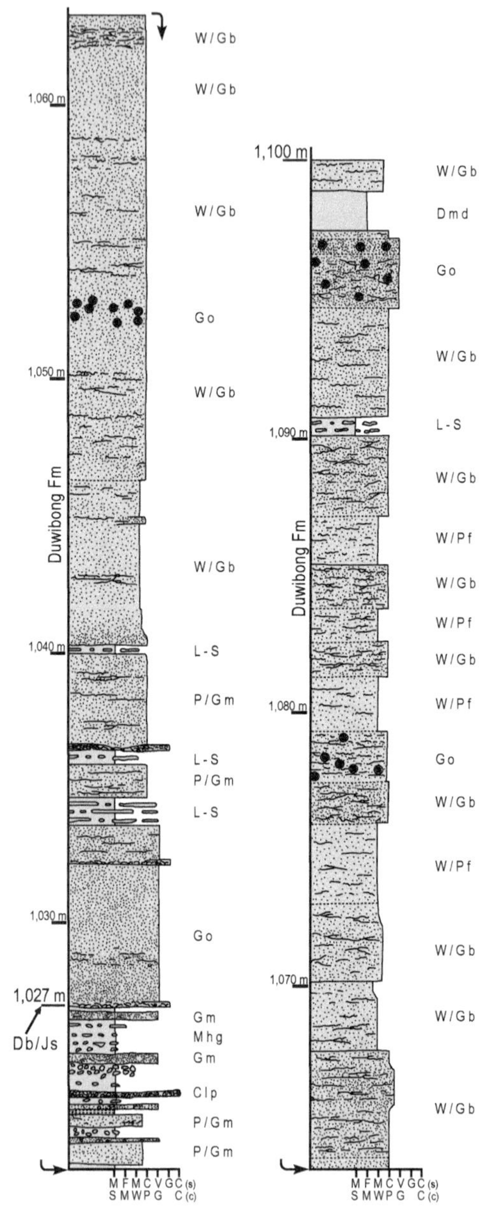

Figure 5.13 (Continued)

5.2.6 Sesong Formation

The Sesong Formation is dominated by gray homogeneous mudstone, nodule-bearing shale, parallel-laminated mudstone, massive sandstone, and crudely stratified sandstone (Table 5.3) (Figure 5.11E). Some sandstone and mudstone beds are normally graded with an erosional base.

The homogeneous mudstone facies formed most likely in outer shelf environments by hemipelagic settling with occasional storm currents. The high proportion of sandstone layers probably indicates deep-water environments where storm currents were prevalent below the fair-weather wave base.

5.2.7 Hwajeol Formation

The Hwajeol Formation is represented by a cyclic succession of meter-scale units, which display a systematic thickening and thinning trend. Each unit consists of lower calcareous shale or nodule-bearing shale and the overlying limestone–shale couplet, flaser wackestone and packstone, or massive grainstone (Table 5.3) (Figure 5.11F, G). It is commonly capped by relatively thin, discontinuous grainstone or limestone pebble conglomerate (Figure 5.11H).

The shale-based meter-scale shallowing-upward cycles are interpreted as shallowing-upward deposits of subtidal environments. The shale facies are indicative of deposition in deep water below fair-weather wave base, whereas the overlying limestone facies are indicative of shallow subtidal deposition. The gradual increase or decrease in cycle thickness reflects an increase or decrease in accommodation due to fluctuations in relative sea level.

5.2.8 Dongjeom Formation

The deposit is characterized by massive and crudely cross-stratified sandstone (quartzite) with slightly inclined pebble grains (Figures. 5.11I, J). The coarse-grained sandstone beds include glauconite and ironstone. They are interbedded with calcareous shale, limestone–shale couplet, and wackestone to packstone in the lower and the upper parts (Table 5.3) (Figure 5.13).

Deposition of sands most likely occurred in shoreface environments (Kwon et al., 2006). The overlying carbonate-dominated formation represents a resurgent ramp facies, followed by intense shelf currents and reworking. Glauconites are indicative of low sedimentation and reworking in shelf environments (Kim and Lee, 2004).

5.2.9 Dumugol Formation

5.2.9.1 Lower Part

The lower part of the Dumugol Formation consists mainly of calcareous shale, nodule-bearing shale, and limestone–shale couplet with subordinate flaser wackestone to packstone and limestone conglomerate (Table 5.3) (Figure 5.11K). The calcareous shale facies commonly includes relatively thin (less than a few centimeters) grainstone layers or limestone pebble conglomerate layers.

The calcareous shale probably formed by hemipelagic settling below the normal wave base. The limestone–shale couplet facies also formed mainly in outer or deep ramp environments. Abundant limestone pebble conglomerate beds resulted from penecontemporaneous deformation, a process of autoconglomeration (Chough et al., 2001; Kwon et al., 2002).

5.2.9.2 Middle Part

The middle part is represented by an alternation of shale- and limestone-dominated beds (Figure 5.11K). It consists of a basal shale-dominated facies of calcareous shale and limestone–shale couplet, and an overlying carbonate-dominated facies of flaser wackestone to packstone, bioturbated wackestone to grainstone, and massive grainstone (Table 5.3) (Figure 5.13). These beds can be grouped into three cycle types: shale–lime-mudstone, shale–wacke/packstone, and shale–grainstone. Each shows a thickening to thinning-upward trend in cycle thickness (Kwon and Chough, 2005).

The meter-scale cycles most likely formed in inner ramp environments where there was an abrupt change in depositional environments. Each sequence is characterized by a thickening to thinning-upward trend in cycle thickness, a fining to coarsening-upward trend in cycle type, and an upward increase to decrease in proportion of shale-dominated facies. The shallowing-upward cycles most likely formed during the high-order sea-level fluctuations.

5.2.9.3 Upper Part

The upper part of the Dumugol Formation comprises a cyclic succession of basal limestone–shale couplet or flaser wackestone to packstone and the overlying bioturbated, wackestone to grainstone, and massive grainstone (Figure 5.11L–N). The grainstone facies are commonly interbedded with normally graded gravelly grainstone, hummocky-stratified gravelly grainstone, sponge reef mounds, and limestone pebble conglomerate, including abundant skeletal fossils (trilobites, echinoderms, brachiopods, and sponges) (Table 5.3; Figure 5.14). Sponge reef mounds are colonized on grainstone or conglomerate beds with an erosional channel base (Figure 5.15).

The meter-scale units capped by grainstone facies probably formed by progradation of shallow subtidal shoals or bars in shallow subtidal environments (Paik, 1987; Woo, 1999). The sponge reef mounds of the lithistid sponge *Archaeoscyphia* and receptaculid *Calathium* are suggestive of hard substrates under high-energy conditions (Kwon et al., 2003) (Figure 5.16). Strong bioturbation is indicative of shallow subtidal environments.

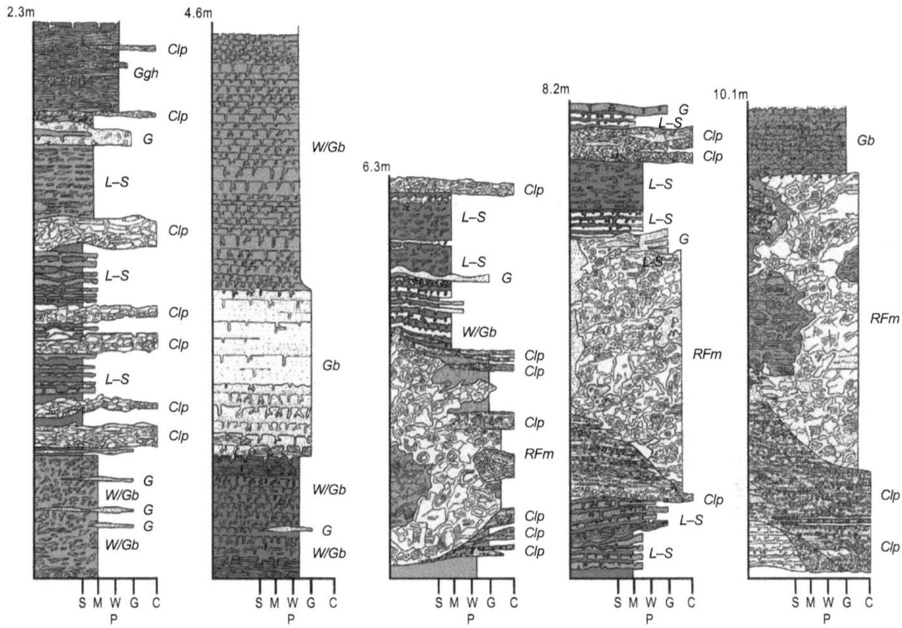

Figure 5.14 Sedimentary log of the Dumugol Formation. Sponge reef mound (RFm) occurs in the uppermost part of the Dumugol Formation. L–S, limestone–shale couplet; W/Gb, bioturbated wackestone to grainstone; G, grainstone; Gb, bioturbated grainstone; Ggh, hummocky-stratified gravelly grainstone; Clp, limestone pebble conglomerate.
Source: Kwon et al. (2003) by permission of the Springer-Verlag.

5.2.10 Makgol Formation

The Makgol Formation comprises bioturbated wackestone to grainstone, massive grainstone, stromatolitic lime-mudstone, wavy and lenticular grainstone–mudstone couplet, finely laminated dolomitic lime-mudstone, massive dolostone, normally graded sandstone, and dolostone breccia (Table 5.3; Figures 5.11O and 5.17). It is characterized by a cyclic succession of peritidal and subtidal carbonates. The peritidal cycle consists of bioturbated wackestone/grainstone and massive grainstone in the lower part, stromatolitic lime-mudstone or wavy and lenticular grainstone–mudstone couplet in the middle part, and finely laminated dolomitic lime-mudstone and dolostone breccia in the upper part. The subtidal cycle comprises basal dolostone or mudstone facies and the overlying bioturbated facies. Each unit is generally 50 cm to about 2 m in thickness. In some cases, bioturbated wackestone to grainstone beds are thick (about 3–4 m thick). In the finely laminated dolomitic lime-mudstone (or dolostone) and stromatolite beds, various sedimentary structures occur such as mudcracks, ripples, and gypsum and anhydrite casts.

Figure 5.15 Photograph and line drawing of the Seokgaejae section, showing sponge bioherms and associated facies. Sponge bioherms occur at the top of meter-scale carbonate cycles, showing laterally abrupt transition to grainstone (G) or pebble-grade limestone conglomerate (PC). Sponge bioherms grew up on the incised channel bottom. BL, bioturbated limestone; L–S, limestone–shale couplet; G, grainstone; PC, pebble-grade limestone conglomerate; SB, sponge bioherm.
Source: Kwon et al. (2003) by permission of the Springer-Verlag.

Stromatolites, mudcracks, ripples, and gypsum and anhydrite casts are suggestive of deposition in arid-climate sabkha-type peritidal environments (Kwon et al., 2006; Paik, 1987). The repetition of peritidal and subtidal cyclic successions was most likely due to higher order fluctuations in relative sea level. The overall regression toward the top of the Makgol Formation culminated in the subaerial exposure of platform carbonates during the earliest Middle Ordovician.

5.2.11 Jigunsan Formation

5.2.11.1 Facies Analysis

The Jigunsan Formation deserves special attention for its depositional environments, since it consists of mudstone and shale which contain abundant trilobite fossils that are useful for worldwide stratigraphic correlation (Choi et al., 2004). In four sections (Sangdong, Sorotgol, Gumunso, and Seokgaejae), eight sedimentary facies have been recognized based on grain size, sedimentary structures, and constituents of carbonates (Table 5.4; Figure 5.18). These facies are grouped into four facies associations, which represent distinct depositional environments (Woo and Chough, 2007; Figure 5.19).

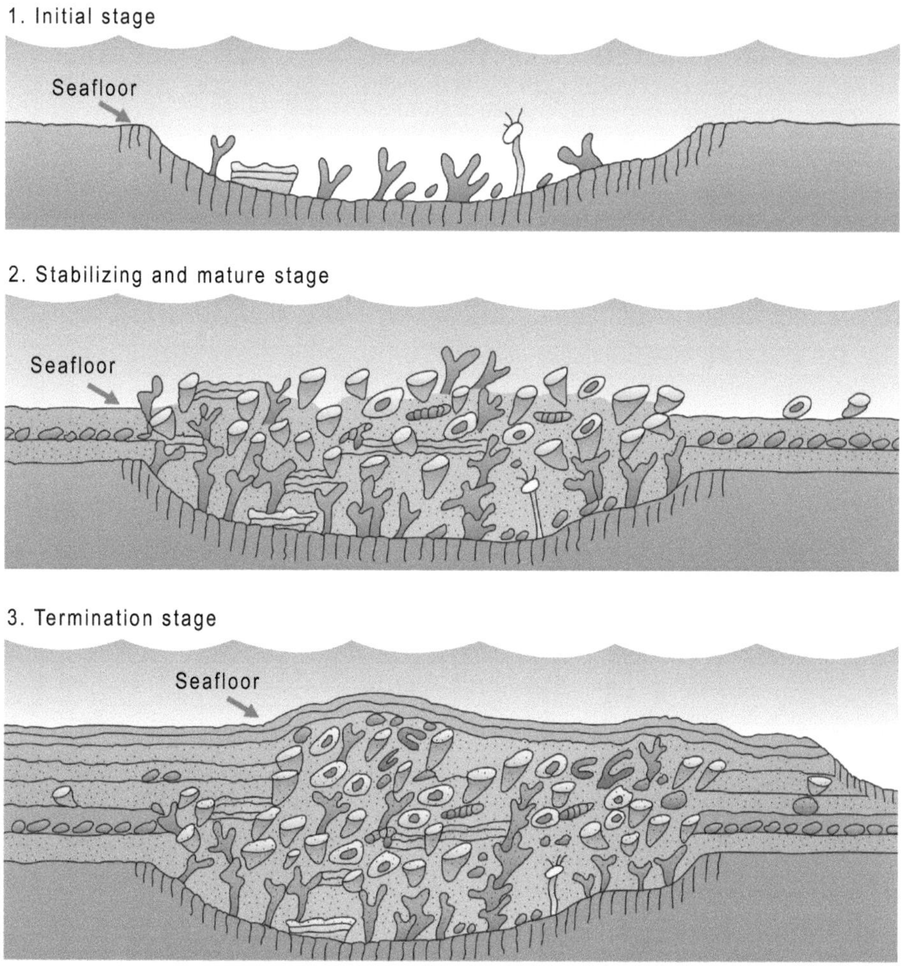

Figure 5.16 Stages of bioherm growth.
Source: Kwon et al. (2003) by permission of the Springer-Verlag.

Facies Association 1: Deep Subtidal to Basin Environments

This facies association is characterized by alternation of laminated lime-mudstone and dark gray mudstone with pyritic calcareous nodules (Table 5.4; Figure 5.20). In the lower part, lime-mudstone and dark gray mudstone beds are decimeters thick (Figure 5.11P). In the upper part, thin nodular lime-mudstone bed is interbedded with thick dark gray mudstone. Small-scale burrows are common (Figure 5.21). The boundary with the underlying oolite of the Makgol Formation is indistinct and transitional. This facies association yields open marine fauna such as cephalopods and graptolites.

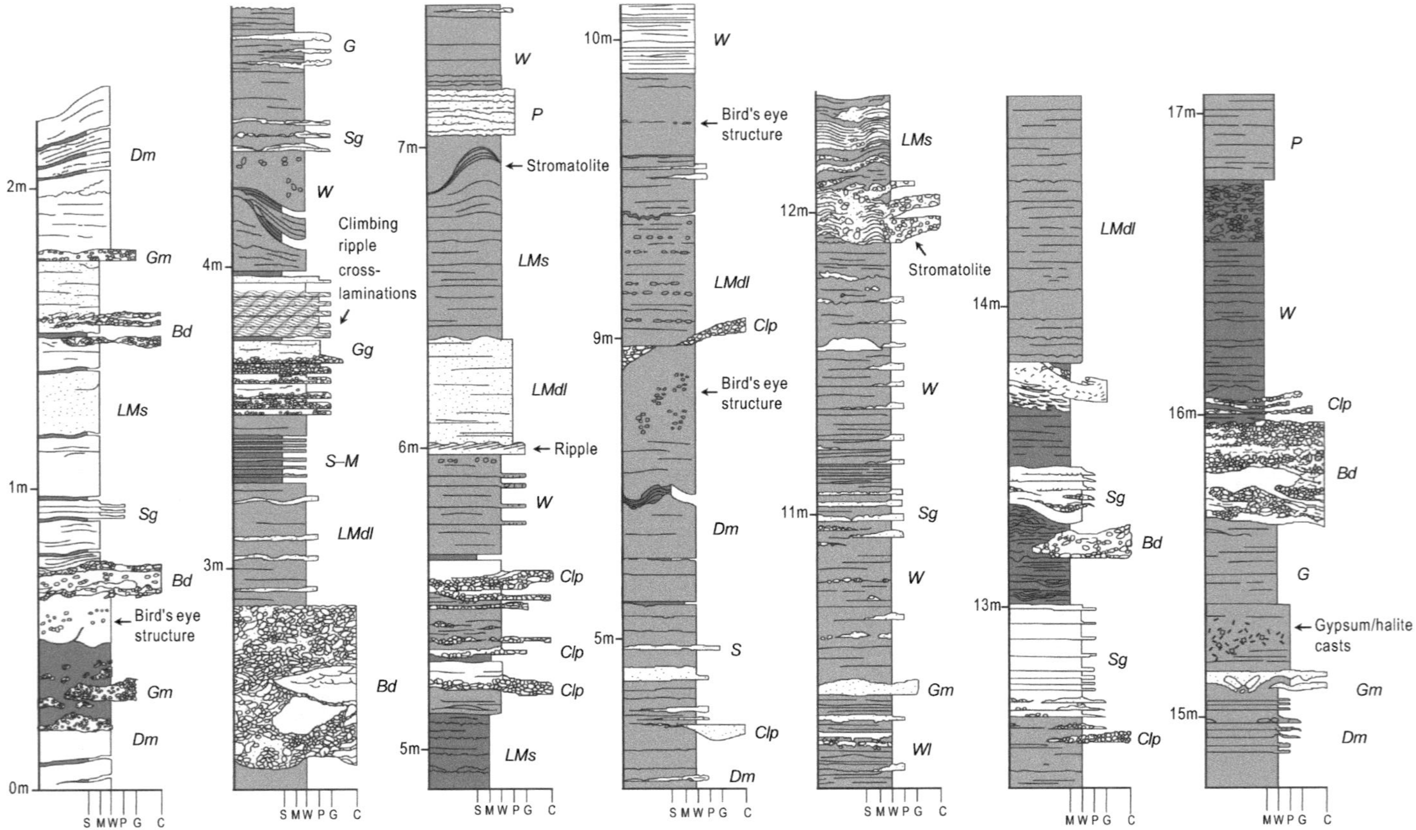

Figure 5.17 Sedimentary log of the Makgol Formation. LMs, stromatolitic lime-mudstone; LMdl, finely laminated dolomitic lime-mudstone; W, wackestone; Wl, laminated wackestone; P, packstone; S–M, sandstone and mudstone couplet; S, sandstone; Sg, normally graded sandstone; G, grainstone; Gg, gravelly grainstone; Gm, massive grainstone; Dm, massive dolostone; Bd, dolomitic breccia.
Source: Kwon et al. (2006) by permission of the Elsevier.

Table 5.4 Description and Interpretation of Sedimentary Facies in the Jigunsan Formation

Facies	Description	Interpretation
Dark gray mudstone (facies Md)	Parallel lamination in some horizons; horizontally elongated micro-burrows; organic-rich (0.75–6% TOC); iron sulfide (pyrite) rich; silt to very fine sand-sized quartz grains in part; illite and chlorite clay mineral assemblage; common pyritic calcareous nodules; abundant fossils (trilobites, brachiopods, gastropods, cephalopods, graptolites, and ostracods) in some horizons	Deep basinal sediment deposited by hemipelagic suspension settling and distal storm current in oxygen-restricted condition (O'Brien, 1996; Schieber, 2003)
Greenish gray siltstone (facies Mg)	Severely bioturbated; partly calcareous; consists of silt-sized quartz grains (0.025–0.05 mm in diameter) and small skeletal grains; a few peloids or mudchips; abundant calcareous nodules; sedimentary structures such as parallel lamination in some nodules	Suspension settling deposit disturbed by bioturbation in well oxygenated deep to shallow subtidal condition
Thickly laminated lime-mudstone (facies LMt)	Crude alternation of lime-mudstone and dolomitic argillaceous (or organic-rich) laminae; less than 5 mm in thickness and indistinct boundaries; a few peloids and intraclasts (0.25–2.5 mm in diameter) in lime-mudstone; pyrites commonly associated with stylolitic structures	Suspension settling under low-energy conditions such as lagoon or deep subtidal; with variation of argillaceous contents
Finely laminated lime-mudstone (facies LMl)	Mainly consist of lime-mudstone laminae which slightly vary in color and occasionally parted by thin argillaceous (organic-rich) laminae; each lamina less than 1 mm in thickness and parallel to each other; slightly wavy and undulatory lamination; thin argillaceous laminae commonly associated with micro-stylolite parallel to bedding plane; about 40% of total volume dolomitized	Intertidal to supratidal cryptalgal laminite (Hardie, 1977; Shinn, 1983)
Bioturbated wackestone (facies Wb)	Lime mud matrix, a few skeletal grains, peloids and dolomite crystals; slightly bioturbated (ichnofacies index-2 or index-3); burrows commonly filled with dolomites and argillaceous materials; crudely laminated; peloids and various shapes of small lime-mudstone clasts in some layers	Shallow subtidal deposit slightly disturbed by bioturbation (Paik, 1987; Overstreet et al., 2003)
Massive pack-grainstone (facies P-Gm)	Massive mixture of skeletal grains (brachiopods, cephalopods, gastropods, and trilobites); various lime mud contents ranging from packstone to grainstone; partly intercalating laminated layers	Deposition in moderately agitating shallow subtidal (Glumac and Walker, 2000)

(*Continued*)

Table 5.4 (Continued)

Facies	Description	Interpretation
Peloidal grainstone (facies Gp)	Light gray to gray in color; various peloids (0.1−10 mm in diameter) and dolomite crystals; well rounded, circular, subcircular, and ellipsoidal peloids; usually massive except stylolite and micro-stylolite swarms parallel to bedding plane; slightly dolomitized; texture selective dolomitization; few fine sand-sized quartz grains	Winnowed deposit in high-energy shoreface (Overstreet et al., 2003)
Oolitic grainstone (facies Go)	Medium sand-sized ooids (~0.4 mm in average diameter) and lesser amounts of peloids and skeletal grains; moderately sorted and rounded, subrounded to ellipsoidal ooids; many ooids with nuclei of skeletal grains of various shape; radial and concentric internal lamina: a few aggregated ooids	High-energy subtidal shoal (Tucker and Wright, 1990)
Massive grainstone (facies Gm)	Predominant sand-sized skeletal grains without lime mud matrix; skeletal grains composed of fragments of echinoderms lesser amounts of brachiopods, trilobites and peloids; thin to medium (1−30 cm) bedded facies unit; laterally continuous with small variation of thickness change; sharp lower and upper boundary; occasional occurrence of normal grading and parallel or wavy lamination	High-energy shallow subtidal deposit such as bioclastic sand sheets; accumulated during full- to waning-storm conditions (Jones and Desrochers, 1992; Lavoie and Asselin, 1998)
Laminated calcisiltite (facies CZl)	Parallel, ripple cross-laminated or low-angle (10−15°) cross-laminated (hummocky or swaley) laminae; lamination of alternating calcareous and shaly materials; calcareous lamina consist of silt-sized particles; darker argillaceous lamina consist of unclear finer grains; distinct, sharp and erosional bases of each unit; faint, gradational, and occasionally wavy upper boundaries; horizontal, oblique and v-shaped burrows; each unit from several millimeters to 15 cm in thickness	Unidirectional current on low-angle platform, with oscillatory movement of water induced by storm (Aigner, 1985; Myrow and Southard, 1996)
Limestone conglomerate (facies LC)	Matrix-supported; grainstone elasts and grainstone matrix; clasts of peloidal and skeletal grainstone; round subcircular and ellipsoidal clasts (about 1−2 cm in diameter); matrix mostly consist of skeletal, peloidal grains, and ooids; sharp erosional base and sharp or diffuse upper boundary; variable in thickness from a few centimeters to decimeter: continuous	Storm lags in a subtidal setting or deposit derived from locally developed gentle slope (Markello and Read, 1981; Séguret et al., 2001)

Source: Woo and Chough (2007) by permission of the Springer.

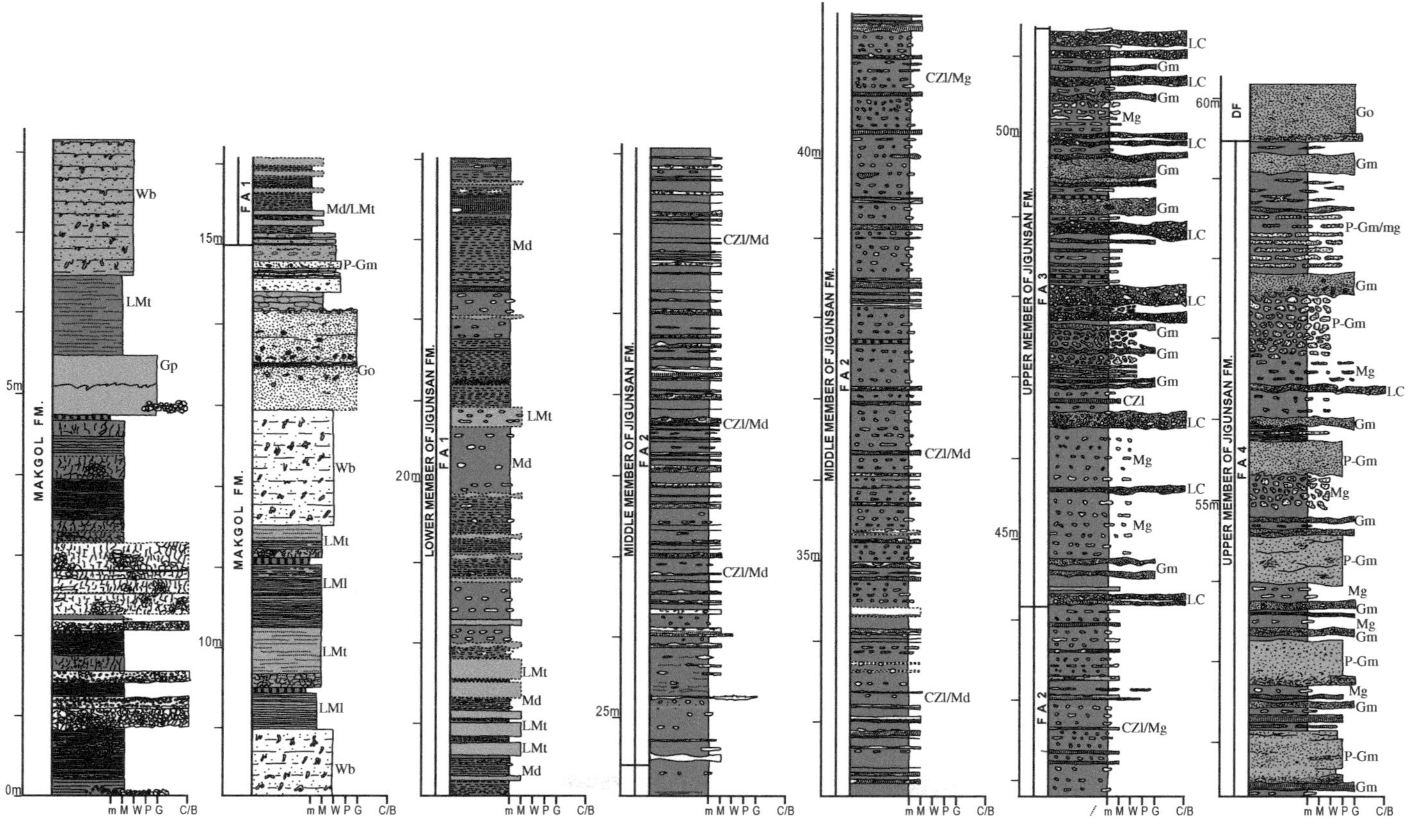

Figure 5.18 Sedimentary log of the upper part of the Makgol and Jigunsan formations in the Seokgaejae section. DF, Duwibong Formation; m, mudstone; M, lime-mudstone; W, wackestone; P, packstone; G, grainstone; C/B, limestone conglomerate/breccia. Arrows in circle indicate direction of paleocurrents. For facies codes, see Table 5.4.

Source: Woo and Chough (2007) by permission of *Geosciences Journal*.

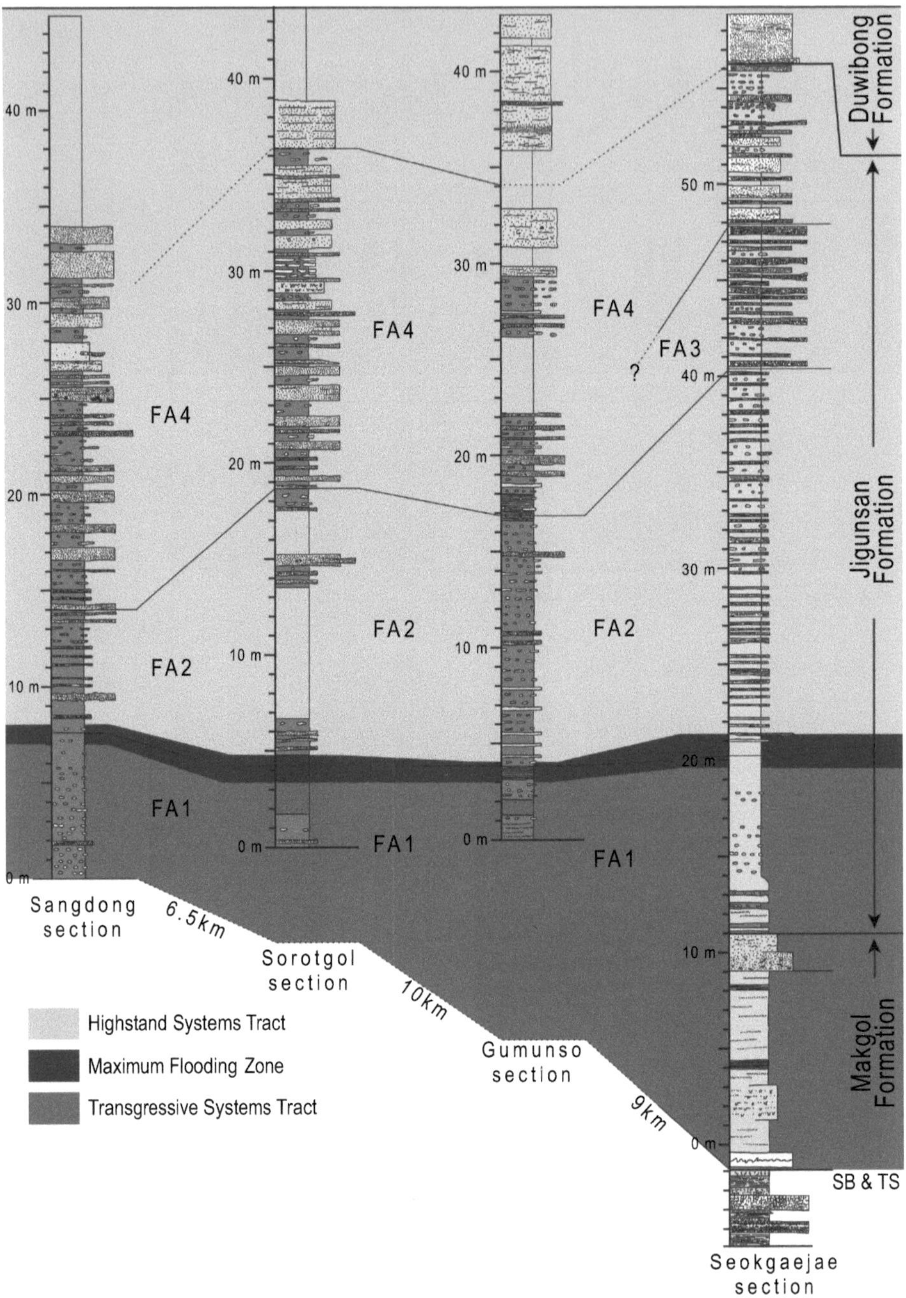

Figure 5.19 Correlation of facies associations (FAs) and bounding surfaces in the Jigunsan Formation. SB, sequence boundary; TS, transgressive surface.
Source: Woo and Chough (2007) by permission of *Geosciences Journal*.

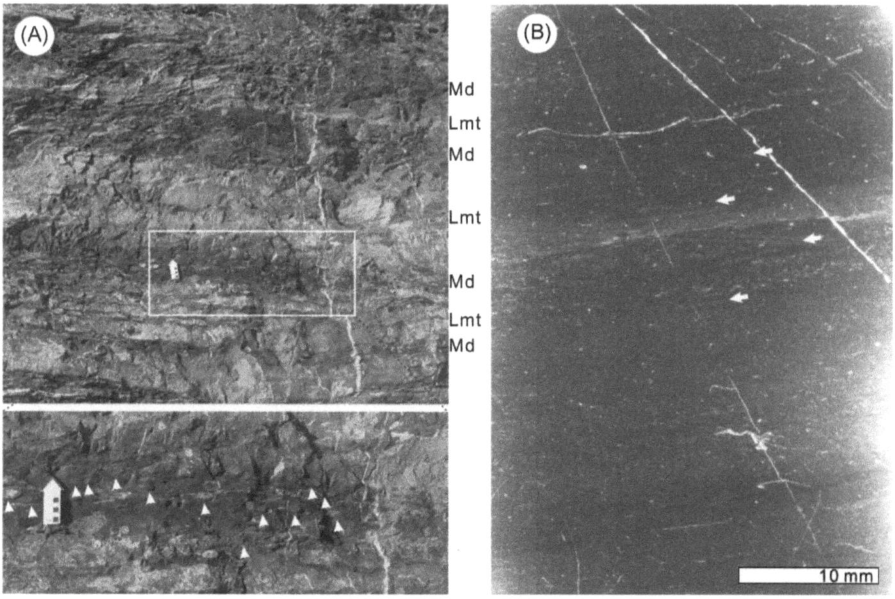

Figure 5.20 (A) Outcrop photograph of facies association 1 in the Seokgaejae Section showing alternation of lime mudstone (LMt) and dark gray mudstone (Md). The lower part is enlarged view of boxed area showing organic rich dark gray mudstone with pyritic calcareous nodules (arrows). Scale is in cm. (B) Photomicrograph of dark gray mudstone (Md) showing horizontal lamination and micro-bioturbations (arrows).

The alternation of dark gray mudstone and lime-mudstone is representative of a deep subtidal cycle. Homogeneous or faintly laminated dark gray mudstone represents hemipelagic settling. Abundant pyrite grains in the dark gray mudstone and relatively high amounts of organic matter suggest oxygen-depleted depositional conditions. The common occurrence of micro-bioturbation is suggestive of dysoxic conditions.

Facies Association 2: Storm-Influenced Deep Subtidal Environments

Facies association 2 consists of thin-bedded laminated calcisiltite, dark gray mudstone, and greenish gray siltstone (Table 5.4; Figure 5.11Q). The lower boundary of laminated calcisiltite bed is very sharp and slightly erosional (Figure 5.22). A few beds contain current ripple laminae and hummocky and swaley cross-laminae (Figure 5.22). The mudstone beds show a coarsening-upward trend from dark gray mudstone to greenish gray siltstone.

The occurrence of hummocky and swaley laminae is suggestive of storm activity. The parallel and cross-laminated calcisiltite beds are interpreted as deposits of dilute density currents induced by storms. Siliciclastic mudstone facies represent background deposition of suspension settling in deep subtidal platform to basinal environments.

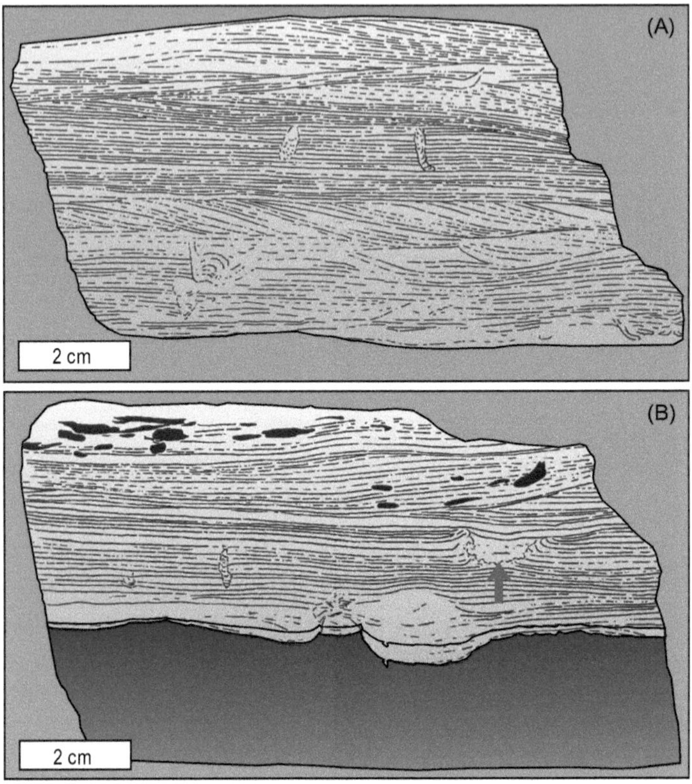

Figure 5.21 (A) Line drawing of laminated calcisiltite, showing parallel lamination, ripple lamination, and internal truncation. (B) Line drawing (lower) of laminated calcisiltite, showing erosional surfaces (arrow). The lower boundary is partly deformed by formation of the calcareous nodules (center of the figure). Note various types of burrows; horizontal elongated burrows of dark gray mudstone and greenish gray siltstone beneath the laminated calcisiltite; elliptical and subhorizontal burrows in the upper part of laminated calcisiltite; and relatively large shallow vertical burrows.
Source: Woo and Chough (2007) by permission of *Geosciences Journal*.

Facies Association 3: Storm-Influenced Shallow Subtidal Environments

Facies association 3 consists of greenish gray siltstone and interbeds of centimeter- to decimeter-thick massive and normally graded pack-grainstone, massive grainstone, and limestone conglomerate (Table 5.4). Clasts are composed of partly laminated argillaceous peloidal/skeletal grainstone consisting of skeletal grains such as crinoids and ostracods as well as quartz grains (Figure 5.23). The matrix consists of skeletal grains (gastropods, trilobites, crinoids, cephalopods, and bryozoans) and peloids (Figure 5.23).

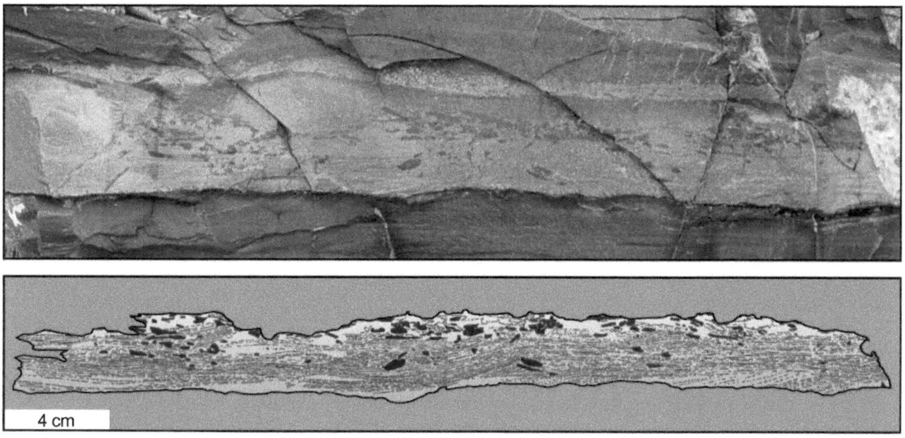

Figure 5.22 Outcrop photograph (upper) and line drawing (lower) of laminated calcisiltite showing swaley cross-laminae and ripple drift. Burrows are abundant in the upper part of the bed.
Source: Woo and Chough (2007) by permission of *Geosciences Journal*.

This facies association is representative of storm-influenced shallow subtidal environments. The sporadic grainstone or conglomerate interbedded with siltstone is suggestive of repetitive event sedimentation. Limestone conglomerate and grainstone interbedded with greenish gray siltstone are indicative of episodic storm activity along the mid-ramp below the fair-weather wave base. Normally graded grainstone beds are suggestive of deposition from waning flows. Bioturbated siltstone is suggestive of deposition in well-oxygenated conditions.

Facies Association 4: Shallow Subtidal Environments

Facies association 4 consists mainly of greenish gray siltstone and massive (or graded) skeletal pack-grainstone (Figure 5.24). Individual beds are tabular and continuous, although thin gravelly skeletal layers are discontinuous and lenticular. The pack-grainstone (massive grainstone) bed is bounded by skeletal base and laminated top.

Strong bioturbation in pack-grainstone and greenish gray siltstone represents a high degree of biological activity under well-oxygenated conditions. Mixed calcareous grains in the siltstone are suggestive of deposition in a moderately agitated platform (Glumac and Walker, 2000). The thin siltstone beds represent deeper-water facies, whereas the thick pack-grainstone beds represent shallower-water facies. The massive or graded pack-grainstone and massive grainstone bed with laminated top are suggestive of waning flows during storm events.

Depositional Model

The occurrence of finely laminated, desiccation-cracked, and brecciated dolomite in the upper part of the Makgol Formation is suggestive of extensive supratidal

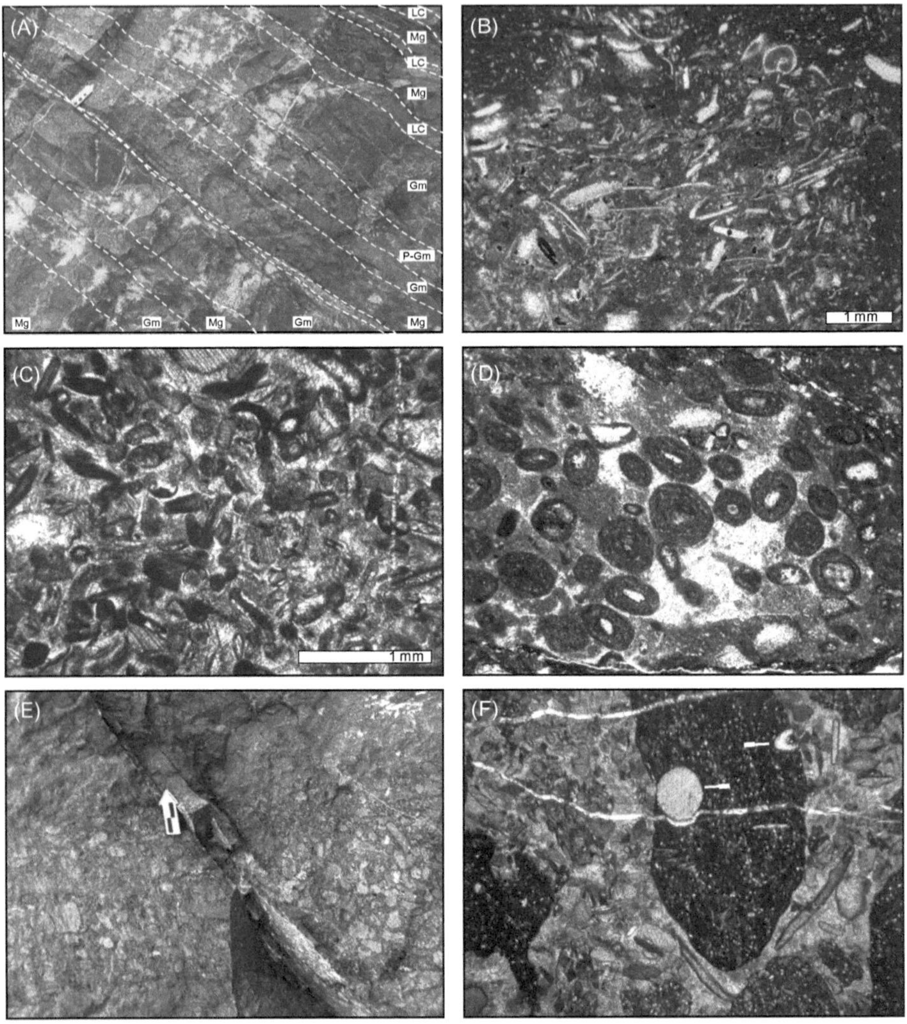

Figure 5.23 Photographs and photomicrographs of facies association 3 and constituent facies. (A) Outcrop photograph of facies codes in the Seokgaejae Section. Each facies unit is tabular and laterally continuous. Scale is in cm. (B) Photomicrograph of massive pack-grainstone (P-Gm). (C) Photomicrograph of massive grainstone (Gm) including recrystalized and micritized grains. (D) Photomicrograph of ooid grains in massive grainstone. (E) Outcrop photograph of limestone conglomerate (LC). Note the sharp lower boundary and matrix-supported texture. Scale is in cm. (F) Photomicrograph of limestone conglomerate. Round or irregular clasts consist of silt-sized grains. Larger skeletal grains form matrix, similar to massive pack-grainstone. For facies codes, see Table 5.4.

Figure 5.24 (A) Outcrop photograph of facies association 4 of the Jigunsan Formation and the overlying oolite bed of the Duwibong Formation in the Seokgaejae section. (B) Photomicrograph of oolitic grainstone (Go). Grains consist of ooids, skeletal grains, and peloids.

environments and subaerial exposure (Figure 5.25A). The dominance of fine-grained siliciclastic units (FA1) is ascribed to rapid regional transgression, which resulted in the drowning of carbonate factory (Figure 5.25B). Rare bioturbation, abundant pyrite, and pyritic calcareous nodules are suggestive of oxygen-restricted (dysoxic) conditions. However, abundant micro-bioturbation and sharp erosional laminae suggest that the basin was neither fully anoxic nor stagnant. Abundant organic matter (about 6% total organic carbon) suggests high surface productivity was associated with upwelling, accompanied by sea-level rise.

The productivity of carbonate system was enhanced due to decrease in the rate of sea-level rise, and the platform continued to prograde (Figure 5.25C). Carbonate sediments in shallow subtidal environments were transported to the deeper part, forming an alternation of laminated calcisiltite and hemipelagic mud (FA2). The erosional base, skeletal lags, and hummocky and swaley cross-laminations are all suggestive of storm-induced flows.

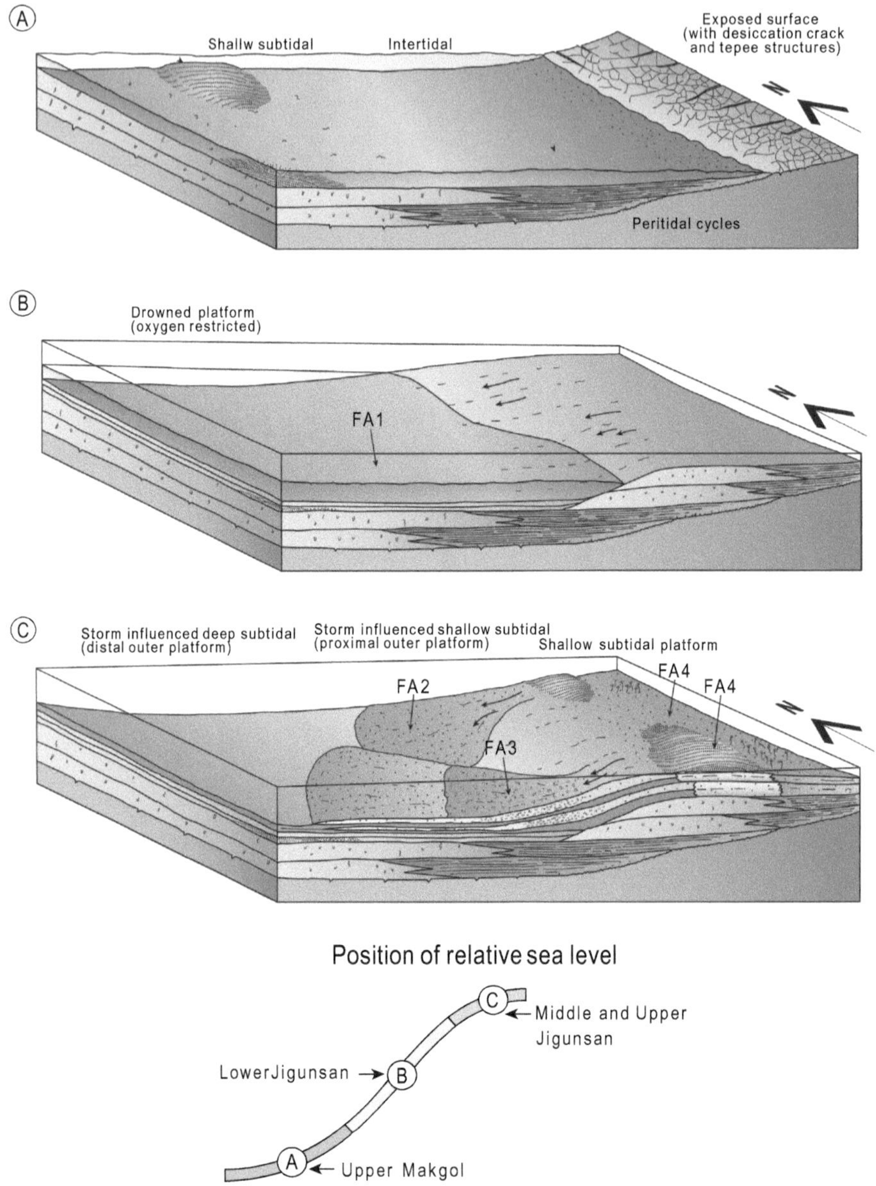

Figure 5.25 Depositional environments of the upper Makgol and Jigunsan formations. (A) In the initial stage of sea-level rise, peritidal environment develops. The peritidal sequences generally retrograde. Oolitic and skeletal shoals form in shallow subtidal environments. (B) Drowning of the platform and retreat of the carbonate factory. During transgressive episode, oxygen-restricted deep water prevails. Local slope forms due to spatial difference in sedimentation rate and oceanic conditions. (C) During slow rise or steady state in sea level, productivity is enhanced and subtidal shoals develop on the platform. Dilute turbidity currents and density currents induced by storms occur. SD, Sangdong section; SR, Sorotgol section; SG, Seokgaejae section.
Source: Woo and Chough (2007) by permission of *Geosciences Journal*.

Massive or graded pack-grainstone and grainstone interbedded with greenish gray siltstone (FA3) probably represent proximal storm deposit. The occurrence of limestone conglomerate in FA3 also suggests deposition during storm events. The outer platform was characterized by discrete carbonate storm beds in fine-grained siliciclastic deposits, which progressively evolved into a shallow subtidal platform with migrating ooid shoals.

5.2.12 Duwibong Formation

The Duwibong Formation comprises massive packstone to grainstone, flaser wackestone to packstone, bioturbated wackestone to grainstone, oncoid/ooid grainstone, limestone–shale couplet, and dark gray massive dolostone (Table 5.3; Figure 5.11R and 5.24B). An oncoid/ooid grainstone bed (about 5 m thick) at the base is overlain by bioturbated wackestone to grainstone and alternation of flaser wackestone to packstone (Figure 5.13). Calcareous shale and limestone–shale couplets occur in the lower part. Gastropods, bivalves, trilobites, crinoids, echinoids, bryozoans, calcareous algae, and brachiopods are abundant (Figure 5.13).

It represents deposition in carbonate shoal environments, varying from subtidal to peritidal environments (Lee et al., 2001). An abundance of gastropods suggests high-energy, highly stressed environments (Woo and Park, 1989). The fining-upward successions are indicative of progradation of shallow subtidal shoal, restricted backshoal, and peritidal environments.

5.2.13 Sequence Stratigraphy

According to the vertical arrangement of sedimentary facies, the mixed carbonate–siliciclastic succession of the Taebaek Group can be grouped into four sequence units (Kwon et al., 2006). Each unit comprises a deepening to shallowing arrangement of facies.

5.2.13.1 Sequence I: Initial Inundation and Transgression

Sequence I consists of siliciclastic-dominated succession (Jangsan-Myeonsan and Myobong formations) and the overlying carbonate-dominated succession (Daegi Formation) (Figure 5.26). The lower boundary is placed at the base of the conglomerate bed in the lowermost part of the Jangsan Formation or the Myeonsan Member. The siliciclastic-dominated succession includes shoreface and inner shelf, restricted shoreface, and inner to outer shelf associations. The shoreface associations occur as coarse-grained quartzose sandstone and generally fine upward. These units are overlain by inner to outer shelf association (Myobong Formation) of marine shale with an occasional intercalation of calcareous sandstone or siltstone bed. The overlying carbonate-dominated succession (Daegi Formation) represents subtidal shoal, backshoal, and shallow subtidal and intertidal environments. The subtidal shoal and backshoal association consists mainly of oncoid and ooid grainstone and the succeeding shale facies. The

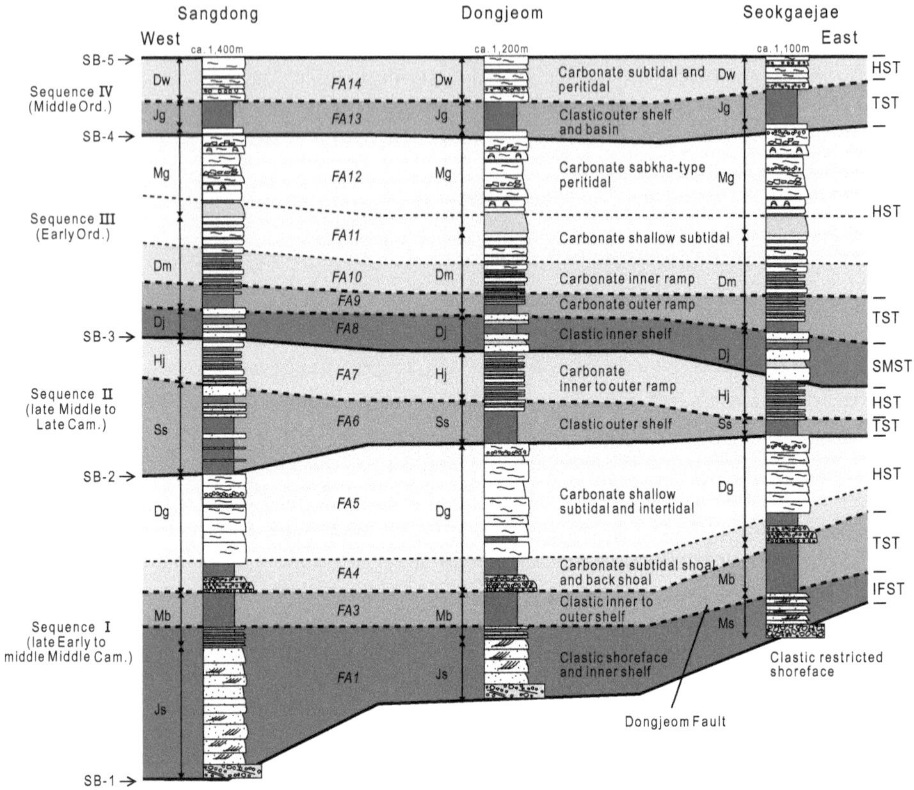

Figure 5.26 Sequence stratigraphy and section correlation of the Taebaek Group. SB-1, base of Sequence I; SB-2, sequence boundary between Sequence I and Sequence II, SB-3, sequence boundary between Sequence II and Sequence III; SB-4, sequence boundary between Sequence III and Sequence IV; SB-5, upper sequence boundary of Sequence IV; IFST, initial flooding systems tract; TST, transgressive systems tract; HST, highstand systems tract; LST, lowstand systems tract; SMST, shelf margin systems tract; Js, Jangsan Formation; Ms, Myeonsan Formation; Mb, Myobong Formation; Dg, Daegi Formation; Ss, Sesong Formation; Hj, Hwajeol Formation; Dj, Dongjeom Formation; Dm, Dumugol Formation; Mg, Makgol Formation; Jg, Jigunsan Formation; Dw, Duwibong Formation. *Source*: Kwon et al. (2006) by permission of the Elsevier.

overlying shallow subtidal and intertidal association comprises an alternation of relatively thick (more than a few meters) wackestone to grainstone.

During initial inundation in the late Early Cambrian, coarse-grained siliciclastics were deposited in shoreface and coastal environments. Siliciclastic sedimentation commonly replaced lowstand or early transgressive deposits in the platform. As the basin was stabilized, fine-grained transgressive mud was deposited in outer shelf or ramp environments, whereas carbonate factory was activated in shallow platform. The occurrence of oncoids, ooids, and grainstone in the overlying shallow subtidal facies is suggestive of continuous progradation in restricted lagoonal and backshoal

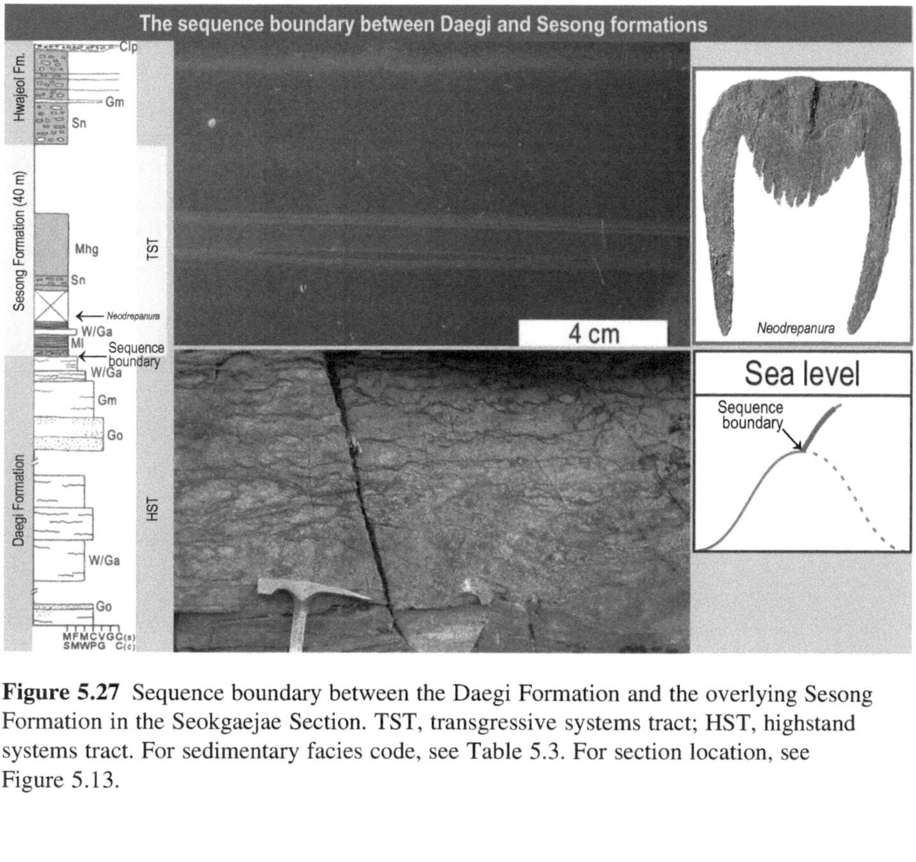

Figure 5.27 Sequence boundary between the Daegi Formation and the overlying Sesong Formation in the Seokgaejae Section. TST, transgressive systems tract; HST, highstand systems tract. For sedimentary facies code, see Table 5.3. For section location, see Figure 5.13.

environments. The high productivity in the Middle Cambrian resulted in the rapid progradation of shallow-marine carbonates.

5.2.13.2 Sequence II: Platform Drowning

Sequence II formed during the late Middle and Late Cambrian (Choi et al., 2004). It started with deposition of shale and siltstone (Sesong Formation), overlying limestone of the Daegi Formation (Figure 5.27). The Sesong Formation is dominated by shale and siltstone with an occasional intercalation of thin sandstone. In the lowermost part of the sequence, calcareous siltstone (or sandstone) is occasionally intercalated with limestone conglomerate and grainstone. The western Sangdong section consists of a relatively thick sandstone-dominated succession, whereas the Dongjeom and Seokgaejae sections are dominated by relatively thin shale with occasional intercalation of sandstone layers (Figure 5.13). The overlying Hwajeol Formation includes meter-scale shallowing-upward subtidal cycles that comprise calcareous shale, limestone–shale couplet, and flaser wackestone and packstone.

The occurrence of deepening facies at the base of the sequence is suggestive of platform drowning (Figure 5.28). The shale beds in the Sesong Formation formed

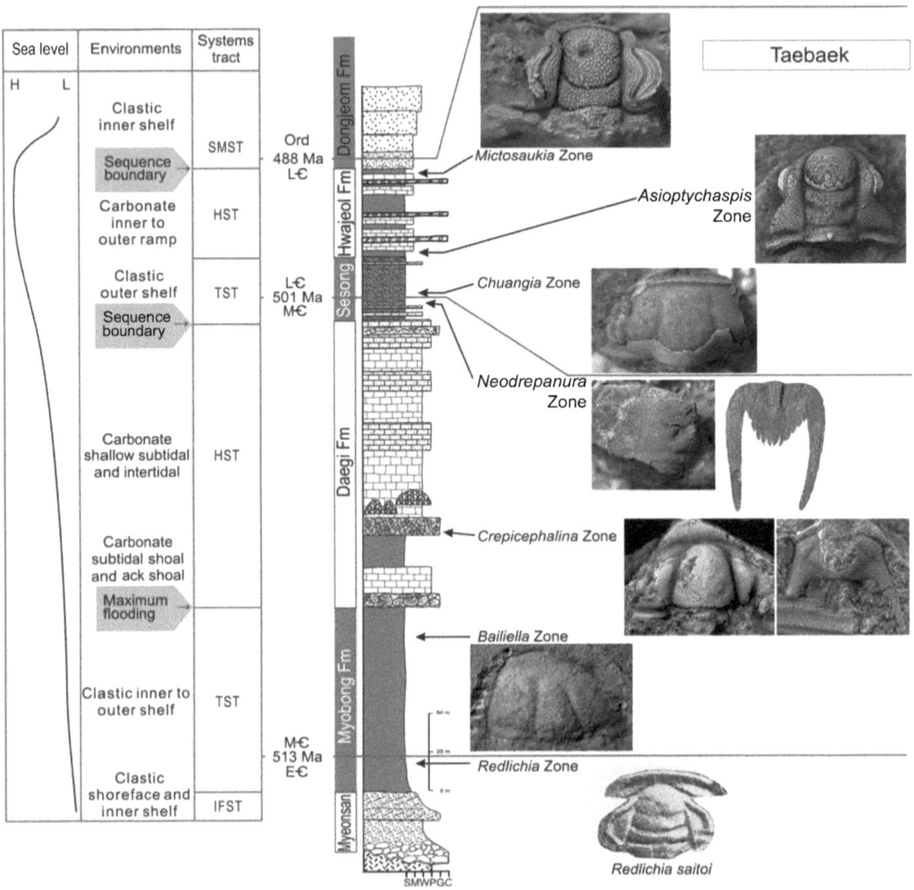

Figure 5.28 Sea-level history and sequence stratigraphy of the Cambrian succession in the Taebaek area. Biostratigraphy is based on trilobites.

during early transgression with an upward decrease in frequency and thickness of the interbedded limestone layers. With the ensuing rise in sea level, the transgressive deposit is overlain by shale and carbonates formed during late transgression and highstand (Hwajeol Formation). Input of subtidal carbonates resulted in meter-scale shallowing-upward cycles (Figure 5.29). In the latest Late Cambrian, progradation of subtidal carbonates in the Hwajeol Formation ceased and the carbonate factory was shut down due to abundant supply of coarse-grained siliciclastics (Figure 5.29).

5.2.13.3 Sequence III: Shallowing-Upward Package

Sequence III comprises quartzose sandstone (Dongjeom Formation) and the overlying cyclic successions of Dumugol and Makgol formations (Figure 5.26).

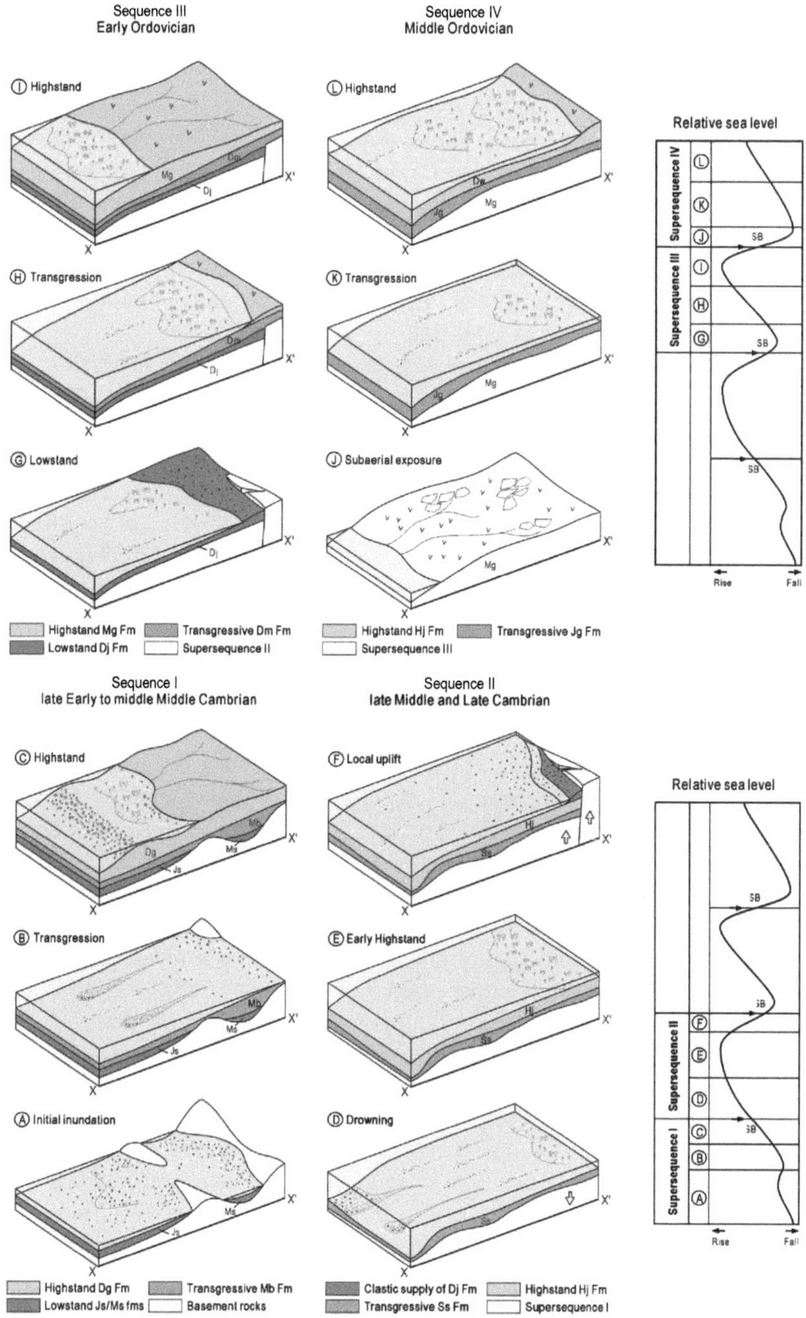

Figure 5.29 Depositional environments and sea-level changes of the Cambrian succession. *Source*: Kwon et al. (2006) by permission of the Elsevier.

According to the trilobite biostratigraphy, it formed in the Early and earliest Middle Ordovician (Choi et al., 2004). The Dongjeom Formation represents shelf environments, ranging from offshore to shoreface. The basal offshore succession of shale and sandstone is overlain by thick massive quartzose sandstone formed in shoreface environments. The overlying Dumugol Formation mostly consists of meter-scale shallowing-upward subtidal cycles of shale and the overlying flaser or bioturbated limestone facies. The succession is overlain by shallow subtidal and sabkha-type peritidal succession of the Makgol Formation. It consists mainly of shallow subtidal carbonates (bioturbated limestone) and reef mounds. Peritidal carbonates of stromatolites, tidal rhythmites, and supratidal dolomites are indicative of progradation of tidal flat.

The glauconite and ironstone beds suggest that the siliciclastic system experienced marine flooding in shallow-marine environments. With the ensuing rise in sea level, the coarse-grained siliciclastic system backstepped, whereas fine-grained siliciclastics (Dumugol Formation) prevailed in outer or inner ramp environments. Approaching the maximum of marine transgression in the Early Ordovician, highstand shallow-marine carbonates formed due to an increase in carbonate production. It was terminated by a relative sea-level fall (Figure 5.29).

5.2.13.4 Sequence IV: Deepening and Shallowing Events

Sequence IV comprises siliciclastics-dominated succession (Jigunsan Formation) and the overlying carbonate-dominated succession (Duwibong Formation) (Figure 5.26). The base of the sequence is placed at the supratidal brecciated dolomite and the overlying subtidal bioturbated limestone (Figure 5.26). The lower part of the Jigunsan Formation (homogeneous or slightly laminated mudstone and siltstone) is gradually transitional to homogeneous or laminated mudstone with intercalated limestone layers in the upper part. The overlying Duwibong Formation formed by progradation of subtidal shoal, shallow subtidal, and peritidal carbonate systems, including subtidal shoal facies of ooid or peloid grainstone. This sequence is unconformably overlain by siliciclastics of the Pyeongan Supergroup (Carboniferous to early Triassic).

Sequence IV formed on a subaerially exposed surface of the underlying peritidal carbonate succession. It is composed of fine-grained siliciclastics deposited during transgression (Jigunsan Formation) and subtidal and peritidal carbonates formed during highstand (Duwibong Formation) (Figure 5.26). Lowstand clastic wedges were probably present further offshore (Figure 5.29). During the early stage of transgression, hemipelagic deposition of mud prevailed over retrograding subtidal facies (Woo and Chough, 2007). As the transgression progressed, the carbonate factory was reactivated in distant shallow subtidal zones. In the late transgressive phase, carbonate materials were deposited by the episodic input of turbidity and storm currents in outer shelf environments (Figure 5.29). During highstand, the carbonate factory expanded, which resulted in an abundant supply of carbonate sediments and progradation of carbonate strata (Figure 5.29) (Lee et al., 2001). The highstand subtidal and peritidal association (FA14) is characterized by fining-upward successions, suggestive of progradation of shallow subtidal shoal, restricted backshoal, and peritidal environments.

5.2.14 Sea-Level Changes

The changes in depositional environments are commonly due to tectonic events (subsidence versus uplift of basin floor), sediment supply, and global sea-level changes. In particular, sea-level changes can be commonly compared among basins worldwide. Figure 5.29 shows sea-level changes for the Taebaeksan Basin during the Cambrian, based on the information provided by the succession of sedimentary facies (Kwon et al., 2006). The geologic time is based on the occurrence of trilobites. In the early Cambrian, sea level was low when initial submergence of the continent began. During this time, the Jangsan and Myeonsan formations formed in the shoreface and inner shelf area (Figure 5.28). It shows a progressive rise in sea level, which was at maximum at the end of deposition of the Myobong Formation. Carbonate sediments began to form at this time, although sea level continued to rise thereafter. The lithologic boundary can be designated as the maximum flooding surface.

Sea level continued to rise in the Middle Cambrian (Figure 5.29). Just prior to the boundary between the Middle and Late Cambrian (501 Ma), the rate of sea-level rise increased. A paleoshoreline delineates the paleogeography of the North China Platform and the Taebaeksan Basin across the Pyeongnam Basin in North Korea (Figure 5.30). The gradual rise in eustatic sea level during the Cambrian was maintained until the Early Ordovician (Figure 5.28). In the earliest

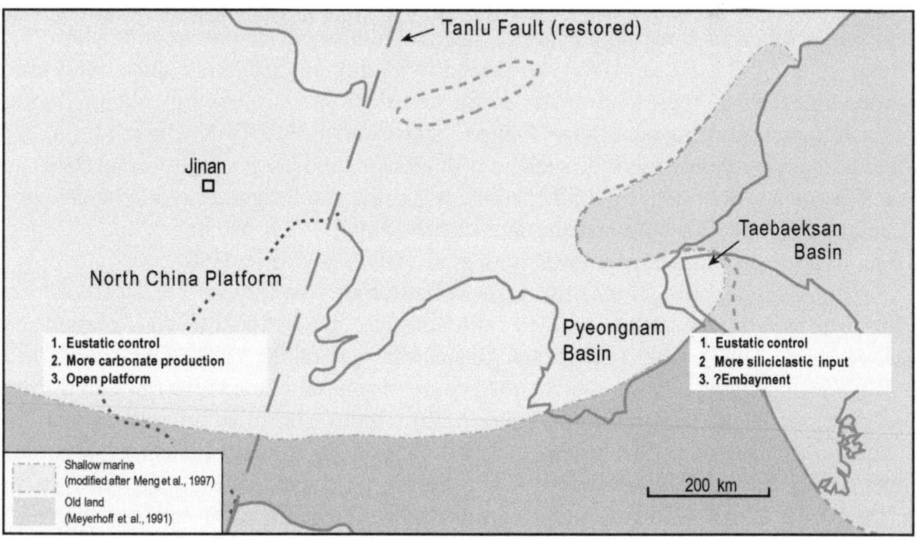

Figure 5.30 The Cambrian paleogeography.

Middle Ordovician, however, eustatic fall occurred, forming the global bounding surface. In the Taebaeksan Basin, this event occurred in the upper part of the Makgol Formation where peritidal carbonates with extensive features of subaerial exposures (desiccation cracks, evaporite mineral casts, stromatolites, and supratidal dolostone) prevailed. The peritidal facies are overlain by a deep subtidal and deep-water facies of the Jigunsan Formation.

5.2.15 Relative Sea Level versus Sediment Supply

In carbonate platforms, supply of clastic sediment plays an important role for the sequence development. On the other hand, carbonate sediment supply is affected principally by changes in relative sea level. As sea level rises, carbonate production catches up with increase in accommodation. On the contrary, sea-level fall causes a decrease in shallow subtidal space for carbonate factory, which results in the reduction of carbonate production.

In the Taebaeksan Basin, deposition of carbonates prevailed during highstand with an increase in supply and active progradation, whereas it was reduced during lowstand or transgression (Figure 5.29). A large input of siliciclastic sediments commonly suppressed production of carbonates and deteriorated the carbonate factory. In the earliest Ordovician, however, the production rate of carbonates was severely reduced, despite the maintenance of a significant subtidal environment (Figure 5.29). This was largely due to the input of siliciclastic sediments caused by regional uplift and lowering of the base level.

5.2.16 Basin Formation and Controls

After the breakup of the Rodinia supercontinent in the late Proterozoic, eustatic rise of sea level resulted in global inundation, forming a number of shallow-marine platforms (Bond et al., 1984; Read, 1989). In the early Cambrian, relatively thick sandstone successions formed, representing tide-dominated shallow-marine sedimentation during basin initiation and transgression (Johnson and Baldwin, 1996; Kwon et al., 2006). During the greenhouse period, sequence development was primarily controlled by basin geometry, subsidence pattern, sediment supply, and long-term sea-level changes (Lee and Chough, 2011; Lehrmann and Goldhammer, 1999; Montanez and Osleger, 1993; Read and Goldhammer, 1988; Sarg et al., 1999; Schlager, 1992).

In the North China Block, the Lower Cambrian flooding of the North China Platform led to the deposition of siliciclastics, phosphorites, and carbonates (Meng et al., 1997). The Taebaeksan Basin was part of the North China Platform across the Pyeonnam Basin, which was progressively inundated and flooded during the long period of transgression (Figure 5.30). Deposition of sediments caused the basin floor to subside, forming an intraplate sag basin. Ensued rise in eustatic sea level during the greenhouse period in the Cambrian caused continuous subsidence. In the North China Platform, more carbonates were produced in an open platform environment. In the Taebaeksan Basin, relatively more siliciclastic sediments were deposited, most likely due to the proximity to the source area in an embayed marginal environment (Figure 5.30). In the latest Cambrian and earliest Ordovician, there was rejuvenation of siliciclastic sediments due to the regional uplift in the entire platform (Meng et al., 1997). Carbonate sedimentation was, however, resumed in the Ordovician. At the end of the Ordovician, the basin was subaerially exposed again and resulted in the absence of Upper Ordovician, Silurian, and Devonian strata. Sedimentation was resumed in the Carboniferous.

5.3 Yeongwol Group

5.3.1 Introduction

The Yeongwol Group comprises a mixed siliciclastic and carbonate succession. It is divided into a number of lithologic units, including the Sambangsan, Machari, Wagok, Mungok, and Yeongheung formations in ascending order (Choi, 1998; Kobayashi, 1966; Yosimura, 1940) (Figure 5.31). According to the trilobite biostratigraphy, the group ranges from the middle Middle Cambrian to the Upper Ordovician (Table 5.1; Choi, 1998). In the northern part of Yeongwol city, Choi (1998) identified eastward-stacking thrust sheets, based on the occurrence of trilobite fauna (Figure 5.31). A brief review of lithologic characteristics and biostratigraphy is presented below. At present, sedimentological data on the entire group are too meager to have a meaningful discussion of sequence stratigraphy.

5.3.1.1 Sambangsan Formation

The Sambangsan Formation (about 400 m thick) consists of homogeneous, purple to greenish gray fine sandstone and siltstone. Sandstones are generally massive, greenish gray, and fine grained, showing small-scale bands and are poorly to moderately sorted (Figure 5.32). They are mainly composed of fine to very fine quartz, feldspar, mica, and opaque minerals, and show foliations. Quartz comprises 70–80% of framework grains. The upper part of the formation yields trilobite fauna (*Metagraulos* and *Megagraulos*) of the middle Middle Cambrian (Kim et al., 1985; Kobayashi, 1961). The occurrence of fine sandstone layers is suggestive of deposition in relatively shallow to deep-water environments.

5.3.1.2 Machari Formation

The overlying Machari Formation consists of alternation of limestone and shale (Figure 5.32). Limestone layers are relatively thicker than shale layers. It yields abundant trilobites (Lee and Choi, 1994, 1995, 1996) of the Middle to Late Cambrian, with lesser amounts of brachiopods and gastropods (Kobayashi, 1962). The base is characterized by dark gray argillaceous limestone and thick-bedded bioclastic grainstone/packstone. It contains Middle Cambrian trilobites such as *Tonkinella*, *Olenoides*, *Dorypyge*, and *Peronopsi* (Lee and Choi, 1994, 1995). The middle part is dominated by laminated dark gray shale with occasional

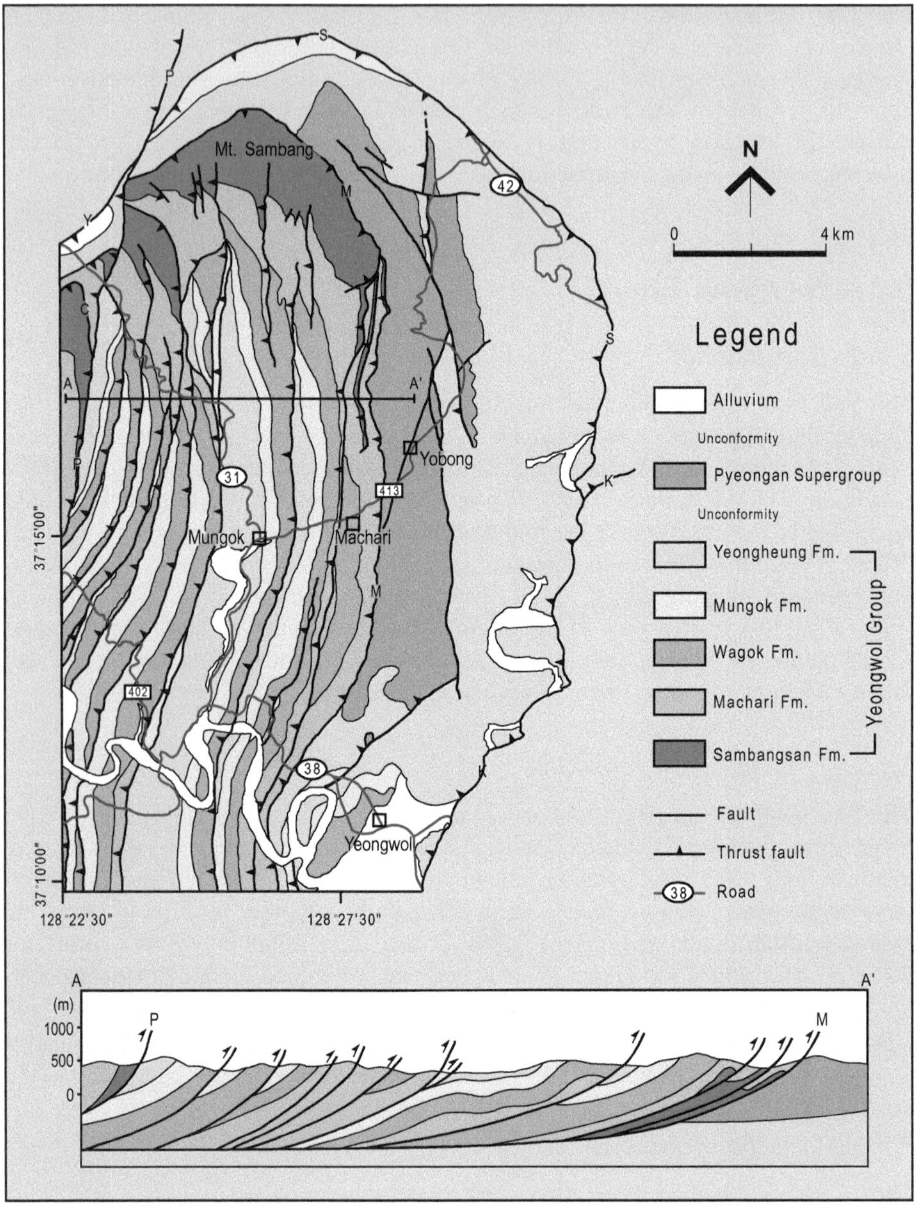

Figure 5.31 Simplified geologic map of the Yeongwol area, showing imbricated stacks of thrust sheets. P, Pyeongchang fault; S, Sangri fault; K, Kakdong (Deokpori) fault; M, Mungok fault.
Source: Choi (1998) by permission of *Geosciences Journal*.

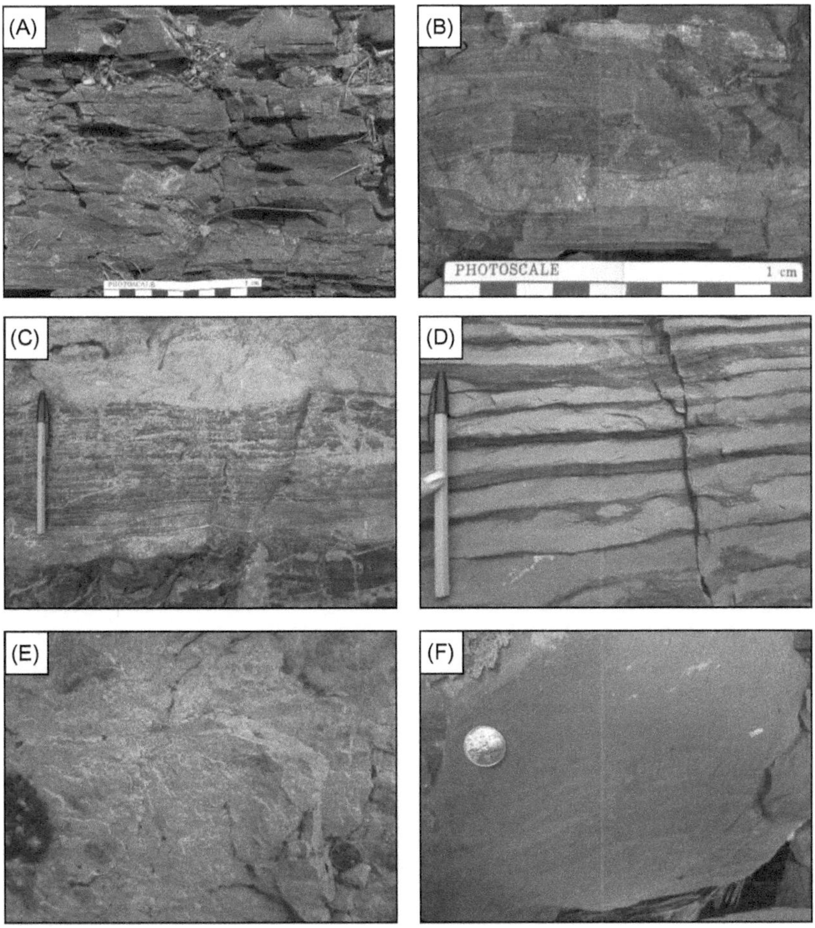

Figure 5.32 (A) Fine-grained sandstone, Sambangsan Formation. (B) Laminated shale with dolomitic limestone interbeds, Machari Formation. (C) Thickly bedded dolomitic limestone and black shale, Machari Formation. (D) Thinly bedded dolomitic limestone and black shale, Machari Formation. (E) Massive dolostone, Wagok Formation. (F) Marlstone to shale, Mungok Formation. (G) Limestone conglomerates, Mungok Formation. (H) Limestone–shale couplet, Mungok Formation. (I) Dolostone, Mungok Formation. (J) Lense-shaped chert interbeds, Mungok Formation. (K and L) Desiccation crack, Yeongheung Formation. Pen for scale, 15 cm, coin for scale, 2.4 cm in diameter.

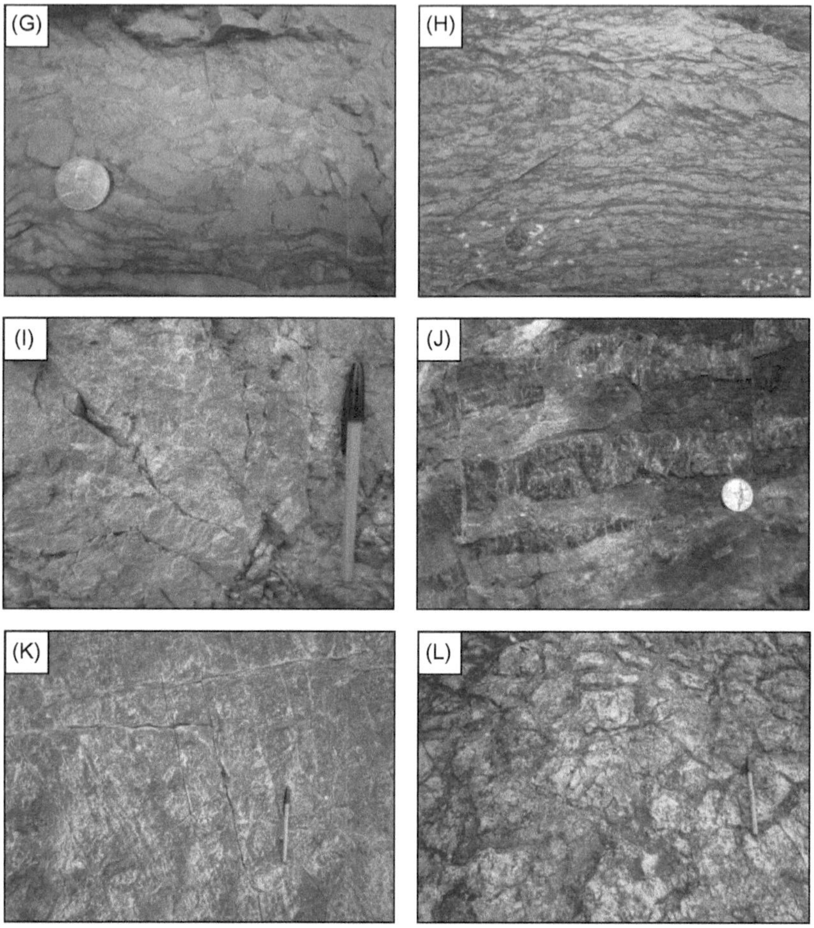

Figure 5.32 (Continued)

intercalation of thin dolomitic limestone bed. The upper part is poorly fossiliferous and characterized by thin-bedded, light gray dolomitic limestone and black shale (Lee, 1995) (Figure 5.33). The predominance of laminated black shale and the abundant cosmopolitan trilobites are suggestive of deposition in dysaerobic deep-water environments (Choi and Kim, 2006; Choi et al., 1999). The lime-mudstone layers are representative of the keep-up phase of carbonates in relatively deep platforms.

5.3.1.3 *Wagok Formation*

The Wagok Formation (200 m thick) (Late Cambrian) is characterized by gray massive dolostone, which is composed of peloids, bioclasts, and ooids

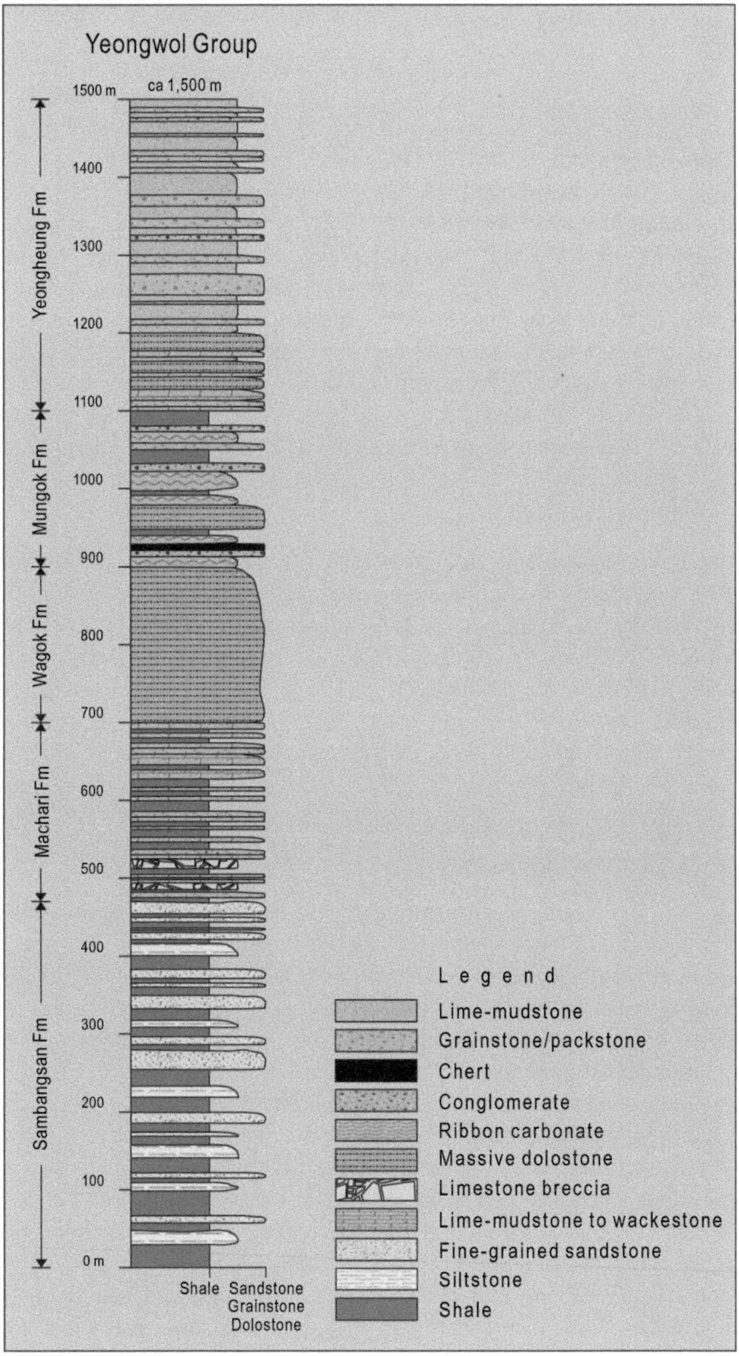

Figure 5.33 Sedimentary log of the Yeongwol Group.

(Figure 5.32). The thick dolostone facies are suggestive of carbonate shoal environments.

5.3.1.4 Mungok Formation

The Mungok Formation (140–210 m thick) consists of limestone and shale (Figure 5.32). It can be divided into four members (Kim and Choi, 2000). The lowermost Garam Member (30–55 m thick) comprises alternation of limestone and shale and grainstone/packstone intercalated with thin chert layers. The overlying Baeiljae Member (30–35 m thick) consists of light gray massive dolostone. The overlying Jeommal Member (30–50 m thick) comprises grainstone/packstone, limestone conglomerate, and marlstone/shale. The limestone/shale beds are commonly transformed into pseudóconglomerate/breccias due to postdepositional deformation (Figure 5.32). The entire succession represents shallow (lagoonal, shoreface, and shoal platform) to deep-water environments (Kim, 2012; Kim and Choi, 2000).

5.3.1.5 Yeongheung Formation

The Yeongheung Formation (400 m thick) consists of massive, thick-bedded, light to dark gray dolostone in the lower part and blue gray limestone in the upper part (Lee, 1987b). It yields trilobites, brachiopods, cephalopods (Kobayashi, 1966), conulariids (Choi and Jeong, 1990), stromatoporoids (Kano et al., 1994; Lee and Yoo, 1993), and conodonts (Lee, 1989). The deposit is characterized by thin laminated lime-mudstone and contains mudcracks and salt pseudomorphs (Figure 5.32) which are indicative of supratidal environments (Choi and Woo, 1993; Yoo et al., 1994).

5.3.2 Depositional Environments

According to the vertical arrangement of sedimentary facies, the mixed siliciclastic–carbonate succession of the Yeongwol Group can be divided into a number of facies association units, representing distinct depositional environments (Figure 5.33). In the lower part of the succession, siliciclastic sediments (Sambangsan Formation) were initially deposited in relatively deep-marine environments. Laminated black shale of the Machari Formation formed in dysaerobic deep-water environments and recorded production of carbonates in deep subtidal realm. According to the trilobite biostratigraphy, the sedimentation rate was very low (D.K. Choi, personal communication). Increased number of carbonate beds in the upper part of the succession (middle Upper Cambrian) is suggestive of an increase in carbonate production that kept pace with the ensuing rise in sea level. In the late Late Cambrian, massive dolostone beds (Wagok Formation) formed, most likely in peritidal environments. In the Early Ordovician (Mungok Formation), water depth increased again in which limestone and shale alternated. In the Middle Ordovician (Yeongheung Formation), the basin was under peritidal environments.

6 North China Platform in Shandong Province

6.1 Geological Setting

The North China Platform was an extensive epeiric sea (more than 1500 km east–west and 1000 km north–south) (Meng et al., 1997). It formed on a tectonically stable Sino-Korean Block in the early Paleozoic. The platform bounds to the north by the Hinggan Fold Belt and to the south by the Dabieshan Belt (Figure 6.1). The Tan-Lu Fault, a major sinistral strike-slip fault, offsets the eastern part of the Sino-Korean Block (Chough et al., 2000) (Figure 6.1). The sedimentary succession in the North China Platform comprises thick (ca. 1800 m in thickness) mixed carbonate and siliciclastic sediments (Blackwelder and Willis, 1907; Lu and Dong, 1953) (Figure 6.2). It is superbly exposed in the western part of Shandong Province (Figures 6.3–6.5).

6.2 Cambrian Stratigraphy and Lithofacies Units

The lithostratigraphy of the Cambrian succession in Shandong Province has been recently refined by detailed sedimentological measurements of major outcrop sections and identification of trilobites (Chough et al., 2010) (Figure 6.6). It consists of 12 siliciclastic facies (Table 6.1; Figure 6.7) and 21 carbonate facies (Table 6.2; Figure 6.8). The succession is divided into six lithologic units (formations) (Figure 6.2).

6.2.1 Liguan Formation

The Liguan Formation occurs in the southeastern part of Shandong Province. It overlies Precambrian metasedimentary rocks (Jinhe section) and granitic gneiss (Anqianzhuang and Zhangjiapo sections) and comprises massive and cross-stratified sandstone intercalated with gray mudstone (Table 6.3). Amalgamated sandstone beds consist predominantly of quartz grains and mudstone chips and show wedge-shaped geometry. In Jinhe section, the sandstone contains unsorted breccia and feldspar grains. Interlayered gray mudstone beds are a few meters thick, showing sharp lower and upper boundaries. The upper part of the formation is characterized by thick dark gray-laminated (and cross-laminated) fine sandstone and mudstone. The fine sandstone changes upward into the bioturbated limestone–shale alternation of the Zhushadong Formation.

Geology and Sedimentology of the Korean Peninsula. DOI: http://dx.doi.org/10.1016/B978-0-12-405518-6.00006-1

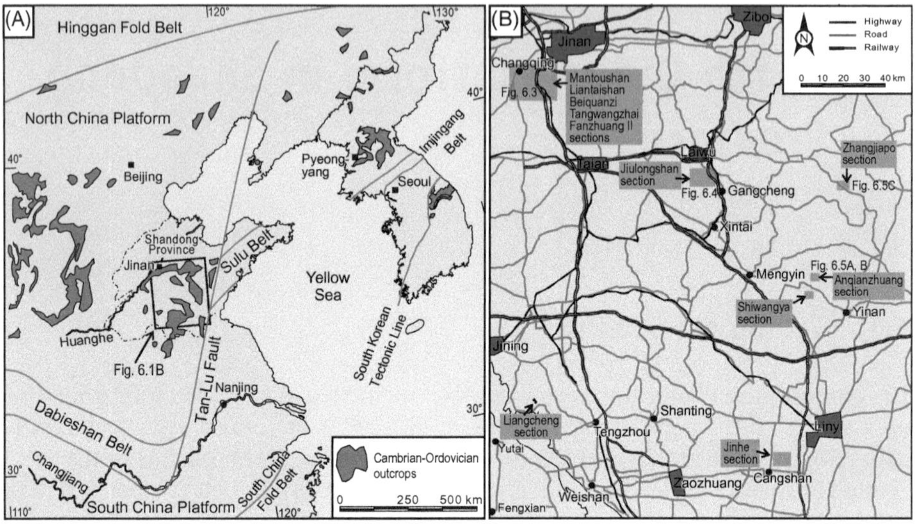

Figure 6.1 (A) Major tectonic boundaries of the Sino-Korean Block and distribution of the Cambrian-Ordovician outcrops of the North China Platform. (B) Location map of the study area in Shandong Province, China.

6.2.2 *Zhushadong and Mantou Formations*

The Zhushadong Formation consists of laminated dolomudstone, stromatolite, bioturbated wackestone, and oolitic packstone to grainstone (Table 6.3). The formation rests conformably on the Liguan Formation in the Jinhe section and unconformably on the Precambrian basement with thick (30 cm to 1 m) paleosol in the Mantoushan, Jiulongshan, and Liangcheng sections (Figure 6.7). The paleosol is crudely stratified and poorly sorted with abundant calcite glaebules, and overlain by laminated and homogeneous dolomudstone, which includes chert concretions. Stromatolites are commonly characterized by domal and irregular geometry. The upper boundary of the Zhushadong Formation is defined by the first occurrence of purple mudstone bed of the Mantou Formation (Figure 6.3).

The Mantou Formation in Shandong Province comprises two members (Table 6.3). The lower member (100–125 m thick) mainly consists of wavy-and lenticular-laminated mudstone, homogeneous purple mudstone, limestone–shale alternation, and microbialite (Table 6.3; Figure 6.9). The upper member (about 100 m thick) is mainly represented by dark purple mudstone and sandstone, cross-stratified sandstone, bioturbated sandstone, hummocky cross-stratified sandstone, cross-stratified oolite, and massive oolitic grainstone (Table 6.3; Figure 6.9). The base of the upper member is defined by the first appearance of massive oolite and hummocky cross-stratified sandstone; the top marked by a thick oolitic grainstone bed (more than 10 m in thickness) of the Zhangxia Formation (Figures 6.4 and 6.9).

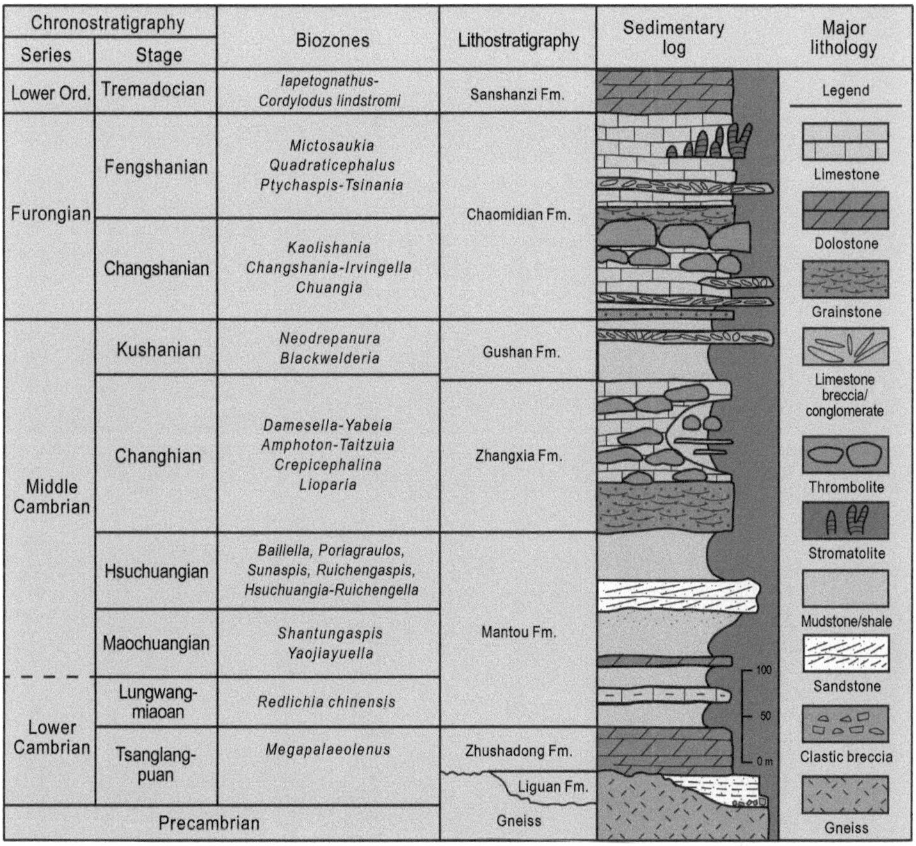

Figure 6.2 Litho- and biostratigraphic summary of the Cambrian succession in Shandong Province, China.
Source: Chough et al. (2010) by permission of *Geosciences Journal*.

6.2.3 Zhangxia Formation

The formation consists of three members (Table 6.3; Figure 6.10). The Lower Member is dominated by oolite and lesser amounts of microbial carbonates. In the Beiquanzi section, it is about 65 m thick and characterized by a thick oolite bed in which microbe-metazoan bioherms occur locally. A finer-grained interval of bioturbated wackestone occurs in the lower part of the member. This member thins eastward to about 40 m in the Jiulongshan section. It consists of a thick oolite bed in the lower part and biohermal thrombolite and oolite in the upper part. A thin and continuous shale layer occurs in the Jiulongshan section between the lowermost oolite and the overlying thrombolite beds. This member is overlain directly by a microbialite-dominated carbonate succession of the Upper Member in the western part (Beiquanzi section) and by shale-dominated Panchegou Member in the eastern part (Jiulongshan section) (Figure 6.10).

Figure 6.3 Type sections of the Cambrian succession of the North China Platform in Jinan, Shandong Province, China. (A) Mantoushan section of the Precambrian basement and the Lower to Middle Cambrian succession. (B) Tangwangzhai section of the Middle to Upper Cambrian succession. Black bars indicate trenches. For location, see Figure 6.1B.

The Panchegou Member, about 100 m thick in the Jiulongshan section, consists predominantly of shale and thin limestone of calcarenite/calcirudite (Table 6.3; Figure 6.8). It pinches out in the western part of Shandong Province. The member contains a few beds of skeletal grainstone and small-scale microbial bioherms (about 50 cm in height) in the lower part. The thickness and grain size of the intercalated calcarenite and calcisiltite beds increase upward. The base of the first continuous thrombolite bed marks the upper boundary of the member.

The Upper Member in the Beiquanzi section (about 120 m thick) is composed mostly of microbial carbonates and lesser amounts of oolite, whereas that of the Jiulongshan section, about 50–60 m thick, is characterized by limestone–shale alternation, microbial carbonates, and oolite (Figure 6.10). The uppermost part of

Figure 6.4 The Jiulongshan section in Laiwu, Shandong Province, China. (A) Paleosol and the overlying stromatolite bed in the Zhushadong Formation. (B) Subsection showing the uppermost part of the Mantou Formation and the lowermost part of the Zhangxia Formation. (C) Subsection showing the upper part of the Gushan Formation and lower part of the Chaomidian Formation. For location, see Figure 6.1B.

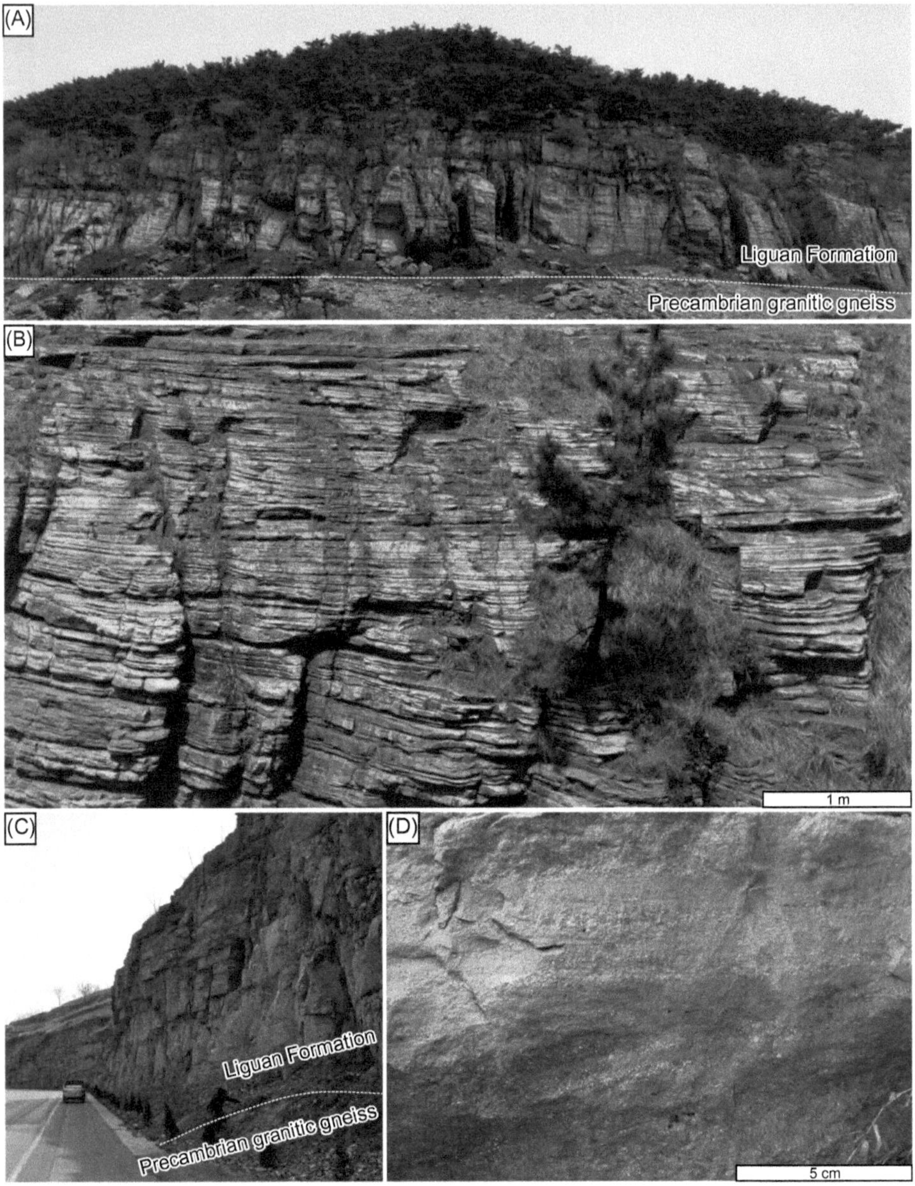

Figure 6.5 (A) Liguan Formation overlying the Precambrian granitic gneiss (Anqianzhuang section). (B) Large-scale cross-stratification of the Liguan Formation (Anqianzhuang section), indicative of shoreface environments. (C) Liguan Formation overlying the Precambrian granitic gneiss (Zhangjiapo section), dominated by immature sandstones. (D) Poorly sorted immature gravelly sandstone, indicative of deposition at a river mouth or fan delta (Zhangjiapo section). For location, see Figure 6.1B.

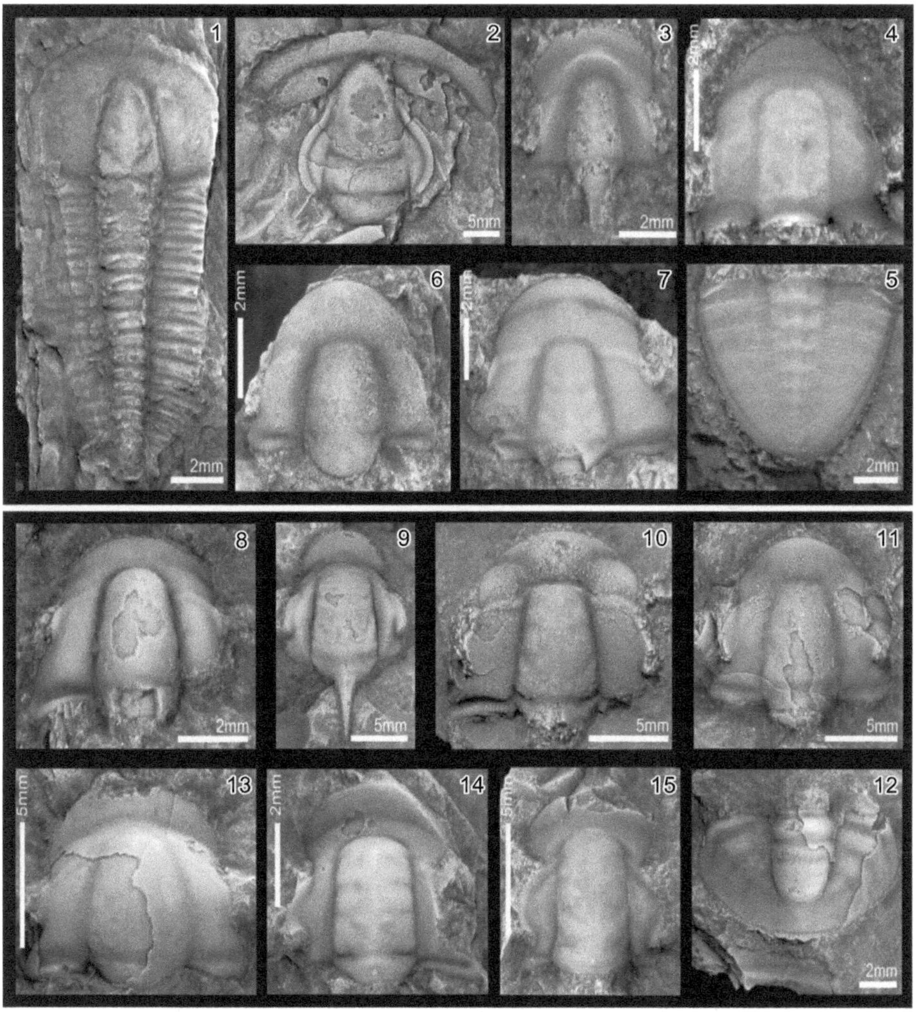

Figure 6.6 Representative trilobites from the Mantou and Zhangxia formations. (1–7) Trilobites from the Mantou Formation. (1) A complete specimen of *Bailiella* sp., Liantaishan section, SNUP5094. (2) *Redlichia chinensis* Walcott, 1905, cranidium, Jiulongshan section, SNUP5095. (3) *Lorenzella* sp., cranidium, Jiulongshan section, SNUP5096. (4–5) *Sunaspis* sp. (4) Cranidium, Liantaishan section, SNUP5097. (5) Pygidium, Jiulongshan section, SNUP5098. (6) *Ruichengaspis* sp., cranidium, Jiulongshan section, SNUP5099. (7) *Yaojiayuella granosa* (Walcott, 1905), cranidium, Jiulongshan section, SNUP5100. (8–15) Trilobites from the Zhangxia Formation. (8) *Yabeia tutia* (Walcott, 1905), cranidium, Jiulongshan section, SNUP5101. (9) *Platylisania* sp., cranidium, Jiulongshan section, SNUP5102. (10) *Poshania* sp., cranidium, Jiulongshan section, SNUP5103. (11–12) *Megagraulos* sp. (11) Cranidium, Jiulongshan section, SNUP5104. (12) Pygidium, Jiulongshan section, SNUP5105. (13) *Liopeishania* sp., cranidium, Jiulongshan section, SNUP5106. (14) *Proasaphiscus* sp., cranidium, Jiulongshan section, SNUP5107. (15) *Anomocarella* sp., cranidium, Jiulongshan section, SNUP5108. Representative trilobites from the Gushan and Chaomidian formations.

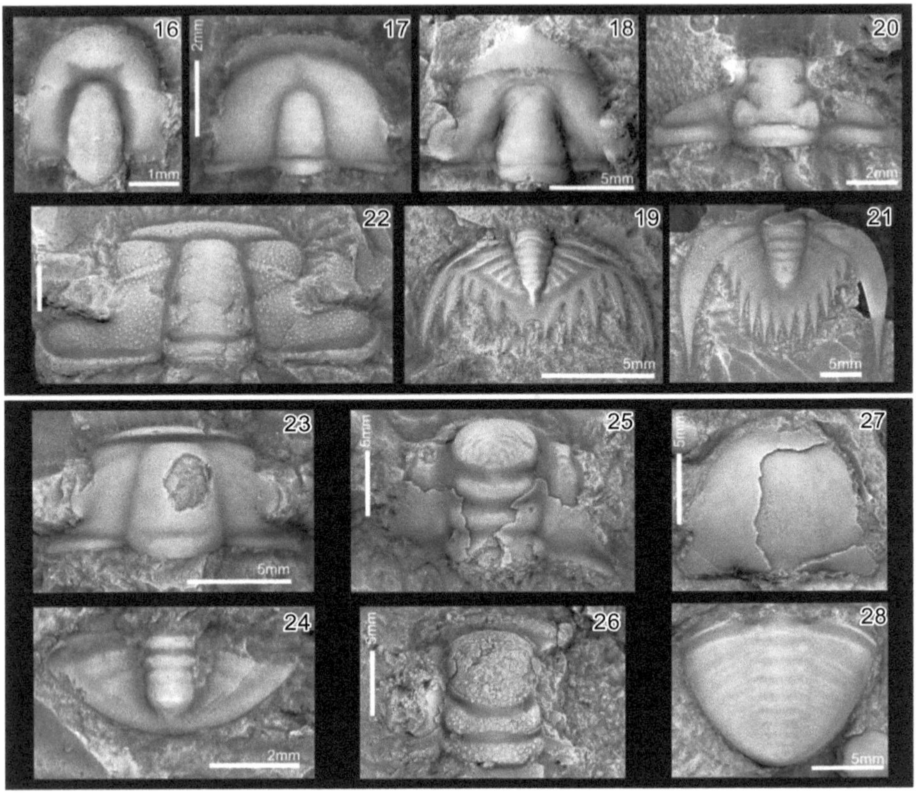

Figure 6.6 (Continued). (16–22) Trilobites from the Gushan Formation. (16) *Jiulongshania acalle* (Walcott, 1905), cranidium, Tangwangzhai section, SNUP5024. (17) *Liostracina* sp., cranidium, Tangwangzhai section, SNUP5109. (18–19) *Shantungia spinifera* (Walcott, 1905). (18) Cranidium, Tangwangzhai section, SNUP5093. (19) Pygidium, Tangwangzhai section, SNUP5110. (20–21) *Neodrepanura premesnili* (Bergeron, 1899). (20) Cranidium, Tangwangzhai section, SNUP5111. (21) Pygidium, Tangwangzhai section, SNUP5112. (22) *Damesella* sp., cranidium, Jiulongshan section, SNUP5113. (23–28) Trilobites from the Chaomidian Formation. (23–24) *Chuangia batia* (Walcott, 1905). (23) Cranidium, Jiulongshan section, SNUP5114. (24) Pygidium, Jiulongshan section, SNUP5115. (25) *Asioptychaspis* sp., cranidium, Tangwangzhai section, SNUP5116. (26) *Mictosaukia* sp., cranidium, Fanzhuang section, SNUP5117. (27–28) *Tsinania canens* (Walcott, 1905). (27) Cranidium, Tangwangzhai section, SNUP5118. (28) Pygidium, Tangwangzhai section, SNUP5119.
Source: Chough et al. (2010) by permission of *Geosciences Journal*.

the Zhangxia Formation in this section is characterized by oolitic and skeletal packstone to grainstone, which is partly rich in glauconite. The upper member of the Jiulongshan section forms distinctive layers of ovoid (or domal) microbial buildups. The uppermost unit of the Zhangxia Formation in the Jiulongshan section is characterized by light-colored thin-bedded skeletal lime mudstone and wackestone intercalated with thin lenses of packstone to grainstone. A sharp boundary between the Zhangxia and Gushan formations is marked by an abrupt lithologic change, i.e., from limestone to shale.

The Zhangxia Formation contains trilobite biozones of Changhian Stage. The Lower Member comprises the *Lioparia* Zone in the Beiquanzi section, whereas the *Lioparia* and *Crepicephalina* zones in the Jiulongshan section. The Panchegou Member yields trilobite fauna of the *Amphoton-Taitzuia* Zone. The Upper Member in Beiquanzi section includes the *Crepicephalina*, *Amphoton-Taitzuia*, and *Damesella-Yabeia* zones, whereas in the Jiulongshan section, it contains only the *Damesella-Yabeia* Zone (Figure 6.5).

6.2.4 Gushan Formation

The lower part of the Gushan Formation consists of purple shale and greenish gray shale with calcareous nodules. These beds are discontinuous and often pinch out in short distances. Some calcarenite beds show flame and loading structures as well as normal grading. The upper part of the Gushan Formation consists of more carbonate deposits than the lower part (Figures 6.4 and 6.5). Cross-stratified fine-pebble conglomerate beds with undulatory bedforms (dunes) are intercalated within calcisiltite facies. A few oolitic grainstone beds occur in the uppermost part. The upper boundary of the Gushan Formation is placed at the base of an extensive bioclastic grainstone bed. The thickness of the Gushan Formation is highly variable—105 m in the Jiulongshan section and 52 m in the Tangwangzhai section (Figure 6.11).

The Gushan Formation contains the *Blackwelderia* and *Neodrepanura* zones of the Kushanian Stage (Zhang and Jell, 1987) (Figure 6.6). The overlying *Chuangia* and *Changshania-Irvingella* zones occur in the upper part of the formation (BGMRSP, 1996). The overlying Chaomidian Formation starts with a thick bioclastic grainstone bed, which yields *Chuangia* (Figure 6.11).

6.2.5 Chaomidian Formation

The Chaomidian Formation comprises various carbonate facies such as limestone–shale alternation, microbialite, grainstone, limestone conglomerate, and wackestone to packstone. It can be divided into two members (Table 6.3; Figure 6.11). The Lower Member (105–120 m thick) consists of thin-bedded limestones such as limestone–shale/marlstone alternation, thin-bedded lime mudstone, laminated calcisiltite, and limestone conglomerates, and thick beds of microbialites (ca. 10 m thick) and grainstones (10–20 m thick).

Table 6.1 Siliciclastic Lithofacies of the Cambrian Succession in Shandong Province

Siliciclastic Facies	Description	Interpretation
Paleosol (P)	Granule-bearing mudstone (Figure 6.7A); overlying the Precambrian granitic gneiss; quartz (45%), calcite and dolomite (20%) as well as organic carbon (0.05%), clay minerals (20%), and hematite (< 6%); amounts of calcite and dolomite increasing upward up to 45%; quartz decreasing to 13%; purple and greenish gray to yellowish gray in color, showing crude stratification; abundant altered feldspar and quartz near the lower boundary, rapidly decreasing upward; calcite glaebules dispersed horizontally	Paleosol above the Precambrian basement
Shale (Sh)	Shale dominant, partly intercalated with fine sandstone, calcareous nodules, or irregular concretions, color variation from gray to purple, partly well-preserved bedding-parallel mold and body of trilobites, brachiopods and other fossils (e.g., hyolith, *Chancelloria*, *Blackwelderia*, and *Neodrepanura*)	Settling of suspended materials under low-energy and either oxidizing or reducing conditions; deep subtidal platform to basin sedimentation
Homogeneous gray mudstone (Fg)	Gray, yellowish gray mudstone; faintly laminated; sharp upper and lower boundaries with cross-stratified medium sandstone; mottled partly; severely weathered	Suspension settling of fine-grained material; reducing condition
Homogeneous purple mudstone (Fh)	Homogeneous purple mudstone (10 cm to a few meters in thickness) (Figure 6.7B); quartz (40%), calcite and dolomite (25%) as well as organic carbon (0.05–0.1%), clay minerals (20%), and hematite (6%); mottled and bioturbated structures made of hematitic and clay rich part; gradational lower boundary and sharp upper boundary	Deposition of suspended mud most likely on uppermost intertidal flat under oxidizing condition

Laminated purple mudstone (Fl)	Wavy- and lenticular-laminated purple mudstone and dolomite as well as fine sandstone and siltstone (Figure 6.7B); normally graded and bidirectional ripple cross-laminated dolomite and calcareous fine sandstone; desiccation cracks and burrows commonly present; upward-thinning or thickening laminae	Tidal deposits in shallow subtidal and intertidal environments
Irregular-laminated mudstone to fine sandstone (Fi)	Irregular-laminated dark purple siltstone to fine sandstone (a few dm to a meter thick); compose of quartz (60%), hematite (8%), and carbonate minerals (7%); less than 3 mm thick laminae dominant; wrinkled and irregular cross- or horizontal-stratified; dark purple, dusky red, alternating dusky red and yellow in color	Intertidal to lowermost shoreface sediments stabilized and modified by microbial mats
Dark purple mudstone and sandstone alternation (Fp)	Alternation of dark purple mudstone and fine calcareous sandstone (Figure 6.7C); discontinuous and wavy sandstone beds (a few mm to 10 cm thick); normally graded or low-angle cross-laminated; abundant fragments of trilobite, echinoderm, and brachiopod; dispersed glauconite grains; mudstone composed of quartz (60%), hematite (8%), carbonate minerals (6%), and clay minerals (18%); sandstone sheet, pot, and gutter cast present; thickening- or thinning-upward trend of sandstone layers	Lower shoreface to offshore environments; lagoonal or subtidal environment with frequently introduced sandy sediments
Horizontally laminated fine sandstone (Shl)	Finely laminated fine sandstone; horizontally laminated laminae, a few mm thick; partly cross-laminated; muscovite-rich; discontinuous fossil-rich layers with bioturbation; calcareous nodules; gray, dark gray, and dark purple in color; mudstone chips and layers frequently recognized	Suspension settling of fine-grained material with frequent input of sandy grains and fossil fragments; alternation of background settling deposits and distal siltstone by storm currents

(*Continued*)

Table 6.1 (Continued)

Siliciclastic Facies	Description	Interpretation
Planar and trough cross-stratified sandstone (Sx)	Planar and trough cross-stratified calcareous medium sandstone (Figure 6.7D); sandstone composed of well-sorted quartz grains and small amounts of ooids and fragments (trilobite and brachiopod); cross-stratification represented by alternation of medium and fine sandstones; 5–25 cm thick cross-set; yellowish gray and gray in color	2D and 3D dune migration on calcareous sand shoal in upper shoreface
Bioturbated sandstone (Sb)	Yellow bioturbated calcareous medium sandstone (ichnofacies index-3 and -4) (Figure 6.7E); yellowish gray and gray in color; crude stratification, represented by alternation of medium and fine sandstones	Shoreface or intershoal deposit with abundant biogenic activities of benthic organisms
Crudely cross-stratified sandstone (Sc)	Crudely cross-stratified and massive medium sandstone; planar cross-stratified partly; partly ripples on the top surface; mudstone pebble clasts aligned; wedge- and sheet-shaped geometry; a few cm to dm thick beds amalgamated; partly parted by thin fine-grained layers; thickening- or thinning-upward trends; abundant quartz	Deposition by 2D and 3D dune migration; rapid sedimentation from heavy sediment-laden flow
Hummocky and swaley cross-stratified sandstone (Shs)	Hummocky and swaley cross-stratified purple medium to fine sandstone (Figure 6.7F); changing upward from swaley cross-bedding to hummocky, trough, and ripple cross-lamination; sand volcanoes (a few dm in thickness and 3–5 cm in diameter) in swaley cross-bedded part; ripples on the top of swaley cross bed; lateral variation from amalgamated to thin-mudstone-parting swaley cross beds	Deposition of sand by storm-induced currents and waves in lower shoreface

Source: Chough et al. (2010) by permission of *Geosciences Journal*.

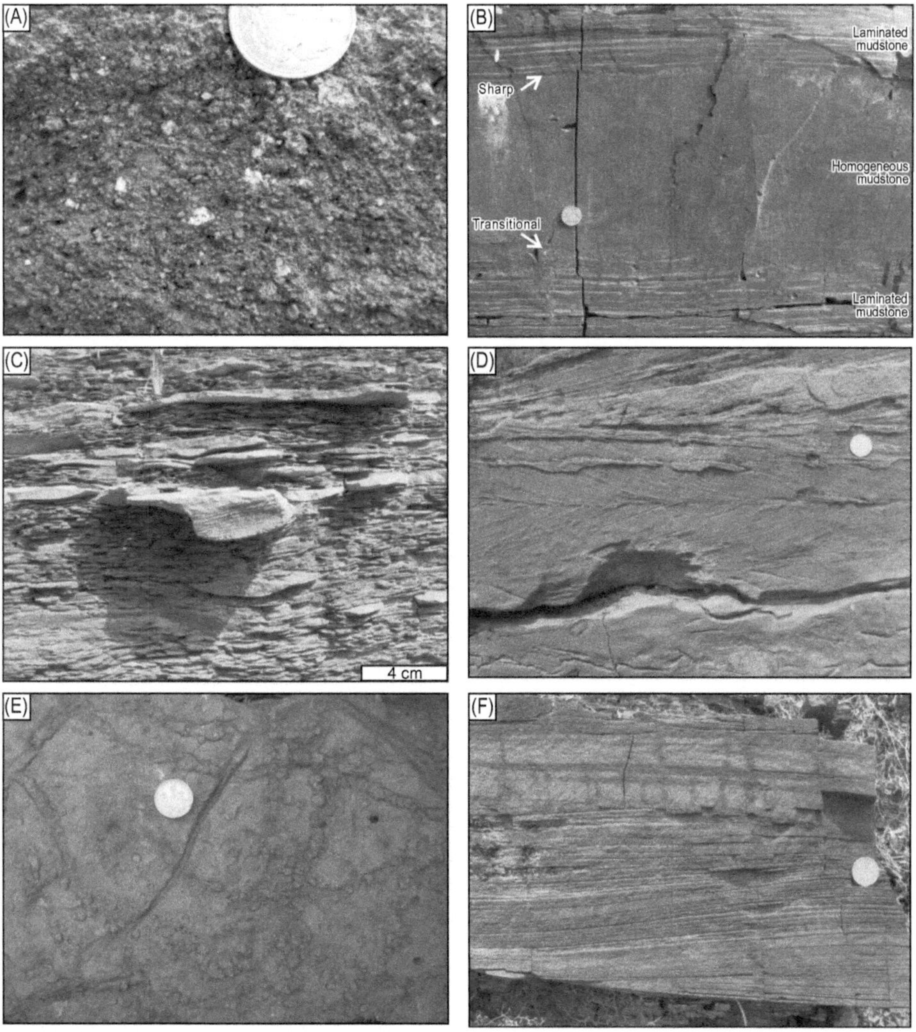

Figure 6.7 Representative outcrop photographs of siliciclastic sedimentary facies of the Cambrian succession in Shandong Province, China. (Coin in (A) and (E) is 25 mm in diameter; coin in (B) and (D) is 20.5 mm in diameter; coin in (F) is 24 mm in diameter.) (A) Paleosol at the base of the Cambrian succession, composed of weathered feldspar and rounded quartz sands (Liangcheng section). (B) Laminated and homogeneous purple mudstone (Shiwangya section). (C) Dark purple mudstone and sandstone alternation. Sandstones show gutter casts and starved ripples in form (Jiulongshan section). (D) Trough cross-stratified calcareous medium sandstone (Jiulongshan section). (E) Bioturbated medium sandstone (Liangcheong section). (F) Hummocky and swaley cross-stratified fine sandstone (Jiulongshan section). For location, see Figure 6.1B.

Table 6.2 Carbonate Lithofacies of the Cambrian Succession in Shandong Province

Carbonate Facies	Description	Interpretation
Limestone–shale alternation (L–S)	Alternation of homogeneous or laminated limestone and dolomitic shale or marlstone (Figure 6.8A); microstylolites commonly present in dolomitic shale layers; intraclasts and desiccation cracks partly present; wedge-shaped grainstone with erosional boundary partly occur; *Planolites*, *Paleophycus*, and *Skolithos*; containing fossiliferous horizons	Cyclic input of carbonates and background sedimentation of argillaceous materials; intertidal to shallow subtidal or deep subtidal environments
Thin-bedded lime mudstone (Ltb)	Thin-bedded lime mudstone (Figure 6.8B); thin shale parting either flaser or flat lime mudstone to wackestone beds; sporadic horizontal to inclined burrows; partly irregular lamination, normal grading, and erosional features	Deposits of relatively low-energy conditions; subtidal to intertidal environments, partly modified by burrows
Laminated dolomudstone (Dl)	Irregularly and discontinuously laminated dolomudstone interlayered with wackestone to peloidal grainstone (Figure 6.8C); dispersed quartz grains, halite pseudomorph, and desiccation cracks in certain laminae; alternation of thick (about 500 μm) and coarse (about 30–50 μm) crystalline lamina and thin (about 70 μm) and fine (<10 μm) lamina	Deposition in shallow intertidal environments with frequent subaerial exposure
Wavy-laminated limestone (Wl)	Wavy-laminated limestone (Figure 6.8D); planar and ripple cross-laminated; interlayered shale from 1 to a few mm in thickness; cross-laminated packstone and grainstone, showing discontinuous wavy bedding; well-sorted peloidal grainstone interlayered; small stromatolite mound and channelized conglomerate with erosional lower boundary	Wave base under relatively high-energy conditions such as sandy tidal flat based on intercalated conglomerate and small stromatolite mounds
Laminated calcisiltite (Cl)	Parallel, ripple, and low-angle cross-laminated calcisiltite intercalated with dolomitic marlstone or shale (Figure 6.8E); composed of silt-sized calcite particles; wavy-bedded, unidirectional and low-angle cross-lamination with truncational boundary; climbing ripples; partly bioturbated with burrows cutting laminae	Subtidal deposits by unidirectional currents (partly combined with oscillatory movement of water) induced by storms, marlstone deposited during fair weather

Bioturbated wackestone facies (Wb)	Moderately to severely bioturbated (Figure 6.8F); gray lime mudstone to wackestone; horizontal to inclined burrows; mottled texture; composed mainly of micrite, and fossil fragments and peloids; partly intercalated with thin bioclastic grainstone with sharp lower boundary; oncoids with distinct laminae and irregular diffuse laminae; brachiopods, spicule of *Chancelloria*, and filaments of *Girvanella* partly present	Low-energy subtidal deposits modified by bioturbation
Wackestone to grainstone facies (WG)	Thin-bedded or flaser-bedded wackestone separated by shale partings; slightly bioturbated; often intercalated with lenses or thin layers of grainstone with sharp lower boundary, partly normally graded and stratified; partial concentration of fossil fragments (Figure 6.8G); lime mudstone clasts derived from the base; partly loading structures	Relatively low-energy subtidal conditions with intermittent higher energy episodes
Thin-bedded wackestone to packstone (WP)	Thin-bedded alternation of irregularly bounded pack-grainstone and bioturbated finer sediments; lime mudstone clasts derived from the base; lenses of grainstone; loading structures	Deposition near the fair-weather wave base in moderately agitating, shallow subtidal environments
Wavy-bedded packstone (Pw)	Wavy-laminated or -bedded calcareous dolopackstone to peloidal grainstone; discontinuous and wavy stratification represented by an alternation of yellow dolomudstone and dolopackstone with sharp or erosional boundary; crude stratification and bidirectional ripple cross-lamination commonly present; sharp lower and upper boundaries	Intertidal deposits modified by tidal currents
Oolitic packstone to grainstone (O)	Massive and crudely stratified oolitic and oncolitic packstone to grainstone; common occurrence of hardground; glauconitic grains and round intraclasts of oolite commonly present	Shallow subtidal shoal with intensive reworking of waves and reduced sedimentation rate
Hummocky and swaley cross-stratified grainstone (Ghsc)	Peloidal grainstone (Figure 6.8H), composed of coarse silt- to very fine sand-grade peloids and small fraction of fossil fragments; laterally discontinuous and undulatory stratum, varying in thickness from a few dm to 2 m; thick beds	Deposits modified by storm-induced combined flows

(*Continued*)

Table 6.2 (Continued)

Carbonate Facies	Description	Interpretation
	amalgamated with internal sharp boundaries; *Skolithos*, 1–5 cm in depth	
Planar and trough cross-stratified grainstone (Gptc)	Planar or trough cross-stratified oolitic or bioclastic grainstone (Figure 6.8I); glauconite grains present; crosssets a few to 10 cm in thickness; dolomudstone clasts present, thrombolite and dendrolite beds, and wedge-shaped conglomerate partly intercalated, partly undulatory bedforms	Migration of 2D and 3D dunes on shoal or foreshore environments
Calcarenite (CA)	About 5–20 cm in thickness; sharp internal bounding surfaces; partly erosional base; containing small lime mudstone clasts and bioclastic grains; faintly laminated or homogeneous; horizontal burrows; loading and flame structures; amalgamated thicker beds	Fast deposition from dilute turbidity currents, modified by sediment loading
Gravelly grainstone (Gg)	Massive or normally graded gravelly packstone to grainstone; composed of bioclasts (trilobite, brachiopod, algae, echinoderm, and cephalopod) and peloids; subangular granules and pebbles of lime mudstone to grainstone commonly at base	Deposition from storm-induced currents
Cross-stratified limestone conglomerate (Cs)	Granule- to pebble-grade polymictic clasts with subangular to rounded corners (Figure 6.8J); clasts of lime mudstone, dolomudstone, grainstone, purple mudstone, and limestone conglomerate; coarse-grained matrix composed of bioclast, peloid, and ooid; mostly flat-lying and imbricated clasts; normally graded or cross-stratified	Deposition by bedload transportation; generation of clasts by reworking of previous deposit by strong currents or waves
Limestone pseudoconglomerate (Cp)	Oligomictic to polymictic clasts in a matrix of marlstone and/or grainstone (Figure 6.8K); flat to irregular sheet-, disk-, or blade-shaped clasts; random internal fabrics (intact, thrusted, mosaic, and disorganized); transitional boundaries from underlying bed	Diagenetic conglomerate due to differential early cementation and deformation

Stromatolite (ST)	Finely to thickly laminated convex-up mat encrusting various sediments (Figure 6.8L); finely laminated by alternation of different-colored micrite; occasional presence of filamentous calcimicrobe, *Girvanella* in laminae; thick laminae with various sand-sized skeletal and glauconitic grains trapped between laminae; either flat, domal, LLH, or columnar in shape	Encrustation of microbial film or mat over substrates; trapping and binding sediments with contribution of microbially induced calcite cementation; intertidal to subtidal
Thrombolite (TH)	Clots of dark-colored micritic carbonate in light-colored lime mudstone and dolomitic matrix (Figure 6.8M); very faint internal structures in some cases; scarce fossil fragments; biostromal and biohermal; calcimicrobes such as *Epiphyton* and *Girvanella*; sessile metazoans like sponges	Formed by blotchy microbial masses; trapping and baffling of carbonate grains; low- to intermediate-energy subtidal environment
Dendrolite (DE)	Columnar and digitate dense dark micritic masses; mixture of shaly material and coarse skeletal peloidal grains among the columns; concave-up stratification in pack-grainstone among the columns	Rapid growth of microbial colony with branching morphology which baffles coarse-grained sediments; higher-energy shallow subtidal conditions
Epiphyton framestone (EP)	Forming meter-scale buildups (Figure 6.8N); stacked or coalescent thick buildup and isolated in muddy sediment; association with coarse debris from the buildup	Shallow marine platform margin setting
Maceriate microbialite (MM)	Biohermal/biostromal megastructure; consists of maceriae and intermacerial sediment with a ragged, obscure boundary (Figure 6.8O); internal structure of maceriae are chaotic; intermacerial sediments composed of lime mud and bioclasts; maceriae composed of micrite, *Girvanella*, and *Renalcis*-like microbes	Low-energy environments near normal wave base

Source: Chough et al. (2010) by permission of *Geosciences Journal*.

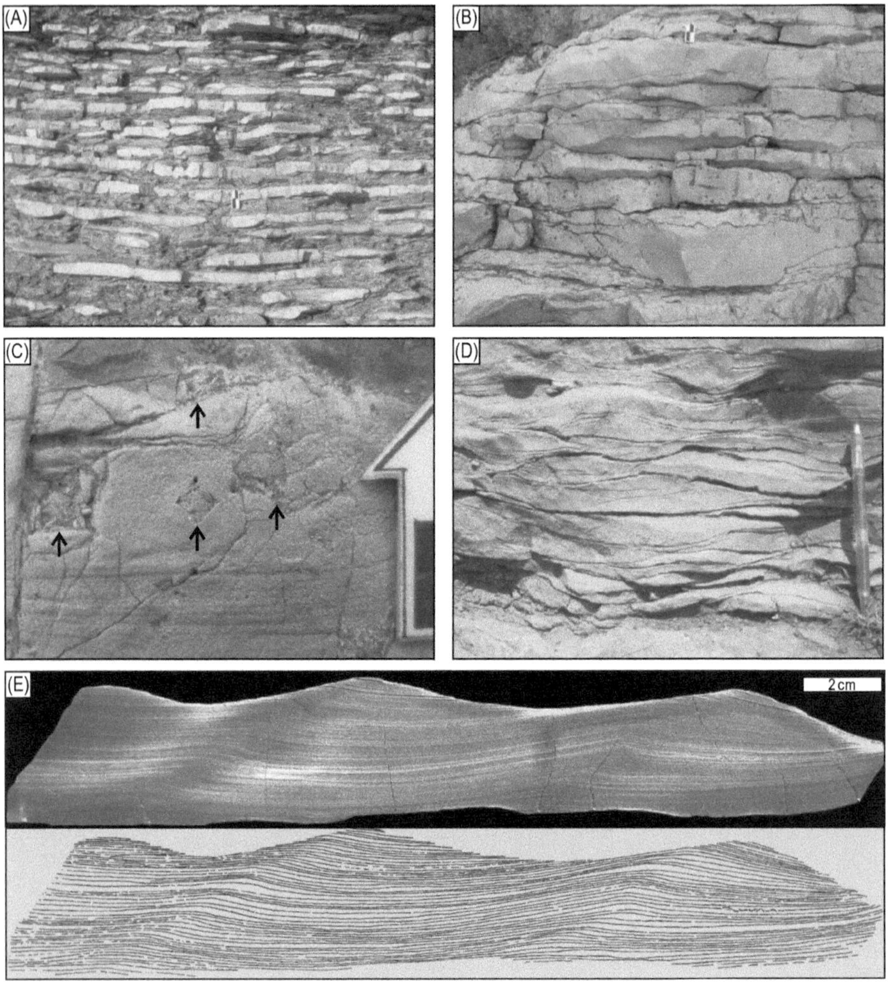

Figure 6.8 Representative photographs of carbonate lithofacies of the Cambrian succession in Shandong Province, China. (A) Limestone and shale alternation (Jiulongshan section). (B) Thin-bedded lime mudstone with orange burrows (Jiulongshan section). (C) Laminated dolomudstone with halite pseudomorphs (arrows) (Mantoushan section). (D) Wavy-laminated wackestone (Jiulongshan section). (E) Climbing ripple cross-laminated calcisiltite (Jiulongshan section). (F) Bioturbated wackestone (Laopozhuang section). (G) Bioclastic wackestone to packstone with rich concentration of bivalve shells showing geopetal structures (arrows) (Fanzhuang section). (H) Hummocky cross-stratified grainstone (Jiulongshan section). (I) Trough cross-stratified grainstone (Jiulongshan section). (J) Cross-stratified limestone conglomerate (Jiulongshan section). (K) Limestone pseudoconglomerate (Jiulongshan section). (L) Domal stromatolite (Jiulongshan section). (M) Thrombolite (Beiquanzi section). (N) *Epiphyton* bioherm (Jiulongshan section). (O) Columnar microbialite (Tangwangzhai section). (Scale bar in (A), (B), (C), (G), and (K) is 2 cm long; pencil in (D), (M), and (O) is 14 cm long; coin in (F), (I), (J), and (L) is 2 cm in diameter; scale in (H) is 15 cm long; hammer head in (N) is 17 cm long.) (For interpretation of the references to color in this figure legend, the reader is referred to the web version of this book.)

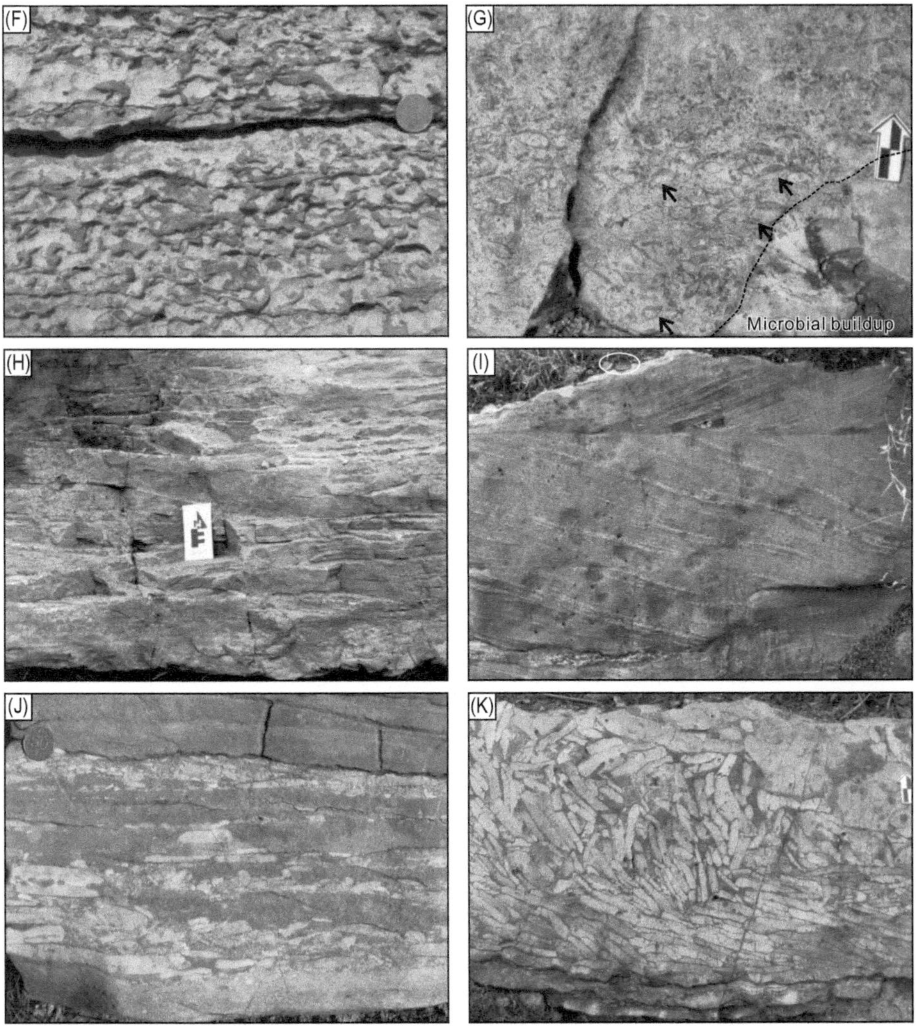

Figure 6.8 (Continued)

The Upper Member (ca. 89–120 m thick) is characterized by a thick monotonous succession of bioturbated wackestone and wackestone to grainstone with lesser amounts of limestone–shale alternation, gravelly grainstone, planar and trough cross-stratified grainstone, and limestone conglomerate. An 11-m-thick interval of columnar stromatolites occurs in the middle part of the Upper Member in the Jiulongshan section (Figure 6.11), whereas it is absent in the Jinan area. The Chaomidian Formation changes gradationally into the Ordovician Sanshanzi Formation, which is characterized by massive dolostone.

The Chaomidian Formation includes six trilobite biozones: the *Chuangia*, *Changshania-Irvingella*, *Kaolishania*, *Ptychaspis-Tsinania*, *Quadraticephalus*, and

Figure 6.8 (Continued)

Mictosaukia zones in ascending order (Figure 6.2) (Zhang and Jell, 1987). The former three represent the Changshanian Stage, whereas the latter three constitute the Fengshanian Stage. The base of the Changshanian Stage in the North China Platform is correlated with the base of the Furongian Series based on the stable isotope excursion (Chen et al., 2011; Zhu et al., 2004).

6.3 Depositional History and Sequence Stratigraphy

Deposition in the North China Platform was initiated in the Cambrian with the ensuing rise in sea level, and the entire platform was submerged except for several highlands in the northwestern part of the platform (the Ordos area) (Meng et al., 1997). The succession represents a long-term transgression of second-order sea-level rise that lasted for about 70 million years without a significant break in sedimentary records (Meng et al., 1997). The lower part of the Cambrian succession (Liguan, Zhushadong, and Mantou formations) (mixed siliciclastic and carbonate deposits) represents initial marine flooding and stabilization of shallow marine environments (Lee and Chough, 2011) (Figure 6.12).

Siliciclastic sediments were derived from unsubmerged archipelago and deposited in coastal floodplains under strong tidal regime, whereas carbonates were produced in flooded peritidal environments (Figure 6.13). With the ensuing rise in sea

Table 6.3 Constituent Facies of Lithostratigraphic Units and Trilobite Biozones

Fm.[a]	Mbr.[b]	Constituent lithofacies: Jinan area[e]	Constituent lithofacies: Jiulongshan section	Trilobite biozones	GA[c]
Chaomidian	Upper	Bioturbated wackestone (Wb) Wackestone to grainstone (WG) Gravelly grainstone (Gg) Limestone and shale alternation (LS)	Bioturbated wackestone (Wb) Wackestone to grainstone (WG) Gravelly grainstone (Gg) Columnar stromatolite (ST) Limestone and shale alternation (LS)	*Mictosaukia* *Quadraticephalus*	Furongian
Chaomidian	Lower	Planar and trough cross-stratified grainstone (Gptc) Hummocky and swaley cross-stratified grainstone (Ghsc) Maceriate microbialites (MM) Thin-bedded lime mudstone (Ltb) Limestone and shale alternation (LS) Shale (Sh)	Planar and trough cross-stratified grainstone (Gptc) Hummocky and swaley cross-stratified grainstone (Ghsc) Maceriate microbialites (MM) Thin-bedded lime mudstone (Ltb) Limestone and shale alternation (LS) Shale (Sh)	*Ptychaspis-Tsinania* *Kaolishania* *Changshania-Irvingella* *Chuangia*	Furongian
Gushan		Planar and trough cross-stratified grainstone (Gptc) Cross-stratified limestone conglomerate (Cs) Laminated calcisiltite (Cl) Limestone pseudoconglomerates (Cp) Limestone and shale alternation (LS) Shale (Sh)	Cross-stratified limestone conglomerate (Cs) Laminated calcisiltite (Cl) Limestone and shale alternation (LS) Calcarenite (CA) Limestone and shale alternation (LS) Shale (Sh)	*Neodrepanura* *Blackwelderia*	Cambrian Series 3
Zhangxia[d]	Upper	Thrombolite (TH) Dendrolite (DN) Planar and trough cross-stratified grainstone (Gptc) *Epiphyton* framestone (EP) Oolitic packstone to grainstone (O) Stromatolite (ST)	Thrombolite (TH) *Epiphyton* framestone (EP) Limestone and shale alternation (LS) Planar and trough cross-stratified grainstone (Gptc)	*Damesella-Yabeia* *Amphotom-Taitzuia* *Crepicephalina* *(Damesella-Yabeia)* *(Amphotom-Taitzuia)*	Cambrian Series 3
Zhangxia[d]	Pan-chegou	-	Shale (Sh) Calcarenite (CA) Limestone and shale alternation (LS)	*Lioparia* *(Crepicephalina* *Lioparia)*	Cambrian Series 3
Zhangxia[d]	Lower	Thrombolite (TH) Oolitic packstone to grainstone (O) Bioturbated wackestone facies (Wb)	Thrombolite (TH) Planar and trough cross-stratified grainstone (Gptc) Dendrolite (DN) Oolitic packstone to grainstone (O)	*Bailiella* *Poriagraulos*	Cambrian Series 3
Mantou	Upper	Homogeneous purple mudstone (Fh) Dark purple mudstone and sandstone alternation (Fp) Planar and trough cross-stratified sandstone (Sx) Limestone pseudoconglomerates (Cp)	Oolitic packstone to grainstone (O) Dark purple mudstone and sandstone alternation (Fp) Hummocky and swaley cross-stratified sandstone (Shs) Planar and trough cross-stratified sandstone (Sx) Shale (Sh) Planar and trough cross-stratified grainstone (Gptc) Thrombolite (TH) Oolitic packstone to grainstone (O)	*Sunaspis* *Ruichengaspis* *Hsuchuangia-Ruichenpella* *Shantungaspis* *Yaojiayuella*	Cambrian Series 3
Mantou	Lower	Laminated purple mudstone (Fl) Laminated dolomudstone (Dl) Stromatolite (ST) Homogeneous purple mudstone (Fh) Irregular laminated mudstone to fine sandstone (Fi)	Homogeneous purple mudstone (Fh) Laminated purple mudstone (Fl) Oolitic packstone to grainstone (O) Limestone and shale alternation (LS) Hummocky and swaley cross-stratified sandstone (Shs) Irregular laminated mudstone to fine sandstone (Fi)	*Redlichia Chinensis*	Cambrian Series 2
Zhushadong		Laminated dolomudstone (Dl) Stromatolite (ST) Paleosol (P) **Liangcheng section** Bioturbated wackestone (Wb) Laminated dolomudstone (Dl) Stromatolite (ST) Paleosol (P)	Oolitic packstone to grainstone (O) Laminated dolomudstone (Dl) Stromatolite (ST) Paleosol (P)		Cambrian Series 2
Liguan		**Jinhe section** Homogeneous gray mudstone (Fg) Horizontally laminated fine sandstone (Sh) Crudely cross-stratified sandstone (Sc)			Cambrian Series 2

[a]Fm.=Formation, [b]Mbr.=Member, [c]GA=Geologic Age.
[d]Biozones of the Zhangxia Formation in parenthesis are for the Jiulngshan section.
[e]Jinan area includes Mantoushan, Liantaishan, Beiquanzi, Tangwangzhai, and Fanzhuang II sections.

Source: Chough et al. (2010) by permission of *Geosciences Journal*.

level, carbonate production kept pace and formed numerous depositional cycles composed of lime mudstone, dolomite, stromatolite, and purple mudstone, showing a general shallowing-upward trend. These cycles formed in a keeping-up phase of carbonate production in response to an increase in accommodation but were affected by the input of fine-grained siliciclastic sediments from the emergent archipelago, displaying autogenic controls. The islands also played an important role in the generation of tidal currents on the vast platform. With the further rise in sea level, sediments were largely deposited in wave-dominant environments above the storm-wave base, forming microbial mounds and flats.

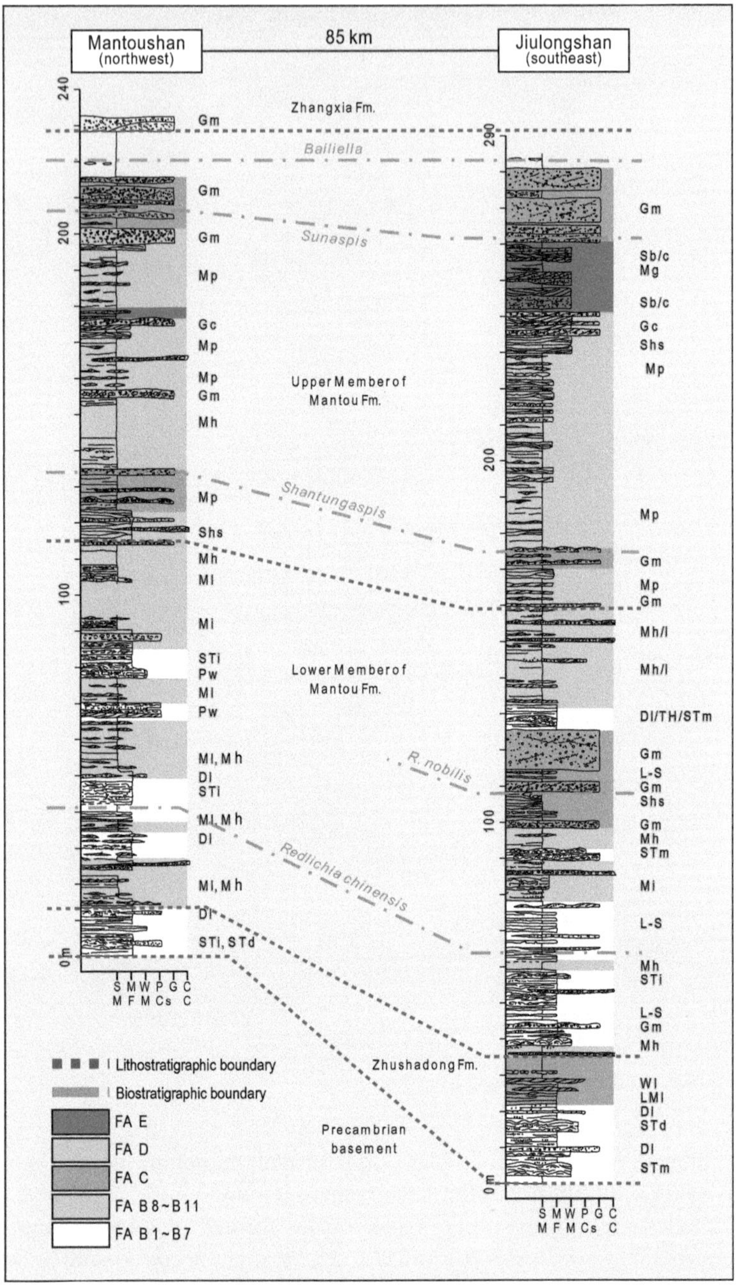

Figure 6.9 Schematic sedimentary logs of the representative sections in the Jinan and Laiwu areas. For sedimentary facies codes, see Tables 6.1 and 6.2. For section locations, see Figure 6.1.
Source: Lee and Chough (2011) by permission of the Wiley-Blackwell.

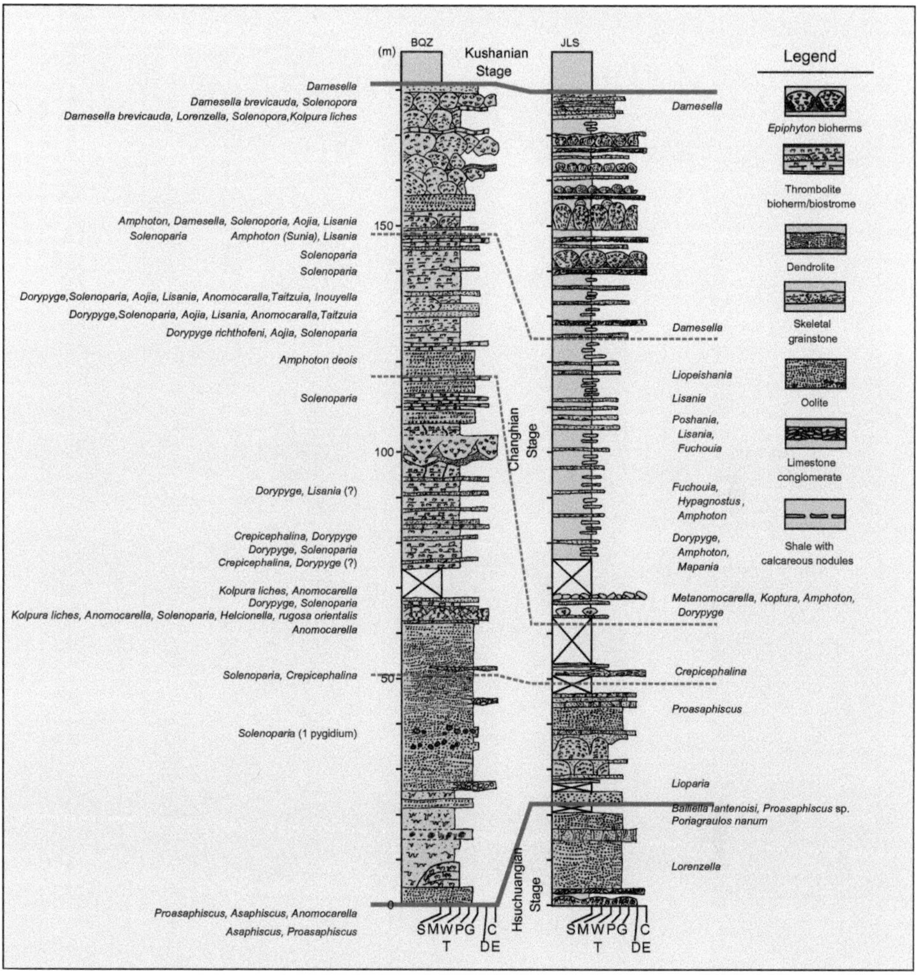

Figure 6.10 Representative sedimentary logs of the Zhangxia Formation with stage boundaries. Probable timelines are drawn based on the occurrence of trilobites. Trilobite data of Beiquanzi and Jiulongshan sections are from Lu and Dong (1953) and Zhang and Jell (1987), respectively. BQZ, Beiquanzi section; JLS, Jiulongshan section; S, shale; M, lime mudstone; W, wackestone; P, packstone; G, grainstone; C, conglomerate; T, thrombolite; D, dendrolite; E, *Epiphyton* bioherm.
Source: Woo (2009).

Production of carbonates kept up with the rise in sea level in the overlying Zhangxia Formation (Middle Cambrian), dominated by oolite and various microbialite facies, representing optimum conditions for carbonate production and microbial activities during highstand in sea level (Woo and Chough, 2010; Woo et al., 2008) (Figure 6.14). With the ensuing rise in sea level, the platform (shoal oolites and microbialites) was continuously aggraded, whereas the inter-platform

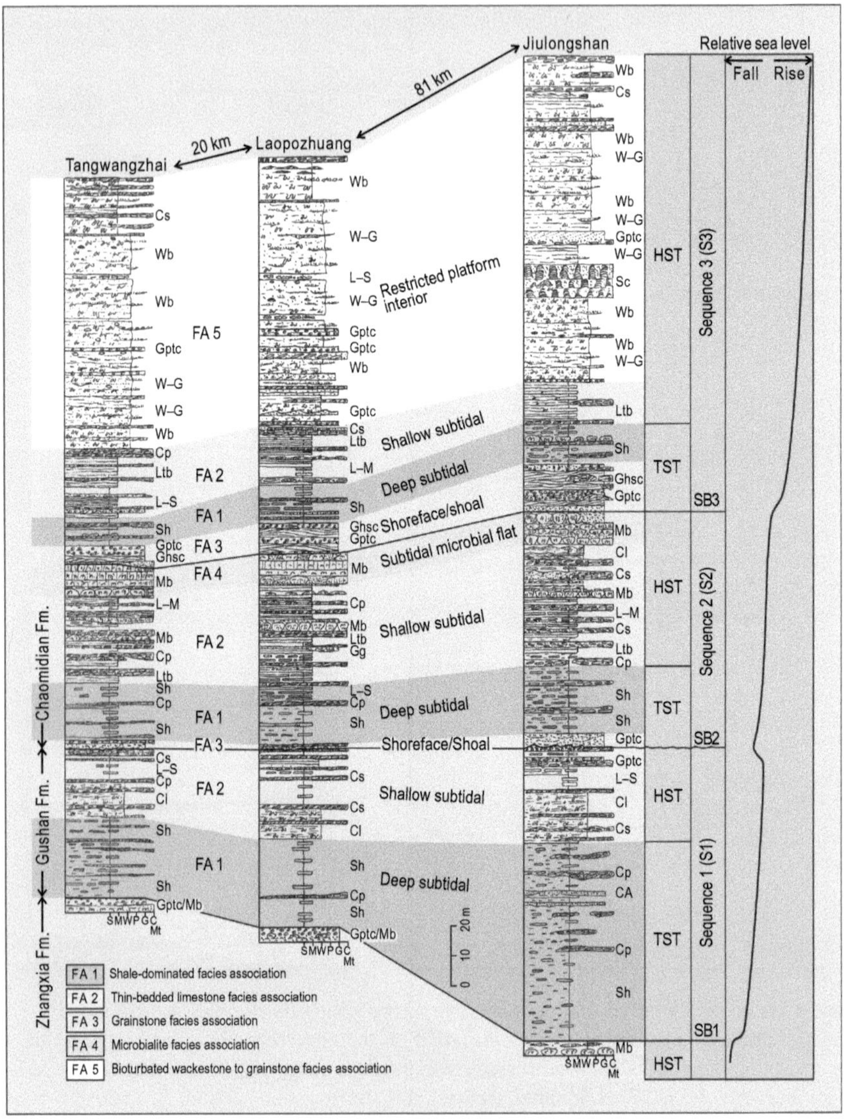

Figure 6.11 Facies associations and depositional sequences of three major sections (for section location, see Figure 6.1B). S, shale; M, lime mudstone; W, wackestone; P, packstone; G, grainstone; C, limestone conglomerate; Mt, microbialite. Sh, shale; L–S, limestone/shale alternation; L–M, limestone/marlstone alternation; Ltb, thin-bedded lime mudstone; Cl, laminated calcisiltite; Wb, bioturbated wackestone; W–G, wackestone to grainstone; Ghsc, hummocky and swaley cross-stratified grainstone; Gptc, planar and trough cross-stratified grainstone; Gg, gravelly grainstone; CA, calcarenite; Cs, stratified limestone conglomerate; Cp, limestone breccia; Mb, biostromal microbialite; Sc, columnar stromatolite. HST, highstand systems tract; TST, transgressive systems tract. SB, sequence boundary.
Source: Chen et al. (2011) by permission of the Elsevier.

areas were under a hemipelagic realm with low sedimentation rates, which resulted in the development of local slopes with deposition from suspension and episodic sediment gravity flows (Woo, 2009) (Figure 6.14).

The microbialite-dominated carbonate platform in the upper part of the Zhangxia Formation was subsequently drowned as a result of eustatic sea-level rise (Figure 6.15) (Kwon et al., 2006; Meng et al., 1997). Bounded by a drowning unconformity, the overlying succession consists mainly of shale (Gushan Formation) and various carbonate facies (Chaomidian Formation) (Chen et al., 2011). Due to a rapid rise in sea level, the increased rate of accommodation overcame that of sediment supply, resulting in sediment starvation (Figure 6.15). Sedimentation was confined merely to background settling of terrigenous fines. Toward the upper part of the Gushan Formation, carbonate production caught up with sea-level rise, accumulating more carbonates. The shallow-subtidal carbonate deposits were deformed by increased pore pressure induced by the rapid fall in sea level, and the platform was subsequently exposed in the Shandong area, forming a cryptic subaerial unconformity (Chen et al., 2011). Microbialites resurged and flourished as the platform was submerged again during the Furongian, and an extensive microbialite bed occurs in the middle part of the Chaomidian Formation, indicating a broad microbial flat (Lee et al., 2010a). The epeiric platform was eventually formed and maintained through the late Furongian to the Early Ordovician.

6.4 Comparison with the Taebaeksan Basin

The lithologic units of the Cambrian succession are well correlated in formation scale between the Shandong region and the Taebaek area, based on the occurrence of trilobites (Figure 6.16). The occurrence of quartzite at the base of the succession is identical in both areas. The overlying lithologic unit of siliciclastics and carbonates (Zhushadong and Mantou formations) is equivalent to that of shale-dominant Myobong Formation. The co-occurrence of *Redlichia chinensis* in the Mantou and Myobong formations is indicative of the Early and Middle Cambrian in age (Figure 6.16). The *Bailiella* Zone is also well correlated. The overlying carbonate-dominant succession (Zhangxia Formation) is very similar to the Daegi Formation. The mixed shale and limestone (Gushan and Chaomidian formations versus Sesong and Hwajeol formations) are generally well correlated in both the lithology and the biozone of trilobites (Figure 6.16).

A comparison of relative sea-level curve between the Shandong region and the Taebaek area during the Cambrian suggests that both areas underwent initial flooding at the same time. In the Shandong region, sediments were delivered from local river mouths and dispersed along the coastline, forming an extensive shoreface facies. In the Taebaek area, sediments deposited in alluvial fans, and in shoreface and tide-dominated environments. During the ensuing transgression, water depth increased slowly in the Shandong region but abruptly in the Tabaek area. As a result, the sedimentary facies of both regions are distinctive from one another

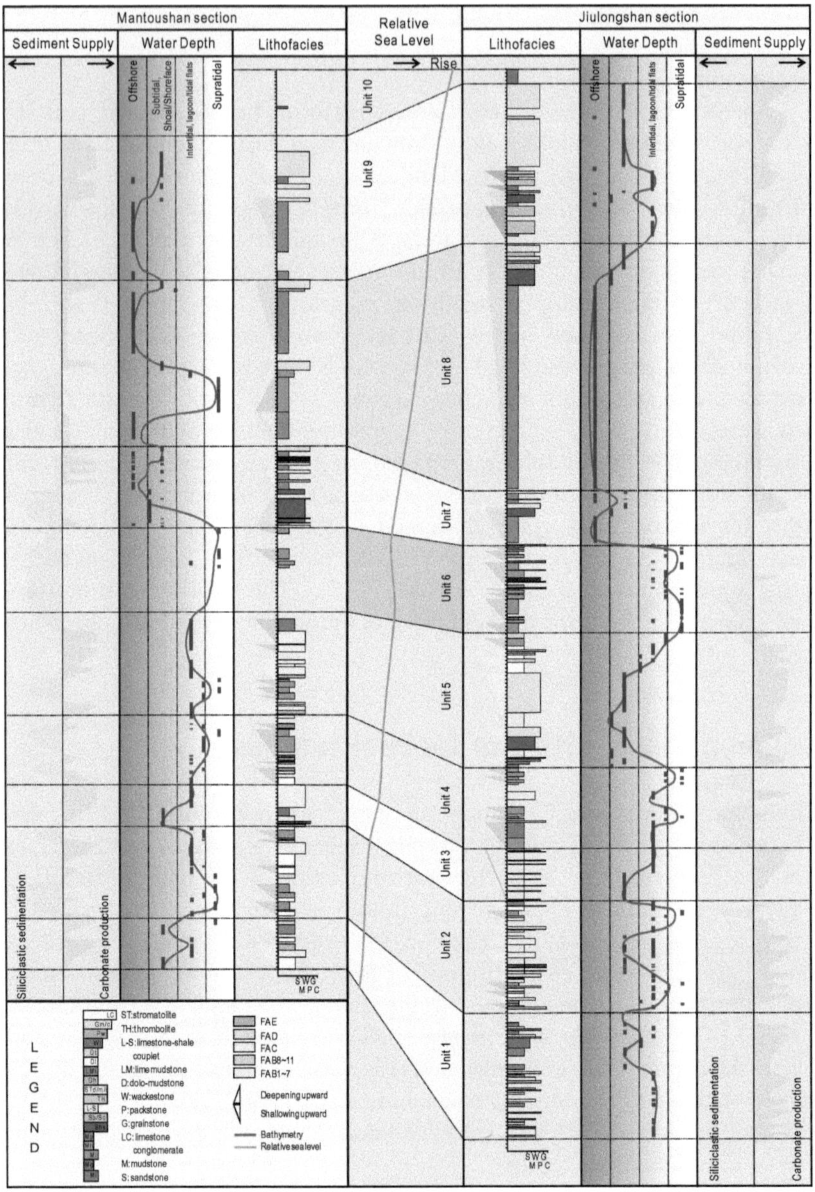

Figure 6.12 Columnar summary of sediment supply, water depth, and relative sea-level changes for the Zhushadong and Mantou formations showing variations in water depth and sediment supply between the Mantoushan and Jiulongshan sections. Red and black lines in water depth indicate smooth bathymetries and depositional environments based on lithofacies and facies associations. (For interpretation of the references to color in this figure legend, the reader is referred to the web version of this book.)
Source: Lee and Chough (2011) by permission of the Wiley-Blackwell.

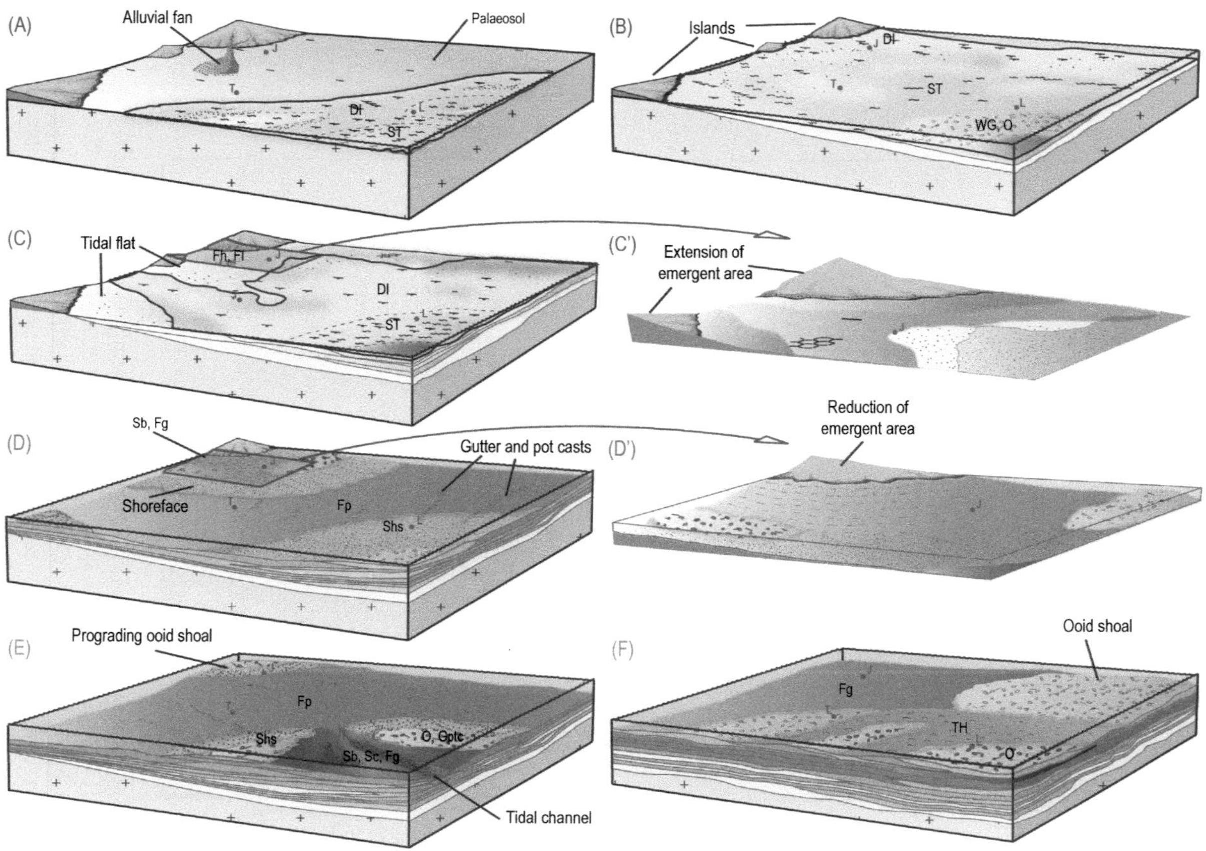

Figure 6.13 Depositional model of the Zhushandong and Mantou formations. J, Jinan; T, Taian; L, Laiwu. For facies codes, see Tables 6.1 and 6.2. *Source*: Lee and Chough (2011) by permission of the Wiley-Blackwell.

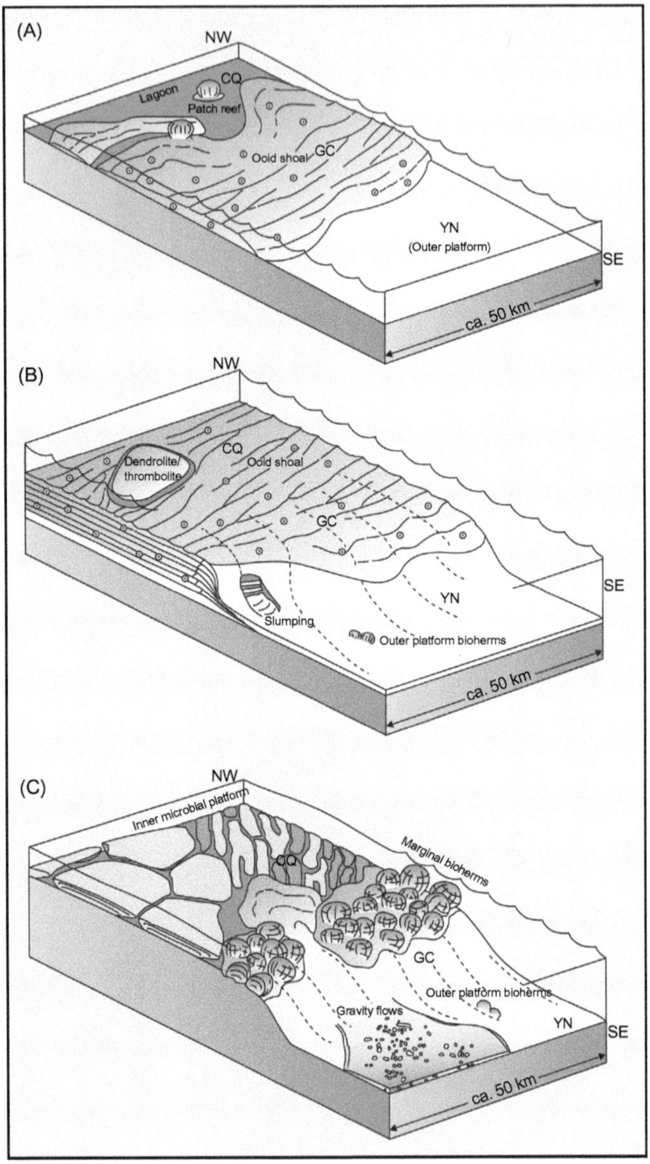

Figure 6.14 Development history of the Zhangxia Formation. (A) The first stage is characterized by low-gradient shoal-lagoon system. (B) As transgression continued, ooid shoal retrograded and aggraded. The platform then had significant relief. The steepening of the platform margin resulted in instability of slope sediments and small-scale failures occurred. Dendrolite or thrombolite biostromes grew, associated with ooids. (C) The last stage is characterized by dominance of microbial sediments in the platform interior and margin. As the rate of sea-level rise decreased, the platform prograded by deposition of active microbial carbonates. The outer platform was influenced by gravity flows. CQ, Changqing; GC, Gangcheng; YN, Yinan. For location, see Figure 6.1B.
Source: Woo (2009).

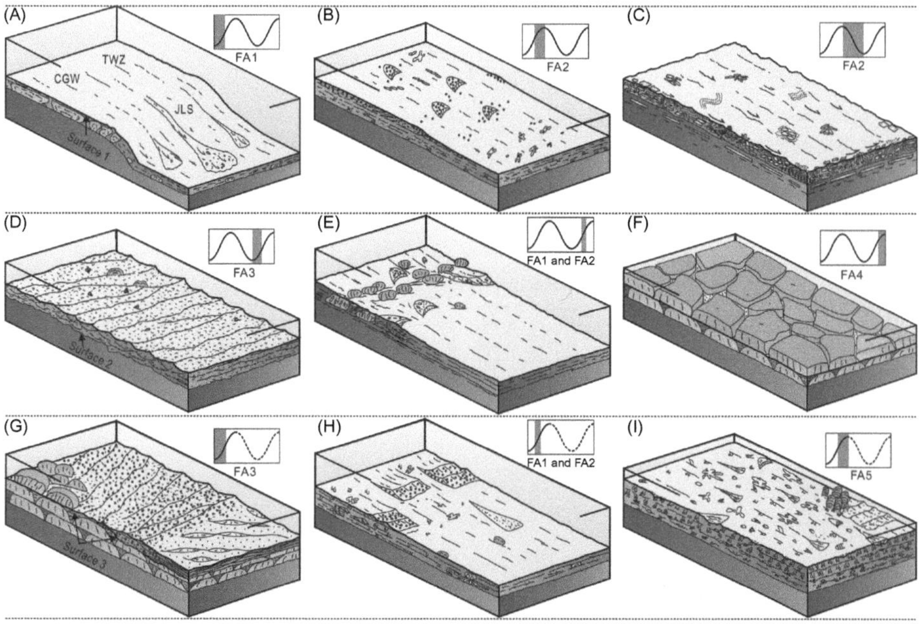

Figure 6.15 Depositional model of the Gushan and Chaomidian formations (for facies association code, see Figure 6.11). (A) Drowning of the Middle Cambrian carbonate platform (Zhangxia Formation) with significant relief, forming a drowning surface (surface 1). (B) Progressive sedimentation of carbonates during highstand base level. (C) Subaerial exposure and erosion of a strongly deformed limestone bed, caused by rapid sea-level fall. (D) Resumed carbonate sedimentation upon the subaerial unconformity (surface 2) during base-level rise, with first occurrence of microbialites after drowning of the Zhangxia carbonate platform. (E) Gradual resurgence of microbialites during slowed rise in base level. (F) Flourish of microbialites during highstand base level. (G) Termination of flat-bedded microbialites by resumed rise in base level, forming a marine flooding surface (surface 3). Microbialites either resurged in the topographic highs or terminated by storm-induced erosion and subsequent deposition of grainstone in the topographic lows. (H) Progradation of carbonate sediments during the early stage of highstand base level. (I) Deposition of thick bioturbated wackestone during highstand base level of the epeiric carbonate platform. CGW, Chengouwan section; TWZ, Tangwangzhai section; JLS, Jiulongshan section.
Source: Chen et al. (2012) by permission of *Geosciences Journal*.

(Figure 6.16). In the Zhushadong and Mantou formations, a facies transition occurred from an initially inundated tide-dominated carbonate platform to a wave-dominated marine environment (Lee and Chough, 2011). In the Taebaek area, deposition of mud prevailed in deep-water environments. The ensuing rise in sea level at slow rate caused deposition of carbonates in both regions, forming prolific biogenic sedimentary facies in both the Daegi and Zhangxia formations (Chough et al., 2010; Hong et al., 2012; Kang and Choi, 2007; Woo, 2009; Woo and Chough, 2010).

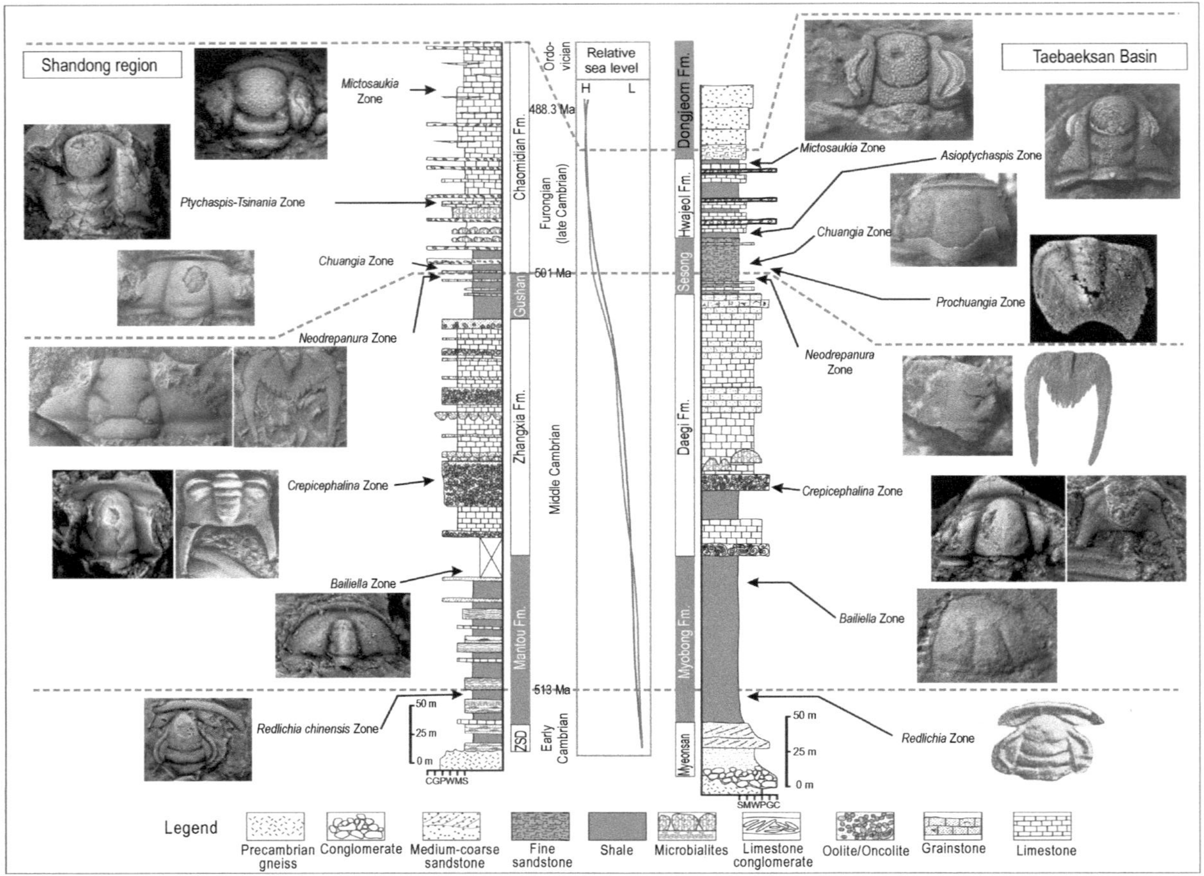

Figure 6.16 Comparison of the Cambrian stratigraphy between Shandong Province and the Taebaek area, based on trilobite biostratigraphy.
Source: Choi et al. (2004) and Chough et al. (2010).

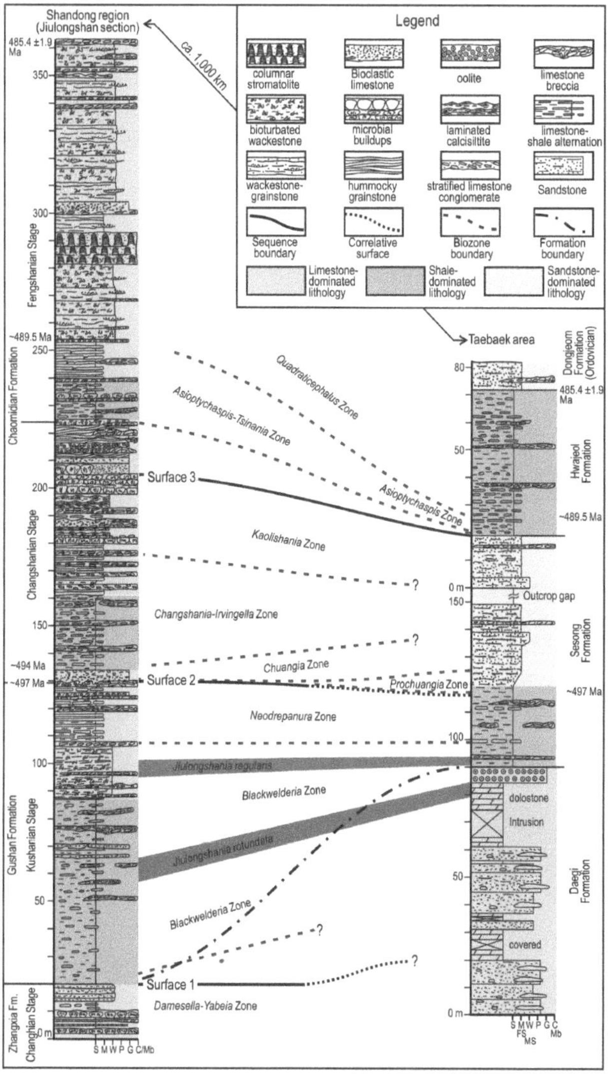

Figure 6.17 Intraplatform correlation of late Middle Cambrian to Furongian succession in the North China Platform between the Shandong region (North China) and the Taebaeksan Basin (South Korea). The sedimentary column of the Taebaeksan Basin and the biostratigraphic data are from Choi and Chough (2005), Kwon et al. (2006), Sim and Lee (2006), Kang and Choi (2007), Park et al. (2008), Chough et al. (2010), and Park and Choi (2011). S, shale; M, lime mudstone; W, wackestone; P, packstone; G, grainstone; C, limestone conglomerate; Mb, microbialite; FS, fine sandstone; MS, medium sandstone. *Source*: Chen et al. (2012) by permission of *Geosciences Journal*.

There are, however, high variabilities of sequence bounding surfaces within the comparable units. In Shandong Province, the sandstone facies at the base of the succession are localized, whereas those in the Taebaeksan Basin underlie the entire succession. In the overlying Zhushadong and Mantou formations, in the Shandong Province, input of siliciclastic fine-grained sediments and production of carbonate formed numerous depositional cycles of generally shallowing-upward trend is due to an increase in accommodation. They contain a few potential third-order sequence boundaries. In the Taebaeksan Basin, however, these cycles are absent in the equivalent Myobong Formation. Production of carbonates kept up with the rise in sea level in the Zhangxia and Daegi formations. The succession is dominated by oolite and various microbialite facies formed under the optimum conditions during highstand in sea level (Figure 6.14). In the Shandong region, however, the intra-platform local slopes formed in the Jiulongshan area during this period in which sedimentation was dominated by hemipelagic settling and episodic gravity flows (Woo, 2009). In the Taebaeksan Basin, these local slopes were absent. As a result, the bounding surfaces for sequence stratigraphic correlation are invalid.

In the late Middle Cambrian and Furongian, a bounding surface shows a significant diachroneity over a 1000 km distance (Figure 6.17) (Chen et al., 2012; Lee et al., 2012a). The distinctive lithologic characteristics of the deformed conglomerate and its erosional surface of the bounding surface 2 in the Chaomidian Formation at about 499 Ma are absent in the Sesong Formation of the Taebaeksan Basin (Figure 6.17). The lithologic boundary between the microbialite bed and the overlying oolite bed (surface 3) at about 493 Ma in the Chaomidian Formation is equivalent to the sandstone–shale boundary in the Taebaeksan Basin (Figure 6.17). To sum up, the correlations of the sequence bounding surfaces are invalid for Cambrian successions more than 1000 km apart due to changes in accommodation and sediment supply.

7 Pyeongan Supergroup

7.1 Introduction

The Pyeongan Supergroup (Carboniferous–early Triassic) comprises a thick (ca. 1700 m) siliciclastic succession of conglomerate, sandstone, purple siltstone, and gray/black shale, unconformably overlying the carbonate–siliciclastic succession of the Joseon Supergroup (Figures 7.1 and 7.2). This unconformity prevails in the entire Paleozoic successions of the northeast Asian continent, including the North China Platform. The absence of Silurian and Devonian strata resulted from tectonic uplift (Meng et al., 1997; Meyerhoff et al., 1991). The Pyeongan Supergroup succession contains abundant plant fossils (such as *Neuropteris* sp.) and marine animal fossils such as crinoids, gastropods, corals, brachiopods, foraminfera, and fusulinids. The entire succession was highly deformed and moderately metamorphosed in the Mesozoic. It contains abundant coal deposits, which were extensively mined during the period 1920–1990.

An identical succession occurs in the Pyeongnam Basin, North Korea, where the lithologic units were initially identified (Kodaira, 1924) (Table 7.1). In the Taebaeksan Basin, Shiraki (1930) established a lithostratigraphic scheme, including the Hongjeom, Sadong, Gobangsan, and Nokam series (Table 7.1). Cheong (1969) reclassified the lithologic units in the Samcheok area, including the Manhang, Geumcheon, Jangseong, Hambaeksan, Dosagok, Gohan, and Donggo formations, based on the occurrence of fusulinids in discontinuous limestone layers. This new biolithologic scheme proved invalid elsewhere in the peninsula (Reedman and Um, 1975).

7.2 Lithologic Units

A detailed sedimentological measurement of the road-cut outcrop sections was made in the Taebaek area (Lee and Chough, 2006a, b) (Figure 7.3). It reveals that the lowermost Hongjeom (Manhang) Formation comprises pebble-grade conglomerate, coarse sandstone, cross-stratified sandstone, and purple and gray shale. Three laterally discontinuous limestone layers (about 20 cm thick) occur in the Jangseong section. The overlying Sadong Formation consists of dark gray sandstone/shale with patchy gray limestone layers (Figures 7.3 and 7.4). A few meter-thick bioturbated limestone layers are interbedded with dark gray shale in

Geology and Sedimentology of the Korean Peninsula. DOI: http://dx.doi.org/10.1016/B978-0-12-405518-6.00007-3

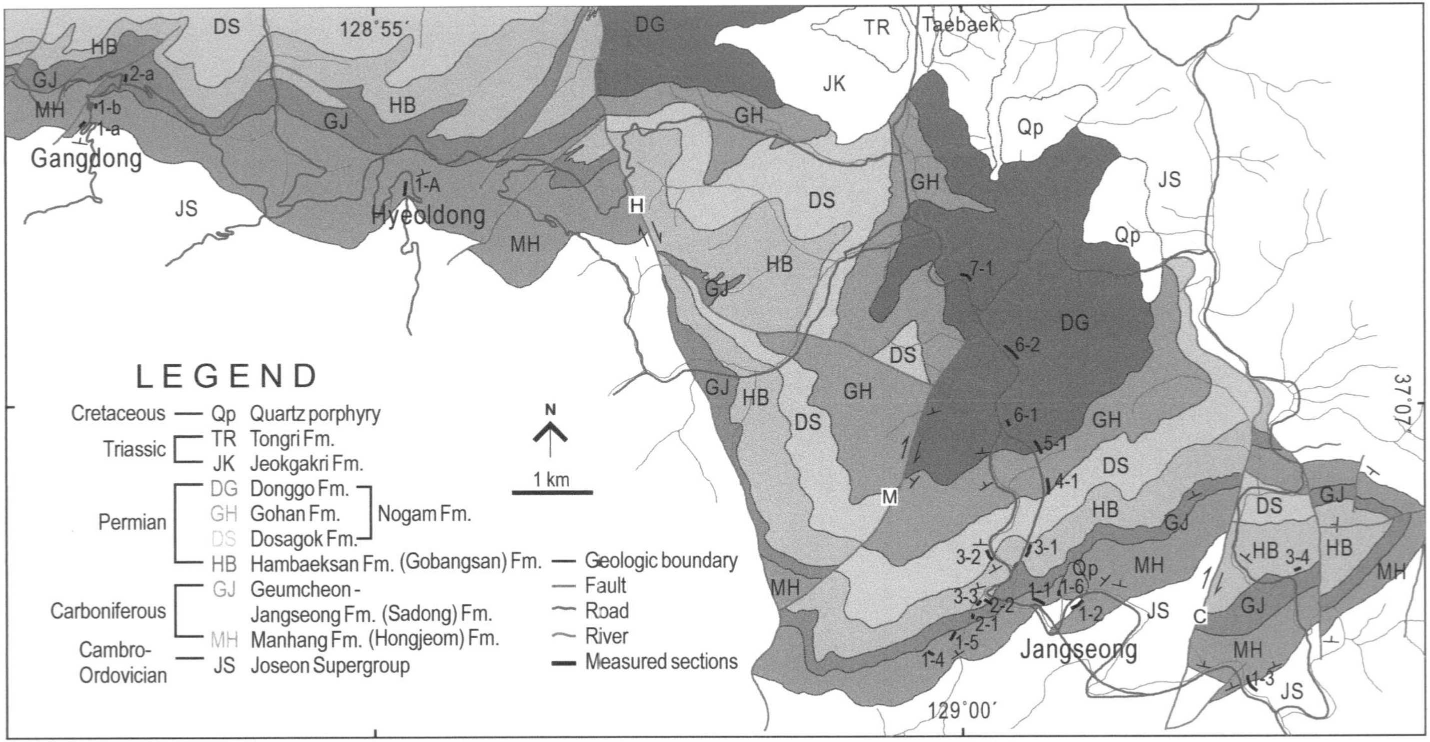

Figure 7.1 Geologic map of the Hwangji Group (Pyeongan Supergroup) in the Taebaek area. H, Hambaeksan fault; M, Mungok fault; C, Cheolam fault.

Source: Lee and Chough (2006a) by permission of the Elsevier.

Figure 7.2 (A) Outcrop photograph and (B) interpretation of the unconformable boundary (disconformity) between the Joseon Supergroup and the Pyeongan Supergroup in the Jangseong road-cut section. For location, see Figure 7.1 (section 1-1).

Table 7.1 Stratigraphic Summary of the Pyeongan Supergroup

<table>
<tr><th>Basin</th><th>Workers / Geologic Age</th><th colspan="2">Carboniferous</th><th colspan="8">Permian</th><th colspan="4">Triassic</th></tr>
<tr><td rowspan="5">Pyeongnam Basin</td><td>Kodaira (1924)</td><td colspan="14">Pyeongan System</td></tr>
<tr><td rowspan="2">Kawasaki (1927)</td><td colspan="14">Pyeongan System</td></tr>
<tr><td>Hongjeom Fm.</td><td colspan="3">Sadong Fm.</td><td colspan="6">Gobangsan Ser.</td><td colspan="4">Nogsaegam Fm.</td></tr>
<tr><td rowspan="2">Tateiwa (1931)</td><td colspan="14">Pyeongan System</td></tr>
<tr><td>Hongjeom Ser.</td><td colspan="3">Sadong Ser.</td><td colspan="6">Gobangsan Ser.</td><td colspan="4">Nogam Ser.</td></tr>
<tr><td rowspan="16">Taebaeksan Basin</td><td rowspan="2">Shiraki (1930)</td><td colspan="14">Pyeongan System</td></tr>
<tr><td>Hongjeom Ser.</td><td colspan="3">Sadong Ser.</td><td colspan="6">Gobangsan Ser.</td><td colspan="4">Nogam Ser.</td></tr>
<tr><td>GICTR (1962)</td><td colspan="2">Hongjeom Ser.</td><td colspan="8">Sadong Ser.</td><td colspan="3">Gobangsan Ser.</td><td>Nogam Ser.</td></tr>
<tr><td rowspan="3">Cheong (1969, 1973)</td><td colspan="14">Pyeongan Supergroup</td></tr>
<tr><td colspan="2">Gomog Group</td><td colspan="8">Cheolam Group</td><td colspan="4">Hwangji Group</td></tr>
<tr><td>Manhang Fm.</td><td>Geumcheon Fm.</td><td></td><td>Jangseong Fm.</td><td colspan="6">Hambaeksan Fm.</td><td colspan="2">Dosagok Fm.</td><td>Gohan Fm.</td><td>Donggo Fm.</td></tr>
<tr><td rowspan="2">Reedman and Um (1975)</td><td colspan="14">Pyeongan Group</td></tr>
<tr><td>Hongjeom Fm.</td><td colspan="3">Sadong Fm.</td><td colspan="5">Gobangsan Fm.</td><td colspan="5">Nogam Fm.</td></tr>
<tr><td rowspan="2">KIGAM (1979)</td><td colspan="14">Pyeongan Supergroup</td></tr>
<tr><td>Manhang Fm.</td><td>Geumcheon Fm.</td><td></td><td>Jangseong Fm.</td><td>Hambaeksan Fm.</td><td colspan="2">Dosagok Fm.</td><td colspan="3">Gohan Fm.</td><td colspan="4">Donggo Fm.</td></tr>
<tr><td rowspan="3">Lee and Chough (2006a)</td><td colspan="14">Pyeongan Supergroup</td></tr>
<tr><td colspan="14">Hwangji Group</td></tr>
<tr><td>Manhang Fm.</td><td colspan="3">Geumcheon-Jangseong Fm.</td><td>Hambaeksan Fm.</td><td>Dosagok Fm.</td><td colspan="2">Gohan Fm.</td><td colspan="3">Donggo Fm.</td><td colspan="3"></td></tr>
<tr><td rowspan="3">This study</td><td colspan="14">Pyeongan Supergroup</td></tr>
<tr><td rowspan="2">Hongjeom (Manhang) Fm.</td><td colspan="3" rowspan="2">Sadong (Geumcheon-Jangseong) Fm.</td><td rowspan="2">Gobangsan (Hambaeksan) Fm.</td><td colspan="6">Nogam Fm.</td><td colspan="3" rowspan="2"></td></tr>
<tr><td>(Dosagok Fm.)</td><td colspan="2">(Gohan Fm.)</td><td colspan="3">(Donggo Fm.)</td></tr>
</table>

Dashed Line Represents Unknown Age.

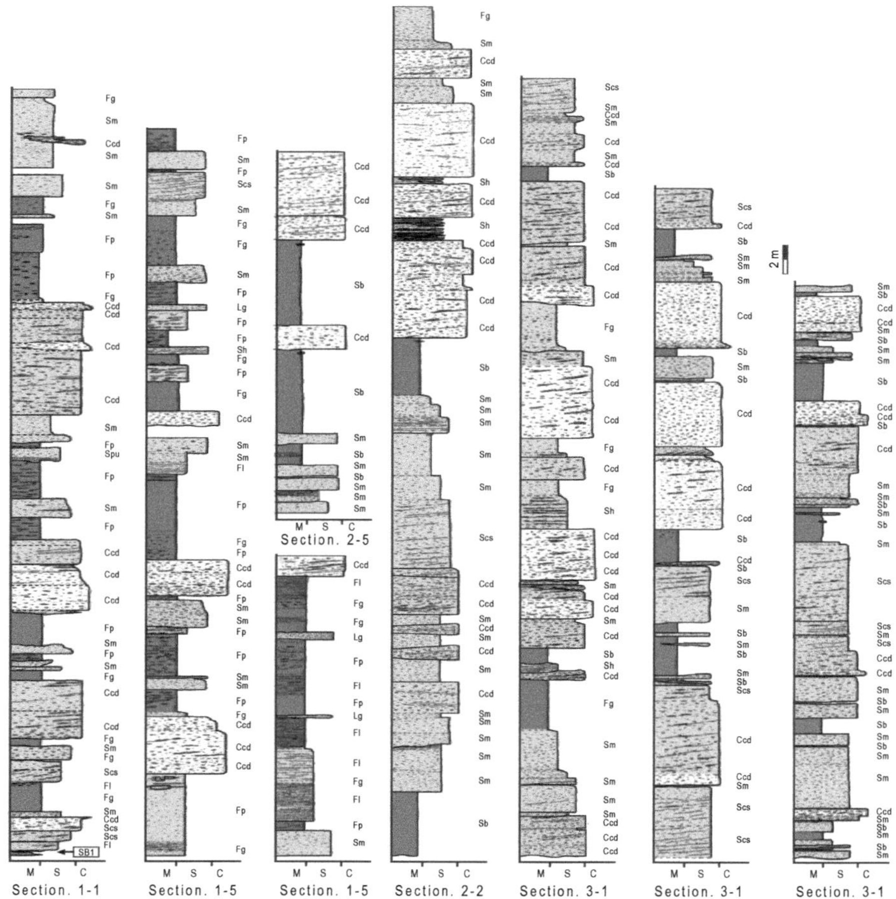

Figure 7.3 Sedimentary log of the Pyeongan Supergroup. M, mudstone; S, sandstone; C, conglomerate. For abbreviation of sedimentary facies codes, see Table 7.2. For section location, see Figure 7.1.
Source: Lee and Chough (2006b) by permission of *Geosciences Journal*.

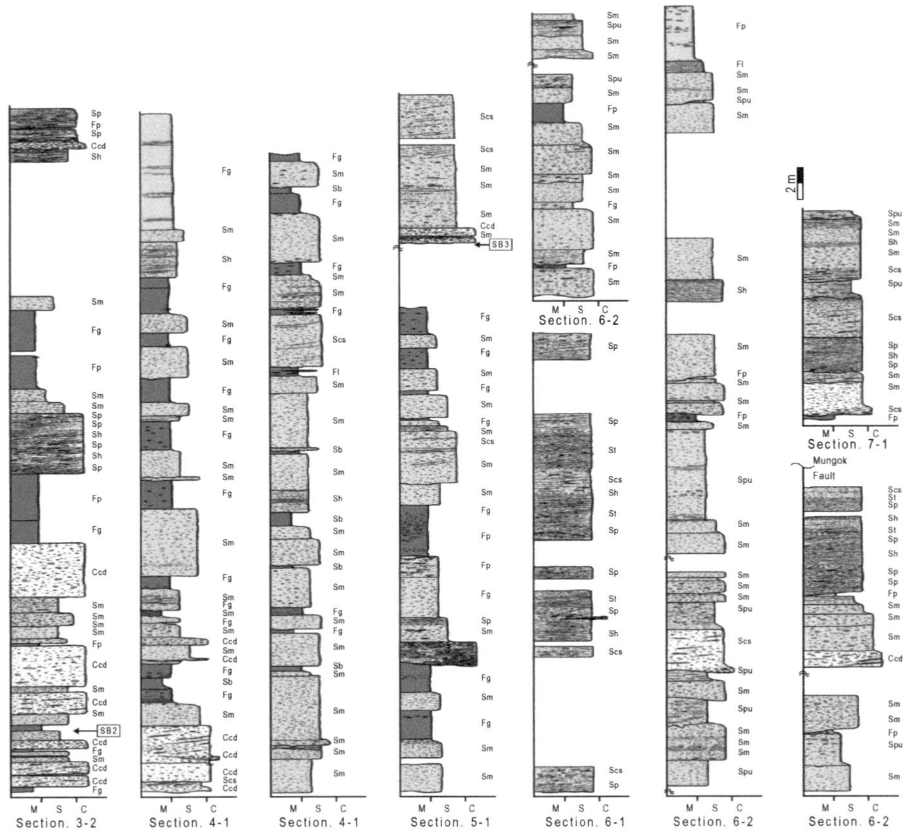

Figure 7.3 (Continued).

Table 7.2 Description and Interpretation of Sedimentary Facies in the Hwangji Group

Facies Code	Description	Interpretation
Crudely cross-stratified conglomerate (Ccd)	This facies is represented by either crudely cross-stratified conglomerate or disorganized clasts in moderately sorted sand to silt matrix. Clasts are well to moderately sorted and granule- to pebble-grade (max. dL = 3 cm). Poorly sorted clasts are scarcely present. The clasts are subrounded to rounded and include dark gray mudstone chips. The matrix is partly abundant (more than 25%). The crude stratifications are represented by discontinuous pebble-concentrated layers or gravel-grade trains and clusters with general parallel orientation of grains. Each facies unit ranges in thickness from 50 to 300 cm, but amalgamated beds reach up to 460 cm. Most units are generally tabular and laterally continuous in outcrop scale, whereas some are lenticular or wedged with a slight change in thickness. The lower bed boundary is commonly sharp and planar, although some beds are bounded by undulatory or erosional boundary. The upper boundary is either gradational or sharp.	Deposition by *en masse* of mixed sand and gravel with diffuse sheets by high-magnitude heavy-laden flows (Miall, 1977; Todd, 1989; Todd and Went, 1991)
Planar cross-stratified sandstone (Sp)	This facies consists of planar cross-stratified medium to very coarse sandstone and pebbly sandstone. Cross-stratification is represented by alternation of either coarser and finer strata or gravelly and sandy strata. Cross-set is 10 cm to a few decimeters thick and commonly solitary set. Each bed ranges in thickness from 30 cm to a few meters.	2D dunes, linguoid dunes, or bar with slipface (Harms et al., 1982; Miall, 1996)

(*Continued*)

Table 7.2 (Continued)

Facies Code	Description	Interpretation
Trough cross-stratified sandstone (St)	The facies is represented by trough cross-stratified medium to very coarse sandstone and pebbly sandstone. Pebbles and/or mudstone intraclasts are recognized at the base of some units. Each set is 1 cm to a few meters thick and set thickness is generally proportional to grain size.	3D sandy dunes and ripples; infills of scour hollow (Collinson and Tompson, 1989; Miall, 1977)
Crudely cross-stratified sandstone (Scs)	The facies consists of poorly to moderately sorted and well rounded to subrounded coarse sandstone to granule-grade conglomerate and is characterized by crude cross-stratification with discontinuous granule trains or streaks. Each set is decimeter to a few meters thick. It shows sharp erosional lower boundary and gradational upper boundary in some beds. Each facies unit ranges in thickness from 150 to 300 cm.	Traction-dominant subaqueous and subaerial stream flow; migration of medium to large-scale bedforms (Bridge, 1993; Miall, 1996, 1977)
Horizontally stratified sandstone (Sh)	This facies consists of moderately to well sorted, horizontally stratified (laminated) fine to coarse sandstone, gray to dark gray in color. Stratifications consist of alternation of coarse sand and fine sand or discontinuous trains of well-sorted granule and sand grains. Each facies unit ranges in thickness from 60 to 380 cm.	Migration of either low-amplitude bedforms or upper flow regime plane bed (Allen, 1984; Bridge, 1981, 1993)
Massive sandstone (Sm)	The massive sandstone is represented by massive medium to coarse sandstone occasionally with some granules or mudstone chips (max. dL = 2 cm). Grains are relatively well sorted and well rounded to subrounded. Each facies unit ranges in thickness from 40 to 500 cm and	Plane bed migration in upper flow regime; bed boundary amalgamated by rapid deposition; rapid deposition from high-concentrated flow (Jo and Chough, 2001)

(Continued)

Table 7.2 (Continued)

Facies Code	Description	Interpretation
	slightly changes in bed thickness in outcrop scale. Occasionally, some beds are wedged or lenticular. The upper and lower bed boundaries are commonly planar and sharp, and often undulatory or irregular.	
Purple sandstone (Spu)	The facies consists of moderately sorted medium sandstone to silstone, showing high variation in grain size. Some beds are horizontally stratified. It shows purple, dark red and chocolate in color. Granule to coarse sand trains or streaks, calcite concretion and burrows are commonly recognized.	Overbank or fluvial plain deposit (Miall, 1996)
Purple homogeneous mudstone (Fp)	The facies mainly consists of massive fine sandstone or homogeneous mudstone platy and blocky structures occur in some beds. In some cases, dark red mudstone chips occur in purple siltstone or fine sandstone and rimmed by gray contact. This facies varies in color from dark red, purple to purple mottled and mingled with gray. The gray parts are commonly coarser than the purple ones. In part, it is characterized by dispersed gray color and platy structures. Each facies unit ranges in thickness from 40 to 550 cm. The upper and lower boundaries are commonly sharp and planar or irregular.	Overbank deposits in coastal plain suggested by its fine grain size and slight bioturbation (Bentham et al., 1993; Walker, 1967)
Gray homogeneous mudstone (Fg)	The facies consists of massive fine sandstone and homogeneous mudstone. Occasionally, dark gray to black mudstone chips (max. dL = 3 cm) or pyrites are dispersed. The mudstone chips are oriented parallel to bedding	Rapid sedimentation from suspended material; slightly bioturbated; post-depositional graying under reducing condition (Miall, 1977; Turner, 1980)

(*Continued*)

Table 7.2 (Continued)

Facies Code	Description	Interpretation
	plane. In some cases, partly faint lamination occurs. The facies shows bright to dark gray, yellowish gray, greenish gray, and milky white in color. The lower bed boundary is commonly gradational, whereas the upper is sharp. Each facies unit ranges in thickness from 10 to 350 cm.	
Laminated mudstone (Fl)	The facies is represented by parallel-laminated siltstone (mudstone) and shale. The lamination is represented by an alternation of mudstone and siltstone. In part, the lamination is discontinuous and lenticular/wavy in centimeter-scale. Occasionally, finer laminae are commonly reddish. The lower laminae boundaries are erosional or sharp. Calcareous nodules or pyrites randomly occur. The lower and upper bed boundaries are generally sharp and planar. Each facies unit ranges in thickness from 40 to 400 cm.	Suspension settling affected by pulses of riverine in flows; deposition from bidirectional or periodic currents (Hjellbakk, 1997)
Black shale (Sb)	The facies is represented by black shale or coaly shale. It contains abundant plant fossils such as *Annularia*, *Lobatannularia*, *Sphenophyllum*, and *Neuropteris*. Parallel discontinuous laminae are well or crudely developed in some part. Black shale beds are commonly squeezed and slipped into the upper or lower beds.	Marsh or swamp deposit; high nutrient supply and anoxic condition (Cabrera and Saez, 1987; McCabe, 1984)
Grainstone (Lg)	This facies is characterized by milky white to gray massive grainstone to packstone. The lime grains consist mainly of fossils such as brachiopods, foraminifera, and algae particles as well as siliciclastic clasts. This facies also contains quartz grains	Storm sedimentation or deposition in coastal pond; deposits of intermediate subtidal zones, i.e., "flaser rocks"; fossil fragments are indicative of subtidal deposition (Aigner, 1985; Allen, 1982)

(Continued)

Table 7.2 (Continued)

Facies Code	Description	Interpretation
	more than 5%. In some cases, shale is intercalated with massive grainstone. The upper and lower boundaries are sharp and planar. These limestone units are a few decimeters to a few meters thick and commonly wedge-shaped or lenticular. This facies is commonly interlayered within purple siltstone (facies Fp) or gray siltstone (facies Fg).	
Bioturbated limestone (Lb)	This facies is represented by dark gray bioturbated packstone to wackestone with bioclastic fragments such as foraminifera and gastropod. Mottled texture is caused by selective dolomitization of burrows (ichnofacies index-2 or -3). Horizontal or subhorizontal and subordinately vertical burrows are recognized. Each facies unit is more than 1 m in thickness. In some cases, wavy chert or shale is interlayered with bioturbated wackestone to packstone.	Intermediate or shallow subtidal deposits modified by bioturbation. The common presence of bioclastic fragments such as gastropod and foraminifera is suggestive of deposition in shallow subtidal with normal marine salinity (Heckel, 1972; Rhoads, 1967).

Source: Lee and Chough (2006b) by permission of *Geosciences Journal*.

the Gangdong section, whereas a few decimeter-scale limestone layers occur in the Jangseong section (Figure 7.5). Cheong (1969) divided the Sadong Formation into the Geumcheon Formation and the parallel unconformable Jangseong Formation (Table 7.1) based on the occurrence of plant fossils and fusulinids in discontinuous limestone lenses. It is, however, a single lithologic unit without a significant time gap.

The overlying Gobangsan (Hambaeksan) Formation (ca. 200 m thick) consists of coarse sandstone and gray shale. The sandstone is partly planar cross-stratified and sparsely intercalated with greenish gray siltstone, coaly shale, and coal (Figure 7.5). The overlying succession (ca. 300 m thick), Nokam Formation, consists of purple/greenish siltstone, pebbly sandstone, and crudely cross-stratified conglomerate. Sandstone beds are either massive or planar cross-stratified. The lower boundary is located where a purple siltstone bed occurs (Figure 7.5), although its correlation in other sections has not been established. The overlying

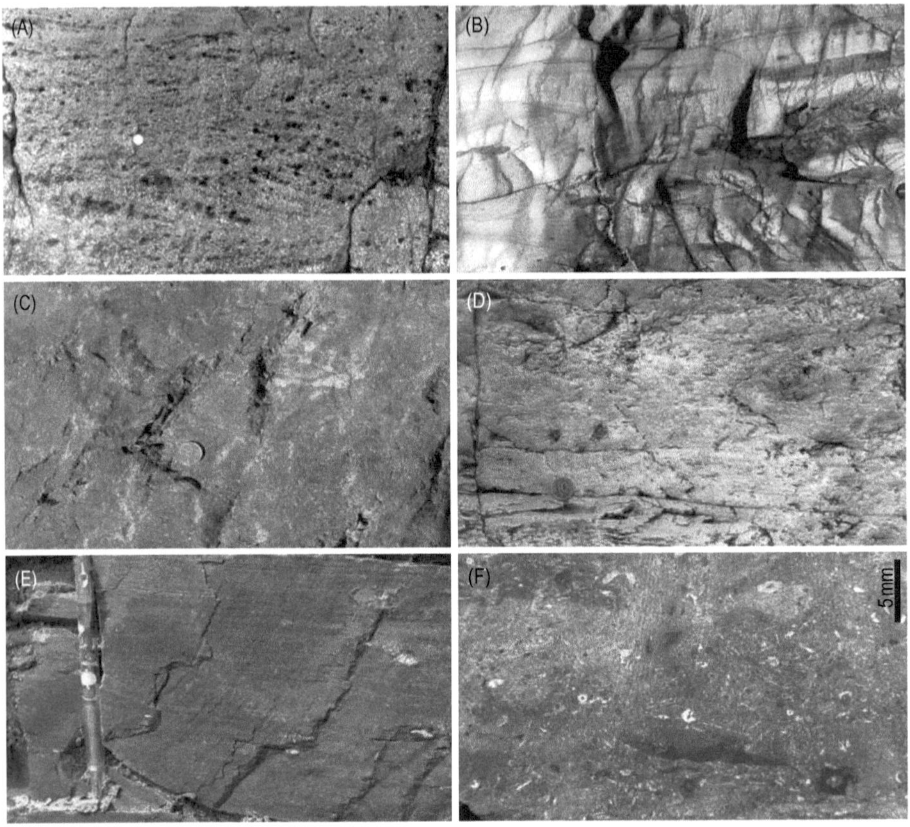

Figure 7.4 Photographs of sedimentary facies in the Hwangji Group, Taebaek area. (A) Crudely cross-stratified conglomerate, section 1-5 (FA2). (B) Planar and trough cross-stratified coarse sandstone (FA7). (C) Bioturbated purple or red sandstone, section 7-1 (FA7). (D) Gray homogeneous mudstone with pyrite and calcareous nodules, section 1-4 (FA1). (E) Gray parallel-laminated mudstone, section 1-A (FA1). (F) Grainstone with purple mudchips, section 1-5 (FA2). Coin is 2.5 cm in diameter, lens cap is 7.5 cm in diameter, and pencil is 14.2 cm long. For section location, see Figure 7.1. (For interpretation of the references to color in this figure legend, the reader is referred to the web version of this book.)

unit is characterized by gray to greenish gray sandstone, massive gray siltstone, and black shale. The common presence of bioclastic fragments such as brachiopods and foraminifera is suggestive of deposition in shallow marine environments.

The overlying Donggo Formation (Table 7.1) consists of massive pebbly sandstone with purple siltstone chips, planar cross-stratified sandstone, trough cross-stratified conglomerate, parallel-laminated sandstone, and massive purple siltstone. The cross-stratified sandstone is commonly interbedded with purple

Figure 7.5 (A) Meter-scale bioturbated packstone interlayered with chert in section 2-a (N15°W/NE35°), Geumcheon-Jangseong Formation. (B) Section 2-1 (N55°E/NW80°) characterized by an alternation of black shale and sandstone in the lower part of the Geumcheon-Jangseong Formation. (C) Photograph of section 3-2 (N30°E/NW75°) showing the boundary (SB2) between the Hambaeksan (FA 4) and Dosagok (FA 5) formations. For sedimentary facies codes, see Table 7.1. For location, see Figure 7.1.

siltstone. The bed boundaries are commonly gradational, whereas the bed geometries change abruptly in both vertical and lateral directions. This unit is age-unknown but probably of the early Triassic (Lee et al., 2012). The upper boundary with the overlying Jeokgakri Formation is characterized by an angular unconformity. The Jeokgakri Formation formed in a post-deformational piggyback basin. Rhyolite bodies, which intruded into the Donggo and Jeokgakri formations, are dated at 88 ± 3 and 67 ± 3 Ma, respectively.

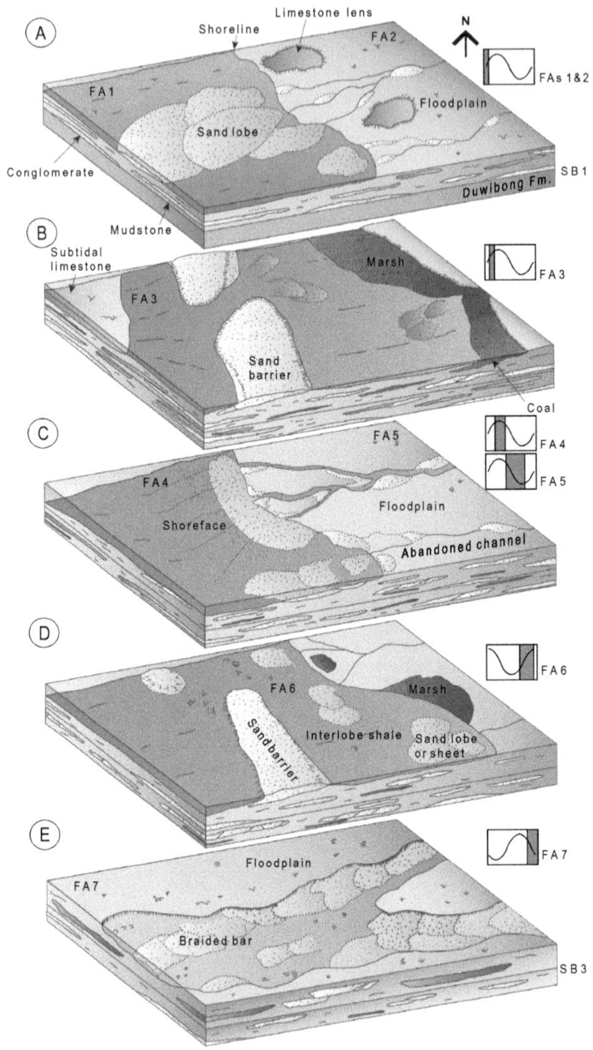

Figure 7.6 Depositional model of the Pyeongan Supergroup in the Taebaek area. (A) Shoreface environments and coastal plain with braided rivers (Manhang Formation). (B) Lagoonal environments (Geumcheon-Jangseong Formation). (C) Regressive shoreface in deltaic environments (Hambaeksan Formation) and fluvial plain environments (Dosagok Formation). (D) Restricted bay (Gohan Formation). (E) Sandy braided river system (Donggo Formation). Shaded curve (inset) represents inferred position of relative sea level.
Source: Lee and Chough (2006b) by permission of the Geosciences Journal.

7.3 Depositional Model

Deposition began with shoreface deposits of gravelly traction flows with small-scale prograding units off local river-mouths (Figure 7.6). The basal conglomerates represent initial subsidence phase, accompanied by an abrupt sea-level rise. Along the coastline, sandstone and conglomerate show changes in thickness. It grades into a coastal plain environment with a bedload-dominated braided river system (Figure 7.6A). Limestone beds with marine fossils indicate continuous sea-level rise. It is transitional to a lagoonal environment with abundant coal and bioturbated limestone (Figure 7.6B). Limestone beds are widely spread across the entire basin. Bioturbated wackestone beds are intercalated in the lower part of the association, whereas coal beds are abundant in the upper part. Well-preserved crinoid stems in the Jangseong section indicate a very calm depositional environment. The base shows coarsening-upward units, representing shoreface progradation, accompanied by a stillstand of relative sea level (Figure 7.6C). The depositional environment changed to a fluvial system with shoreface progradation (Figure 7.6C). Black shale and gray siltstone facies indicate anoxic to suboxic conditions with limited biogenic activities in interdistributary bay where sea level fluctuated (Figure 7.6D). It was followed by trough cross-stratified coarse sandstone and bioturbated purple fine sandstone, which formed in a fluvial system during relative fall in sea level (Figure 7.6E). This fall was most likely due to the regional uplift in the Yeongnam Massif.

Figure 7.7 Thick sandstone beds interlayered with conglomerate, Yobong Formation, Pyeongan Supergroup in the Machari area, Yeongwol. Top to the left.

7.4 Pyeongan Supergroup in the Yeongwol Area

The Pyeongan Supergroup in the Yeongwol area comprises siliciclastic deposits of conglomerate, sandstone, and shale (Figure 7.7). The succession can be divided into the Yobong (sandstone and limestone), Pangyo (sandstone), Bamchi (limestone), and Mitan (sandstone and shale) formations (Cheong and Park, 1977). It unconformably overlies the Yeongwol Group of the Cambrian–Ordovician. Limestone in the lower part of the Yobong Formation contains fusulina *Eostafella* and *Pseudostafella*, which indicates Bashkirian in age (Lee, 1984). Cheong and Park (1977) have suggested that abundant biomicritic limestone beds in the Bamchi Formation, containing fusulinids, formed in a lagoonal environment. Lee (1990) suggests that sandstone of the Yobong Formation consists largely of quartz grains with lithic fragments, which is similar to that of the Manhang Formation in the Taebaek area.

8 Tectonic Evolution

8.1 Active Plate Collision

Major crustal deformation in the Korean Peninsula occurred in the Triassic when the South China Block and the Sino-Korean Block collided against each other along the suture zone of the Imjingang Belt (Figure 8.1) (Ree et al., 1996). The event was most important in the making of the geologic features in the peninsula and eventually transformed the stable platforms and basins into a subduction-related tectonic belt. At the same time, the eastern margin of the northward-moving South China Block was abutted against the Sino-Korean Block, forming a collision belt along the South Korean Tectonic Line (SKTL) (Figure 8.2) (Chough et al., 2000). The Gyeonggi Massif was in dextral shear relative to the Yeongnam Massif (Figure 8.2). Compressional deformation continued in the middle to late Jurassic, caused by orthogonal subduction of the paleo-Pacific Plate across the shear zone, as manifested by the spectacular arrays of faults, folds, and thrusts (Figure 8.3). The subduction also caused extensive emplacement of granite batholiths.

8.2 Imjingang Belt

8.2.1 Introduction

The Imjingang Belt was first mapped in 1962 by North Korean geologists (Ri and Ri, 1994). Thirty years later, it was recognized by South Korean geologists for its tectonic importance (Cho et al., 1995; Ree et al., 1996). It is an east–west trending fold and thrust zone of late Paleozoic metasedimentary successions underlain by Precambrian basement rocks (Ree et al., 1996). In the southern part of the demilitarized zone, the metasedimentary rocks can be observed in the Yeoncheon and Sintanri areas, about 25 km north of Seoul, which consist of siliciclastic (shale, quartzite, and phyllite) and carbonate (marble) successions with volcaniclastics. The Yeoncheon complex consists of pelitic schist, phyllite, quartzite with marble (Silurian–Devonian), and calc-silicates with minor amphibolites (Late Proterozoic) (Ree et al., 1996). These rocks underwent metamorphism under the conditions of 8–13 kbar and 630–790°C (Ree et al., 1996).

The metamorphic event was initiated in the late Permian–Triassic when the Sino-Korean Block collided against the South China Block (Cho et al., 2007).

Geology and Sedimentology of the Korean Peninsula. DOI: http://dx.doi.org/10.1016/B978-0-12-405518-6.00008-5

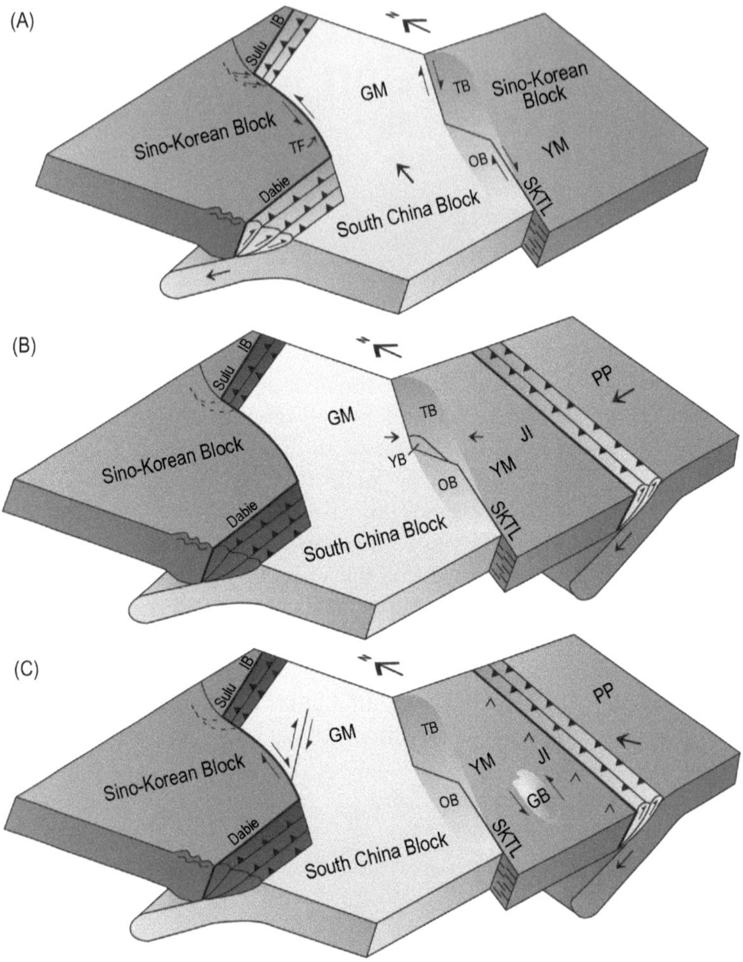

Figure 8.1 Three-stage deformation of the Korean Peninsula in the Mesozoic. (A) In the Triassic–early Jurassic, the South China Block (SCB) moved northward against the Sino-Korean Block (SKB) at the Imjingang Belt (IB), an extension of the Dabie Belt (DB) and the Sulu Belt (SB), offset by the Tanlu Fault (TF). At the eastern end of the Imjingang Belt, the Sino-Korean Block rotated clockwise and moved southward relative to the South China Block along the SKTL. (B) In the middle–late Jurassic, the paleo-Pacific Plate (PP) began to subduct under the Asian continent in the east of the Japanese islands (JI), causing compression in the peninsula across the SKTL, accompanied with extensive emplacement of granite batholiths as well as volcanism. (C) In the Cretaceous, the Pacific plate subducted northward, causing extension of the backarc basins such as the Gyeongsang Basin (GB), forming a backarc-arc-trench system, accompanied with extensive volcanism and magmatism. TB, Taebaeksan Basin; OB, Okcheon Basin; GM, Gyeonggi Massif; YM, Yeongnam Massif; YB, Yeongwol Block. Plutonism and volcanism are omitted.

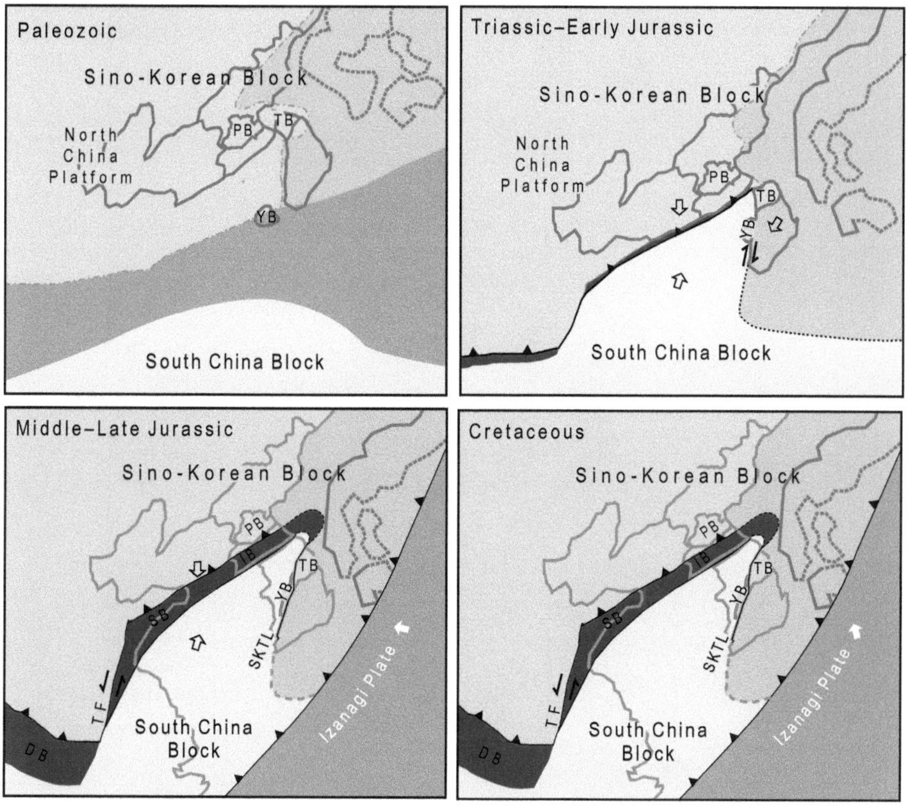

Figure 8.2 Tectonic reconstruction of the Sino-Korean and South China blocks. PB, Pyeongnam Basin; TB, Taebaeksan Basin; YB, Yeongwol Block; DB, Dabieshan Belt; TF, Tanlu Fault; SB, Sulu Belt; IB, Imjingang Belt; SKTL, South Korean Tectonic Line. Note that the indented South China Block collided against the Sino-Korean Block, forming an offset along the Tanlu Fault and rotational offset along the SKTL.

Shearing and fold-thrust structures formed during the collision by a contractional deformation. Although the details of these events warrant further observation in the northern part of the demilitarized zone, the present data on geologic structures, metamorphic rocks, and isotopic dating suggest that the Imjingang Belt is an eastward extension of the Qinling-Dabie-Sulu collision belt of China (Ree et al., 1996; Cho et al., 2007).

8.2.2 Hwangju and Imjin Groups

8.2.2.1 Introduction

The Paleozoic successions (Ordovician–Carboniferous), an affinity of the South China Block, occur in the northern part of the demilitarized zone in scattered areas

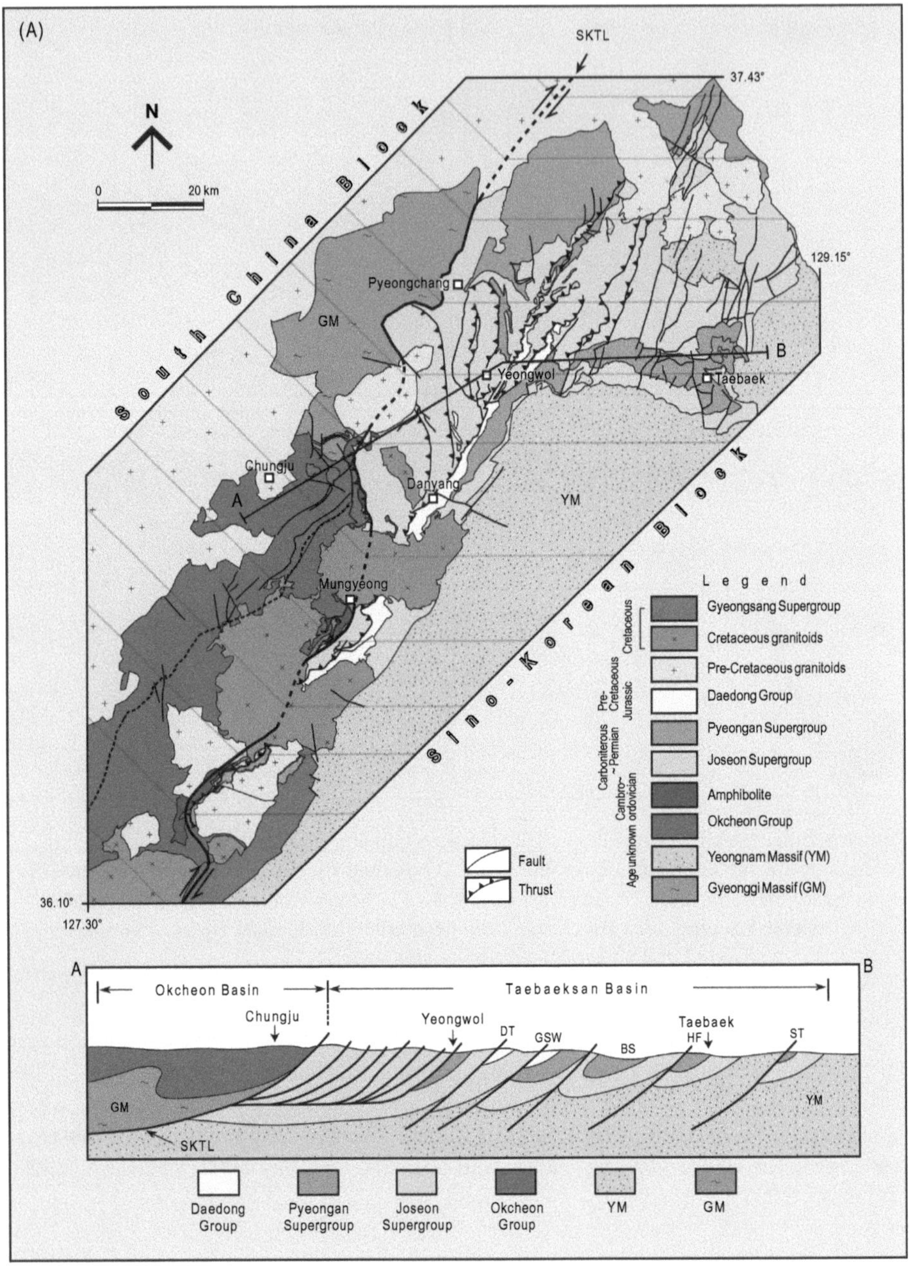

Figure 8.3 (Continued)

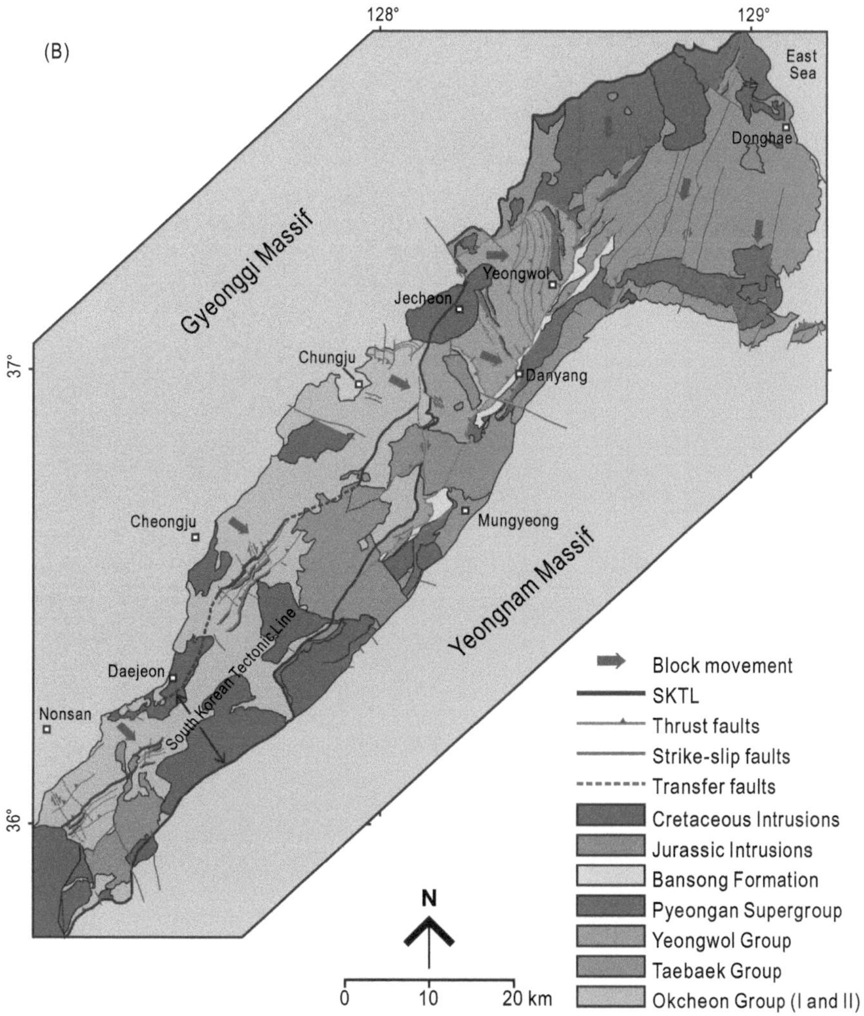

Figure 8.3 (A) Geologic map along the SKTL, a collision boundary between the Gyeonggi Massif (South China Block) and the overlying Okcheon Group and the Yeongnam Massif (Sino-Korean Block) and the overlying Joseon and Pyeongan supergroups. GM, Gyeonggi Massif; YM, Yeongnam Massif; DT, Deokpori thrust; GSW, Gongsuwon thrust; BS, Baekunsan syncline; HF, Hambaeksan fault; ST, Sambang thrust fault. (B) Simplified geologic map, showing the deformation structures of the SKTL.
Source: (A) Chough et al. (2000) by permission of the Elsevier. (B) Hwang et al. (1996), Kim et al. (1998), and Lim et al. (2005, 2006, 2007).

of Hwanghae Province (Figure 8.4). These limestone-dominant successions contain corals, brachiopods, bivalves, crinoids, cephalopods, and plant fossils, collectively designated as the Hwangju and Imjin Groups (Paek et al., 1996). Corals of the Ordovician, in particular, are identical to those of the South China Block, but not found

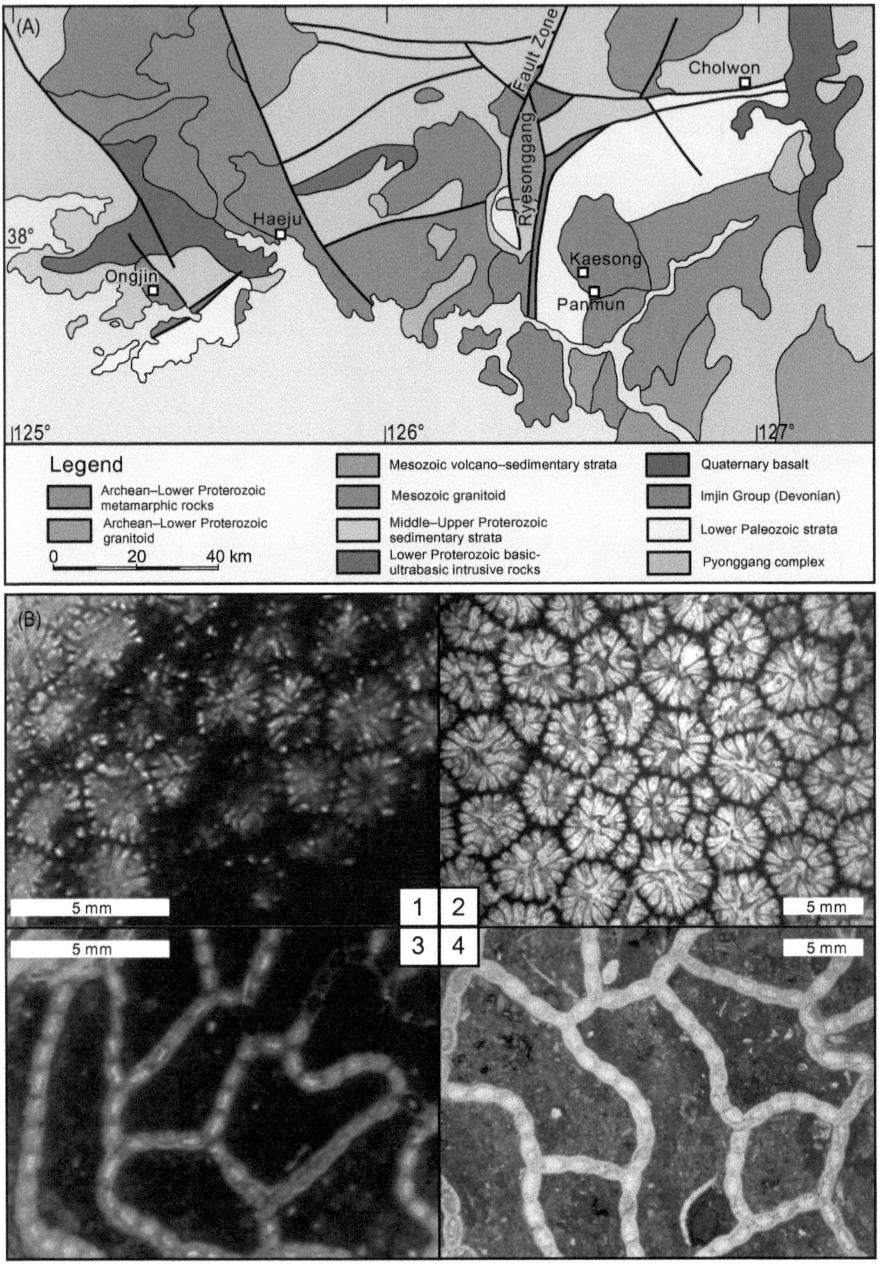

Figure 8.4 (A) Geologic map of Hwanghae Province. (B) Two Upper Ordovician tabulate corals from North Korea and South China.

1. *Agetolites* sp., Koksan, Hwanghae-Do, Korea.
2. *Agetolites* sp., Yushan, Jiangxi Province, China.
3. *Catenipora zhejiangensis*, Hwanghae-Do, Korea.
4. *Catenipora zhejiangensis*, Yushan, Jiangxi Province, China.

Source: (A) Paek et al. (1993) and (B) the courtesy of D.J. Lee.

in the Sino-Korean Block (D.J. Lee, personal communication). Kobayashi (1930) also described limestone pebbles, which contain coral, cephalopods, and others of the Silurian. The patchy deposits of the Ordovician, Silurian, and Devonian represent tectonic windows (thrust blocks) of the South China Block.

8.2.2.2 Lithology and Fossils

The Paleozoic succession (Hwangju Group) in Hwanghae Province comprises 10 formations (Paek et al., 1996). The lower part of the Sangsori Formation (Upper Ordovician) consists mainly of dolostone with intercalation of mudstone, marlstone, and dolomitic limestone, containing corals, gastropods, and crinoids. The upper part of the formation comprises black shale and marly limestone, including crinoids. The Koksan Formation (Lower Silurian) consists of discontinuous mudstone and argillaceous limestone, black shale, bedded limestone, and dolomitic limestone. It contains corals (*Syringopora* sp., *Favosites* sp.), graptolite, gastropods, and crinoids (Paek et al., 1996). The Wolyangri Formation (Upper Silurian) consists of mudstone and argillaceous limestone, including corals, crinoids, gastropods, cephalopods, and brachiopods. These fossils were earlier found in limestone pebbles of the Songrim Formation in the Gyeomipo area (Kobayashi, 1930; Tateiwa, 1976), including corals (*Halysites*, *Heliolites*, *Propora*, *Favosites*, and others) and stromatoporoids.

8.2.2.3 Evidence of Tectonic Collision

An independent evidence for the collision of the two blocks comes from the occurrence of tabulate corals in the Hwangju and Imjin groups (Ordovician–Carboniferous) of Hwanghae Province North Korea (Figure 8.4). These corals are identical to those of Yushan, Jiangxi Province, China, the characteristic Upper Ordovician fossil components in South China (D.J. Lee, personal communication). Earlier, Kobayashi (1930) identified Ordovician trilobites in deformed limestone pebble layers of the Songrim Formation (Triassic). In the Songrim area of Hwanghae Province, Shimizu et al. (1934) identified corals and graptolites of the Silurian in limestone conglomerates. It is most likely that the deposits represent part of an obducted thrust block of the Gyeonggi Massif across the Imjingang Belt. The obducted upper Paleozoic (Silurian and Devonian) successions are, in fact, scattered in the Pyeongan and Hwanghae Provinces, North Korea, and surrounded and bounded by the Joseon and Pyeongan supergroups of the Sino-Korean Block.

However, a possible occurrence of the upper Paleozoic (post-Silurian) succession far from the Imjingang Belt has recently been suggested using a sensitive high-resolution ion microprobe dating on detrital zircons in metasandstones of the Taean Formation within the Gyeonggi Massif. Choi et al. (2008) and Cho et al. (2010) proposed that the greenschist–amphibolite-grade metasandstone of the Taean Formation formed posterior to the Silurian (430–230 Ma), which was mapped previously as a Precambrian–Mesozoic unit (Na et al., 1982). According to Cho et al. (2010), the majority of detrital zircons yield age of the

Archean–Proterozoic (2.5–2.4 Ga, 1.0–0.9 Ga, and 820–750 Ma) and the Silurian (~430 Ma). These ages are identical to those of the Qinling-Dabie Belt. The overall distribution pattern of detrital zircons is similar to that of the Imjingang Belt, which indicates that the Taean Formation represents a southward extension of the collision belt (Cho et al., 2010).

8.3 South Korean Tectonic Line

8.3.1 The Shear Zone

Although the Imjingang Belt demarcates the Nangrim Massif in North Korea from the Gyeonggi Massif in South Korea (Cho et al., 1995; Ree et al., 1996), the SKTL divides the Gyeonggi Massif from the Yeongnam Massif across the southern part of the peninsula (Figure 8.3). The SKTL is the major structural discontinuity, from the mideastern continental margin near Gangneung to the southwestern tip of the peninsula, extending for over 400 km (Chough et al., 2000). This tectonic belt represents a northeast–southwest trending dextral shear zone in which the Gyeonggi Massif, and the overlying metasedimentary rocks of the Okcheon Group in the northwest are in thrust contact with the Yeongnam Massif and the overlying sedimentary successions of the Joseon and Pyeongan supergroups in the southeast (Figure 8.3).

In the mideastern part of the peninsula, the SKTL runs along the boundary between the Okcheon and Yeongwol groups (Figure 8.3). It is different from the fault system of Cluzel et al. (1990, 1991), which placed it at the boundary between the Yeongwol and Taebaek groups. Cluzel et al. (1990, 1991) and Cluzel (1992) have misinterpreted that both the Okcheon and Yeongwol groups formed in a single intracontinental rift basin. In the southwestern part of the peninsula, the SKTL deviates from the anastomosing ductile shear zone, the Honam shear zone (Otoh and Yanai, 1996; Yanai et al., 1985, 1993; Yin and Nie, 1993), although it is difficult to trace the boundary zone between the Gyeonggi and Yeongnam massifs. This shear zone also crosscuts the boundary at a low angle (Ree et al., 2001). The shear zone is largely obliterated by granite, which was intruded in the Jurassic and Cretaceous periods (Figure 8.3).

The thrust deformation along the SKTL spanned approximately 70 million years (from Triassic to Jurassic) during the collision of the two blocks. According to the traditional terminology of geosynclinal evolution and orogeny, the Mesozoic deformation in the Okcheon Fold Belt occurred in three distinct events: the Songrim Orogeny (Late Permian to Early Triassic), followed by the Daebo Orogeny (Late Jurassic), and the Bulguksa event (Cretaceous to Paleogene) (Kim, 1996; Lee, 1987a; Reedman and Um, 1975). In view of the plate tectonic evolution of the colliding lithospheric plates, the Songrim Orogeny most likely corresponds to the collision event at the Imjingang Belt and the clockwise rotation and right-lateral movement of the Sino-Korean Block along the SKTL (Figure 8.1). The Daebo Orogeny would represent continued folding and thrust, caused by compression and adjustment. This orogeny was most likely due to orthogonal subduction of the paleo-Pacific Plate under the Asian continent, associated with extensive emplacement of granite batholiths (Figure 8.1). The right-lateral strike-slip fault (Honam shear zone) activated during this time (mid-late Jurassic). However, the Bulguksa event was dominated by backarc extension and the associated plutonism and volcanism (Figure 8.1).

In the following section, I will briefly outline geological features in a few sections of the shear zone (SKTL) and the associated intrablock deformation. The shear zone includes the Bonwhajae, Mungyeong, and Pibanryeong sections, whereas the associated intrablock deformation includes the Yeongwol, Deokpori, Gongsuwon, and Danyang sections.

8.3.1.1 Bongwhajae Thrust

The Bongwhajae section reveals strongly deformed beds of metamorphosed siliciclastics of the Seochangri Formation (Okcheon Group) in the west and of the carbonate succession of the Yeongwol Group in the east (Chough et al., 2006) (Figures 8.5 and 8.6). It is the major stratigraphic discontinuity formed by thrust deformation along the SKTL. Although the present topography has been subdued by denudation for about 200 million years since the collision event, the abrupt changes in both sedimentary facies and deformational features are suggestive of major discontinuity formed by tectonic deformation. In fact, the Seochangri phyllite beds form a thrust front against the younger limestone and dolostone beds of the Cambrian–Ordovician (Figure 8.7). Both rocks show upright isoclinal-to-tight folds, sheath folds, and boudins (Figure 8.8). The fold axes generally trend north-northeast with a low plunge angle.

The thrust front is represented by a quartzite–shale alternation that occurs at the base of the Seochangri Formation along the boundary. The quartzite is generally massive and coupled with thin shale layer (Figure 8.7). Each couplet is characterized by a sharp boundary, although they are commonly deformed, showing strong foliation, injection, pseudonodules, and flame structures. The couplet beds formed most likely in a marginal deep-water environment (submarine fan). The thickening-upward trend is indicative of progradation of depositional lobes in the middle and outer fans (Chough et al., 2006). However, the overlying phyllites represent hemipelagic sedimentation. This depositional setting was quite in contrast to the underlying limestone beds, which formed in a shallow carbonate platform. In the nearby sections of limestone beds, Ordovician fossils of brachiopods, crinoids, and sponges occur (Chough et al., 2006).

The entire succession underwent two phases of deformation (Figure 8.9). The first compressive deformation formed an isoclinal fold where the phyllite beds thrust over the limestone beds. The second phase shows higher angle thrust. These beds were strongly deformed under brittle–ductile conditions of about 300°C. The phyllites contain two distinct deformational features. One is major regional foliation, aligned by biotite and muscovite; the other is crenulation without mineral growth. However, the intruded amphibolite bodies are undeformed (Chough et al.,

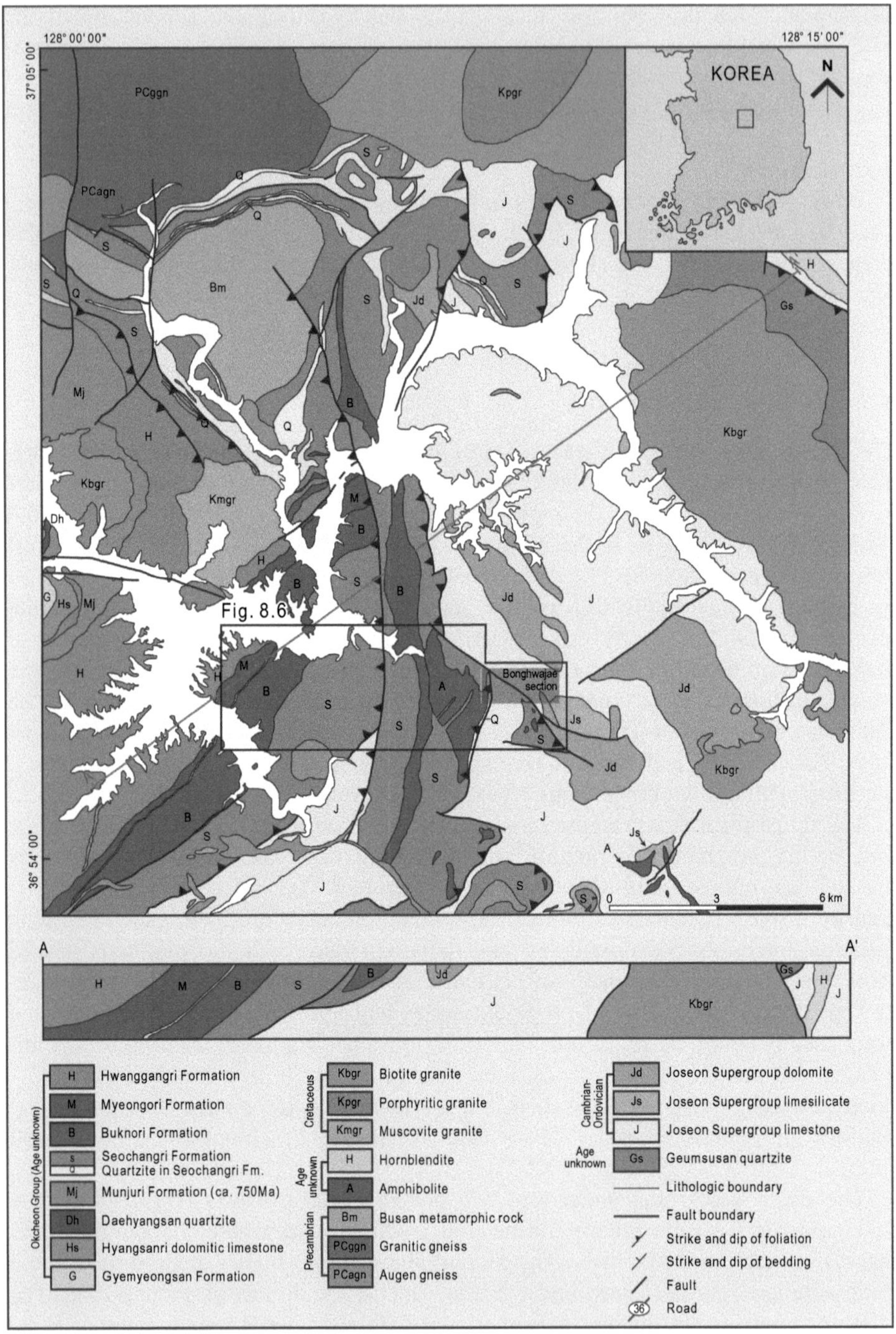

Figure 8.5 Geologic map and major structural features in the vicinity of the Bonghwajae section.
Source: Lee and Park (1965), Kim et al. (1967).

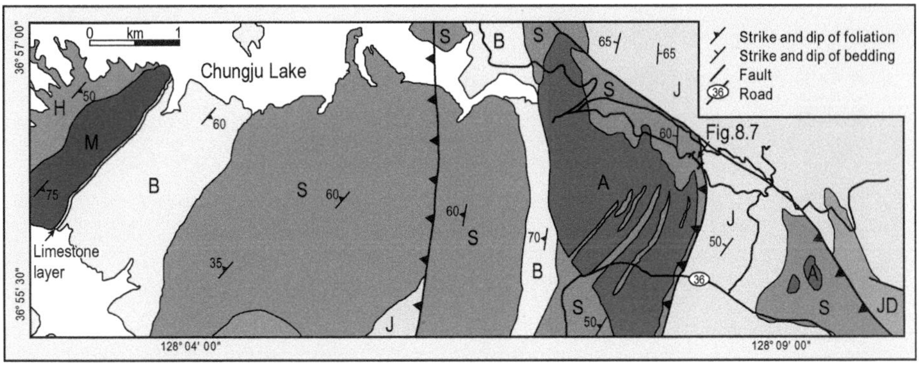

Figure 8.6 Geologic map of the Bonghwajae area. For legend, see Figure 8.5.
Source: Lee and Park (1965).

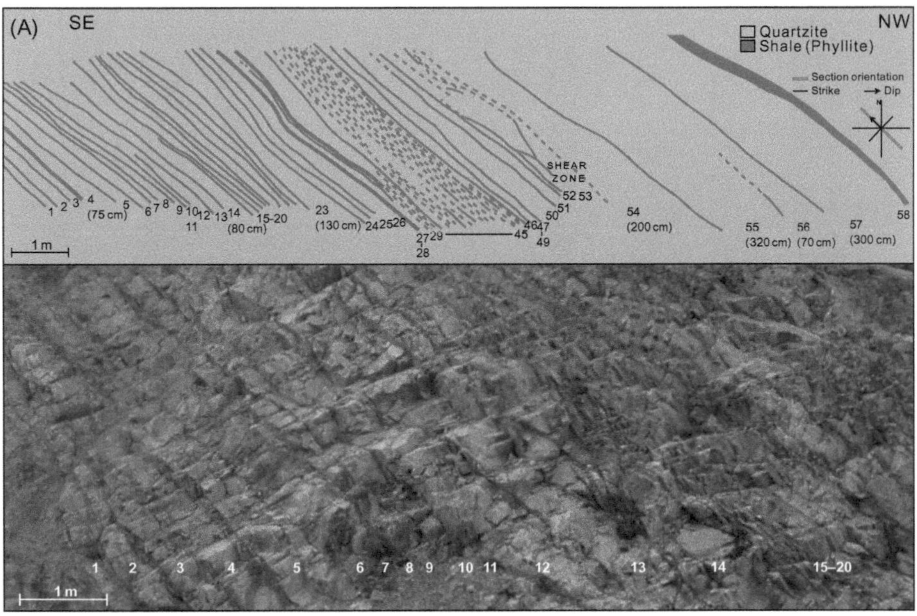

Figure 8.7 (A) Photograph and line-drawing of quartzite-shale interbeds at the base of the Seochangri Formation, Bongwhajae section. For location, see Figure 8.6. (B) Sedimentary log of the quartzite–shale beds in the Bongwhajae section. (C, D) Photographs of quartzite-shale beds, showing Bouma-A and -D/E divisions with an erosional boundary.
Source: Chough et al. (2006) by permission of *Geosciences Journal*.

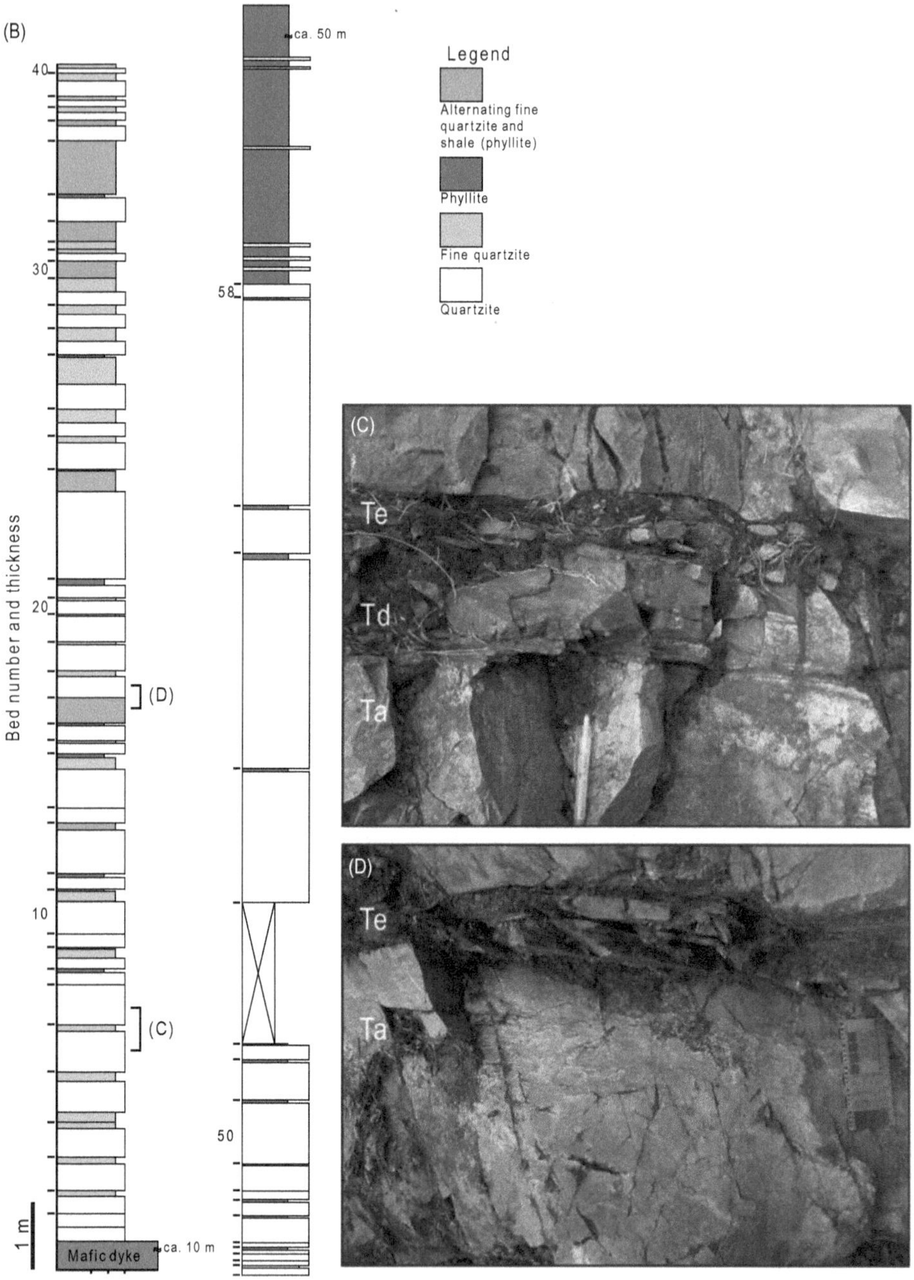

Figure 8.7 (Continued)

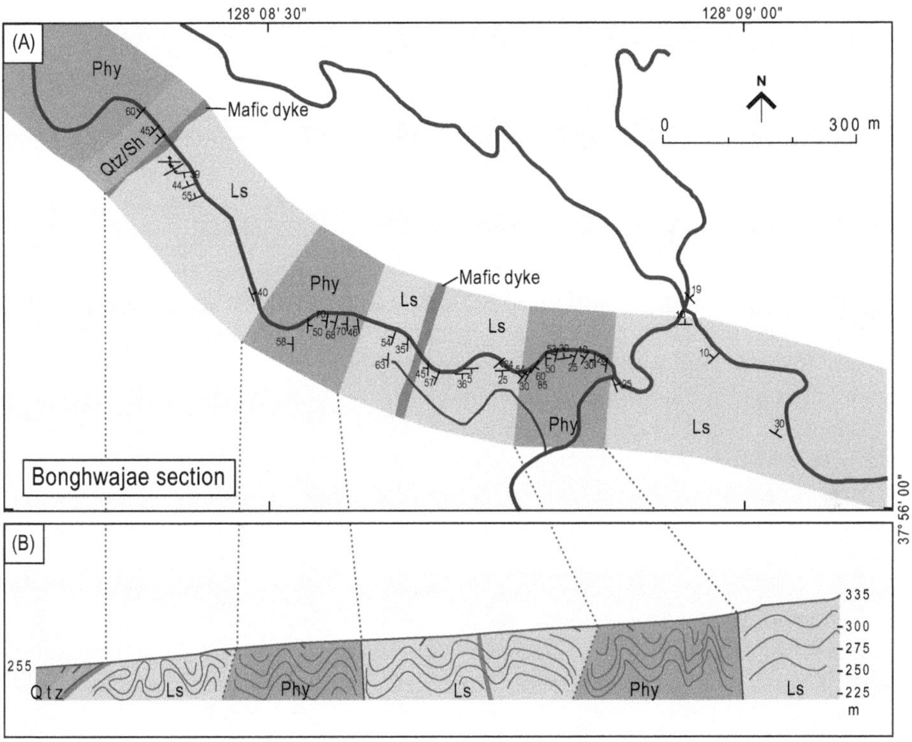

Figure 8.8 Geologic description of the Bonghwajae section. (A) Lithologic distribution, attitudes of foliations, and fold axial planes of the phyllite and limestone units along the roadcut. (B) Inferred cross-section. Qtz, quartzite; Phy, phyllite; Ls, limestone; Sh, shale.
Source: Chough et al. (2006) by permission of *Geosciences Journal*.

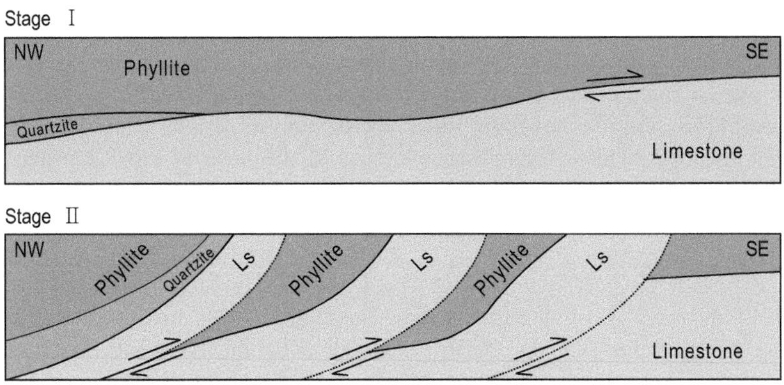

Figure 8.9 Deformational model of the Bonghwajae section. Stage I shows phyllite beds of the Seochangri Formation that thrust over the limestone beds of the Joseon Supergroup. Stage II shows subsequent shortening. Ls, limestone.
Source: Chough et al. (2006) by permission of *Geosciences Journal*.

2006). The deformation front trends northeast–southwest with southeastward tectonic transport (Kihm and Kim, 2003). This trend was due to the collision in a restraining bend of the SKTL. The occurrence of amphibolites most likely suggests reactivation of the tectonic belt in an extensional regime since the collision event.

The Bongwhajae section thus represents part of the restraining bend of the collision belt between the Gyeonggi and Yeongnam massifs. It is a thrust front of the colliding tectonic plates, i.e., the South China Block versus the Sino-Korean Block (Figure 8.1). The collision belt extends from the Qinling-Dabieshan Belt in China, offset by the Tanlu fault, to the Sulu Belt across the Yellow Sea and the Imjingang Belt in the Korean Peninsula (Figure 8.2). Although a paleomagnetic study of the late Paleozoic succession (Pyeongan Supergroup) (Doh and Piper, 1994) suggests that the Korean Peninsula rotated clockwise (about 30°) in the Mesozoic, the amount of rotation was most likely minimal at the eastern end of the Imjingang Belt where the South China Block terminated (Figure 8.2). It is quite similar to the clockwise rotation of the Eurasian plate at the eastern end of the Himalayan Orogenic Belt.

8.3.1.2 Thrust of Yeongwol Group

The Yeongwol Group succession in the northwest of Yeongwol city, approximately 20 km east of the SKTL, is characterized by a series of thrust faults. Individual sheets of intra- and interformational thrust were mapped by a correlation of trilobites (Fig. 5.31) (Choi, 1998). The entire eastward thrust block of the Yeongwol and Pyeongan groups is enclosed by a loop-shaped outer thrust front, the Sangri fault, against the Taebaek and Hwangji groups (Figure 8.3). The outer thrust fault cuts a series of northeast–southwest-bounded strike-slip faults in the latter (Figure 8.3). Although the thrust was due to eastward compression against the Taebaek Group in the final stage of deformation, the block probably traveled for long distances (tens to hundreds of kilometers) toward the northeast from the depositional site, far from that of the Taebaek Group. The thrust deformation occurred within the Sino-Korean Block by compressional force generated by the northeast and eastward movement of the South China Block against the Sino-Korean Block along the SKTL. This far-reaching influence of the collisional thrust front is manifested by a klippe of the Geumsusan quartzite at the base of the Yeongwol thrust block (Figure 8.5).

8.3.1.3 Deokpori Thrust

The Deokpori thrust is a low-angle reverse fault (thrust) bounded by the underlying limestone beds of the Taebaek Group (footwall) and the overlying limestone beds of the Yeongwol Group (hanging wall) (Figure 8.10). The thrust trends northeast (N25°E, 35°NW) and comprises foliated cataclastic rocks and a fault gauge (Figure 8.10). The fault zone is about 100 m wide and the slickenside lineations on the cataclastic foliation in the fault zone trend northwest with a plunge angle of 20°, which indicate southeastward tectonic transport (Han et al., 2006). Because of

Figure 8.10 The Deokpori thrust. TG, Taebaek Group; YG, Yeongwol Group; BF, Bansong Formation. For location, see Figure 8.11.

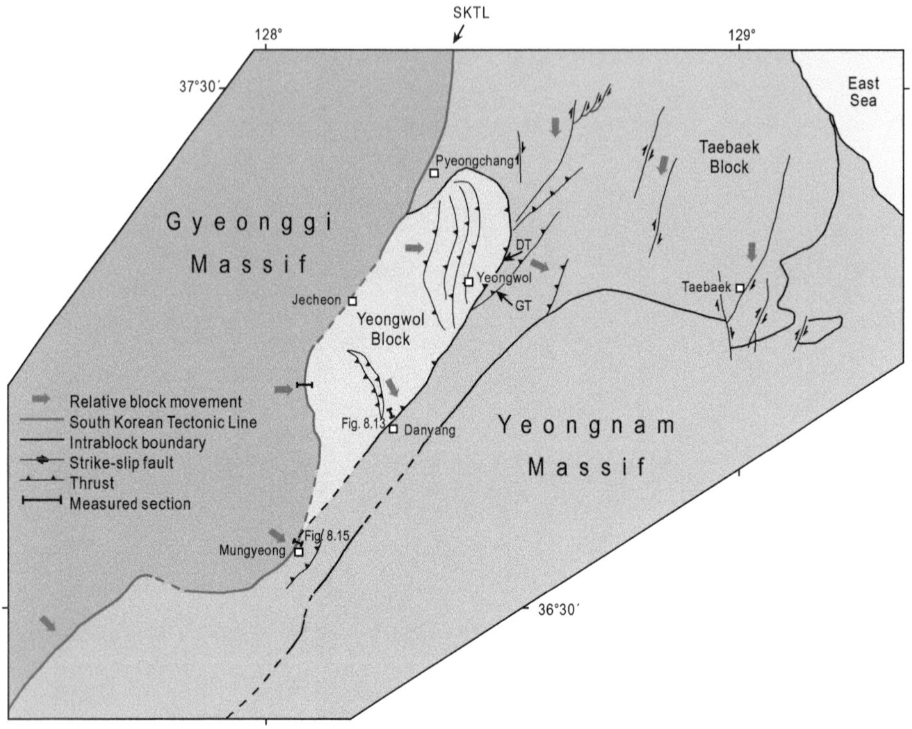

Figure 8.11 Intrablock deformation of the Taebaek and Yeongwol groups. DT, Deokpori thrust; GT, Gongsuwon thrust.

the load (weight) of the overriding limestone thrust blocks, the underlying beds subsided, forming a linear depression (piggyback basin). The deposits (Bansong Formation) comprise conglomerate, sandstone, and shale, which are also largely sheared and deformed (Figure 8.10). The thrust resulted from intrablock compression within the Sino-Korean Block, influenced by eastward movement of the South China Block (Figure 8.11). The formation mechanism of a piggyback basin is similar to that of the foreland basin on a large scale.

8.3.1.4 Gongsuwon Thrust

The Gongsuwon thrust runs northeast across the East River, an offset branch of the Deokpori thrust (Figure 8.11). The thrust constitutes a high-angle reverse fault in which the limestone succession (Taebaek Group) overrides sandstone beds of the Hambaeksan Formation (Pyeongan Supergroup) with a clastic wedge (Bansong Formation) of conglomerate, sandstone, and shale in between (Figure 8.12). The Bansong Formation formed in the fault zone during a thrust movement in which sediments were derived from both blocks. The fault zone consists of cataclasites and generally strikes northeast with a dip angle of 70–80° to the northwest. The

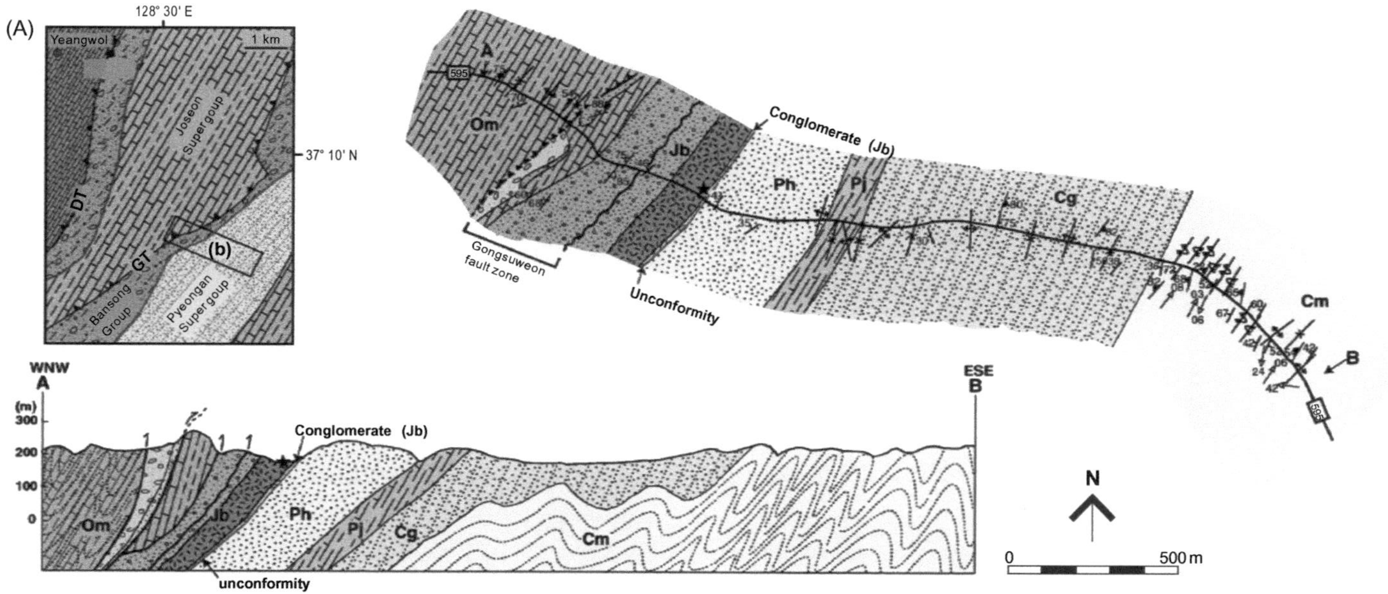

Figure 8.12 (A) Geologic map and cross-section of the eastern Yeongwol area. Om, Makgol Formation; Cm, Manhang Formation; Cg, Geumcheon Formation; Pj, Jangseong Formation; Ph, Hambaeksan Formation; Jb, Bansong Group. For location, see Figure 8.11. (B and C) The Gongsuwon thrust. The Bansong Formation formed between thrust of the Taebaek Group and Pyeongan Supergroup. (D) Details of the erosional boundary between the Hambaeksan and Bansong formations. (E and F) Photographs of metamorphosed conglomerates, Bansong Formation. *Source*: (A) Han et al. (2006) by permission of the Elsevier.

Figure 8.12 (Continued)

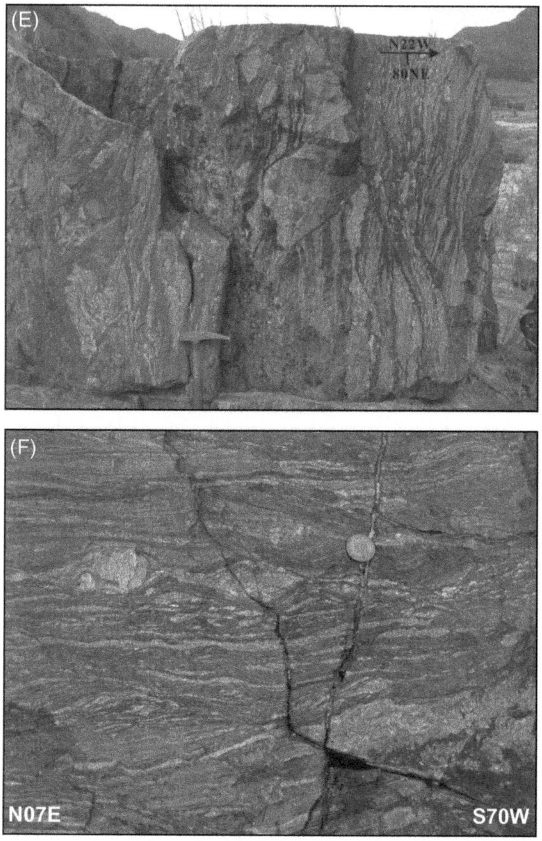

Figure 8.12 (Continued)

boundary between the conglomerate (Bansong Formation) and the underlying sandstone (Hambaeksan Formation), whose clasts originated from debris of fault breccia, is erosional. One of the most interesting features of the fault zone is the occurrence of metasedimentary rocks, i.e., metaconglomerate, metasandstone, and shale in the Bansong Formation (Figure 8.12). These rocks display texture of metamorphic rocks. It suggests that the succession underwent a series of burial deformations and metamorphism since its deposition in a narrow belt of the depression, experiencing lateral displacement for long distances. The metamorphism is evinced by porphyroblasts of andalusite, cordierite, and biotite, which formed posterior to the foliation (Han et al., 2006). The Bansong Formation contains volcanic rocks (lapilli tuff and tuff breccia). Two independent age dates for volcanic rocks indicate that the Bansong Formation formed about 180 Ma (Han et al., 2006; Jeon et al., 2007).

8.3.1.5 Danyang Thrust Complex

Highly deformed limestone beds of the Yeongwol Group succession occur in a road-cut section north of Danyang (Figure 8.13). This section is represented by an

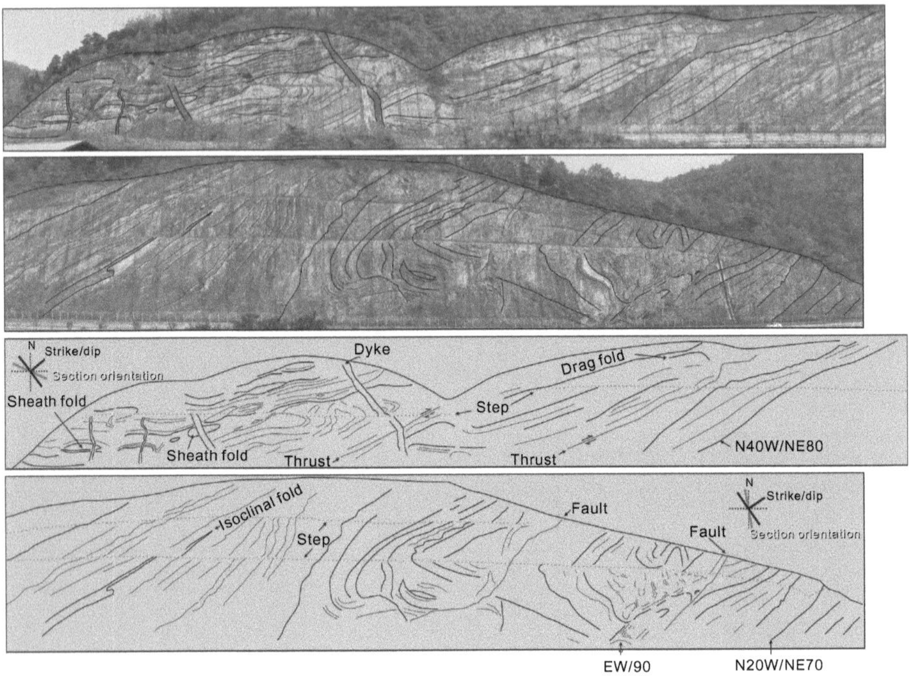

Figure 8.13 Highly deformed limestone beds (Yeongwol Group) north of Danyang (N36°59′16″, E128°20′28″). The section is represented by alternation of dark and light gray limestone beds. Isoclinal tight and sheath folds occur in the dark gray limestones, whereas faults and recumbent folds prevail in the limestone beds. Two north-dipping thrusts formed in the middle part of the section. For location, see Figure 8.11.

alternation of dark and light gray limestone beds, showing tight isoclinal, sheath, and recumbent folds as well as faults. The entire section is indicative of thrust deformation with compression trending approximately north to south.

8.3.1.6 Pibanryeong Thrust

The thrust deformation in the Gyeonggi Massif along the SKTL is also manifested in the Pibanryeong area, southeast of Cheongju (Figure 8.14). The metasedimentary successions of the Okcheon Group are bounded by a series of thrust, directed southeastwards (Lim et al., 2007), which reveals a two-stage deformational history. The first-stage thrust formed a relatively low-angle offset, whereas the second-stage fault retrograded backward to form a relatively higher angle offset under the similar compressional regime. The thrust trends southwestward along the SKTL (Lim et al., 2005, 2006).

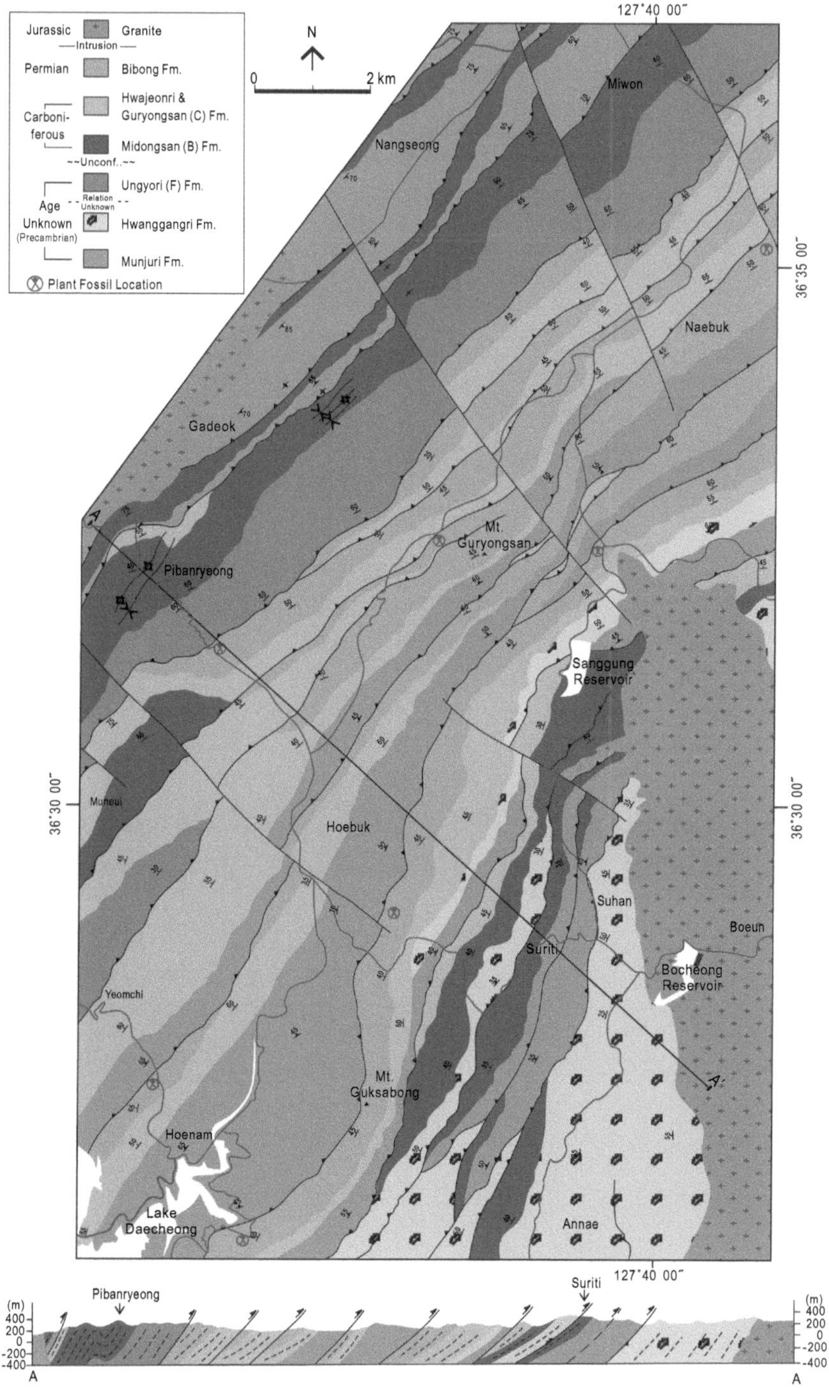

Figure 8.14 A geological map of Pibanryeong area.
Source: Lim et al. (2007).

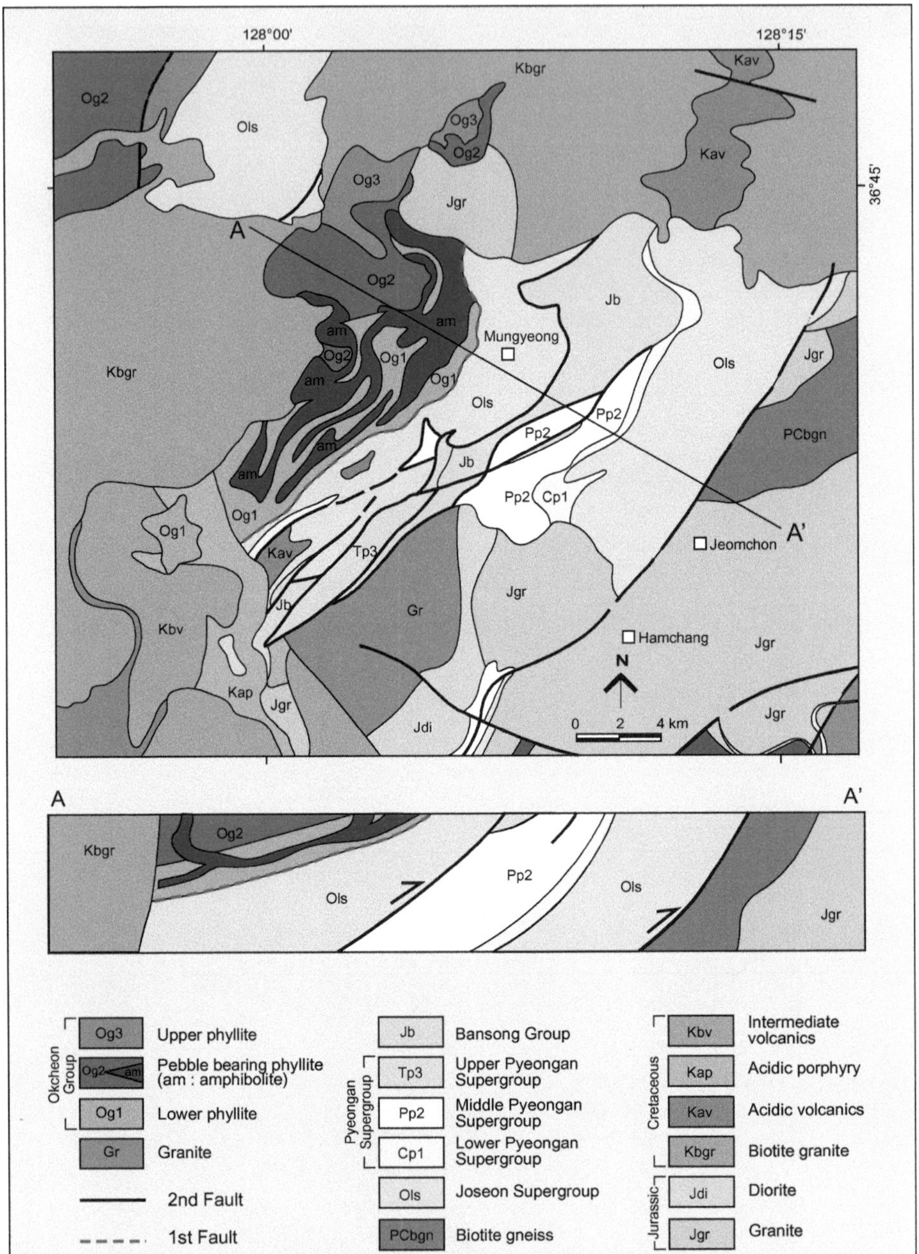

Figure 8.15 Simplified geologic map and cross section (A-A′) of the Mungyeong area, showing the thrust contact between the Okcheon Group succession of the Gyeonggi Massif (South China Block) and the Joseon and Pyeongan supergroups of the Yeongnam Massif (Sino-Korean Block). For location, see Figure 8.11.

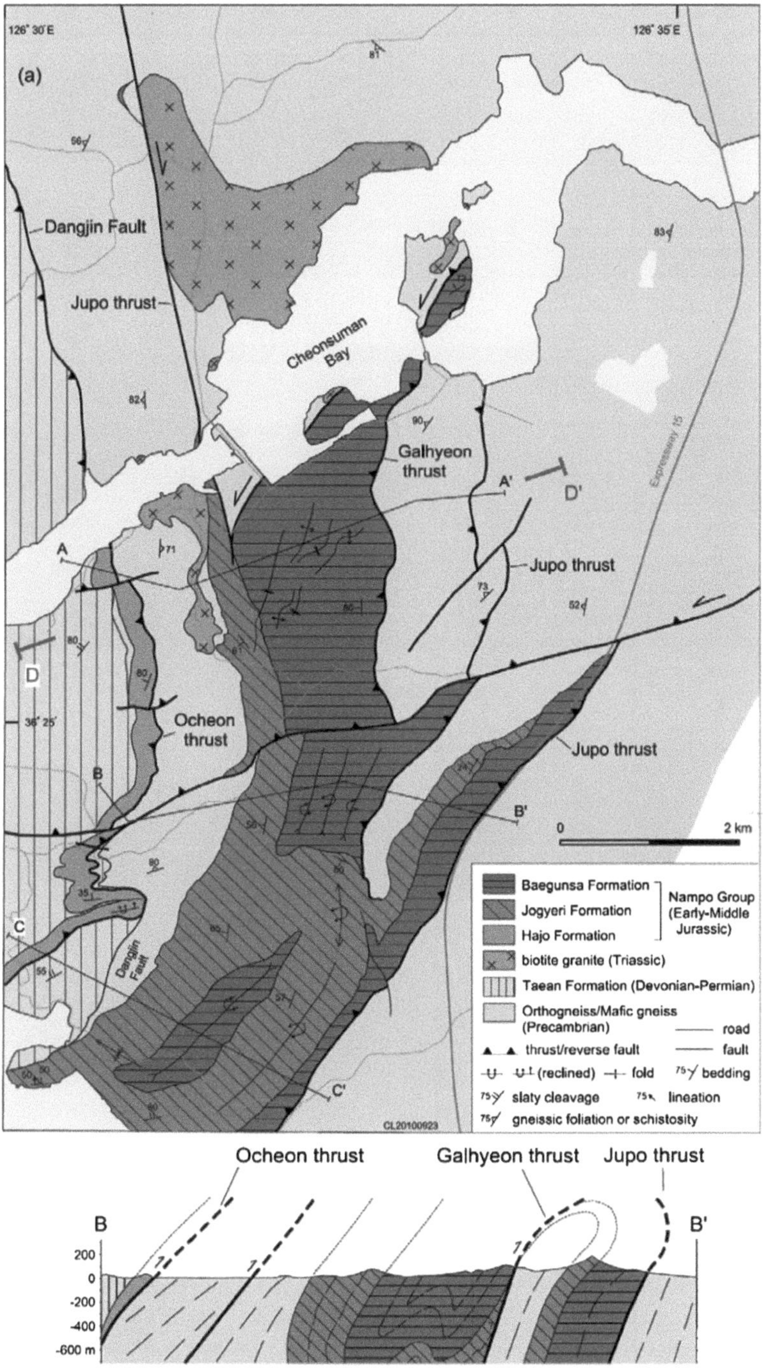

Figure 8.16 (A) Geologic map and (B) cross-section of the Ocheon area, Chungnam Basin. *Source*: Lim and Cho (2012) by permission of the American Geophysical Union.

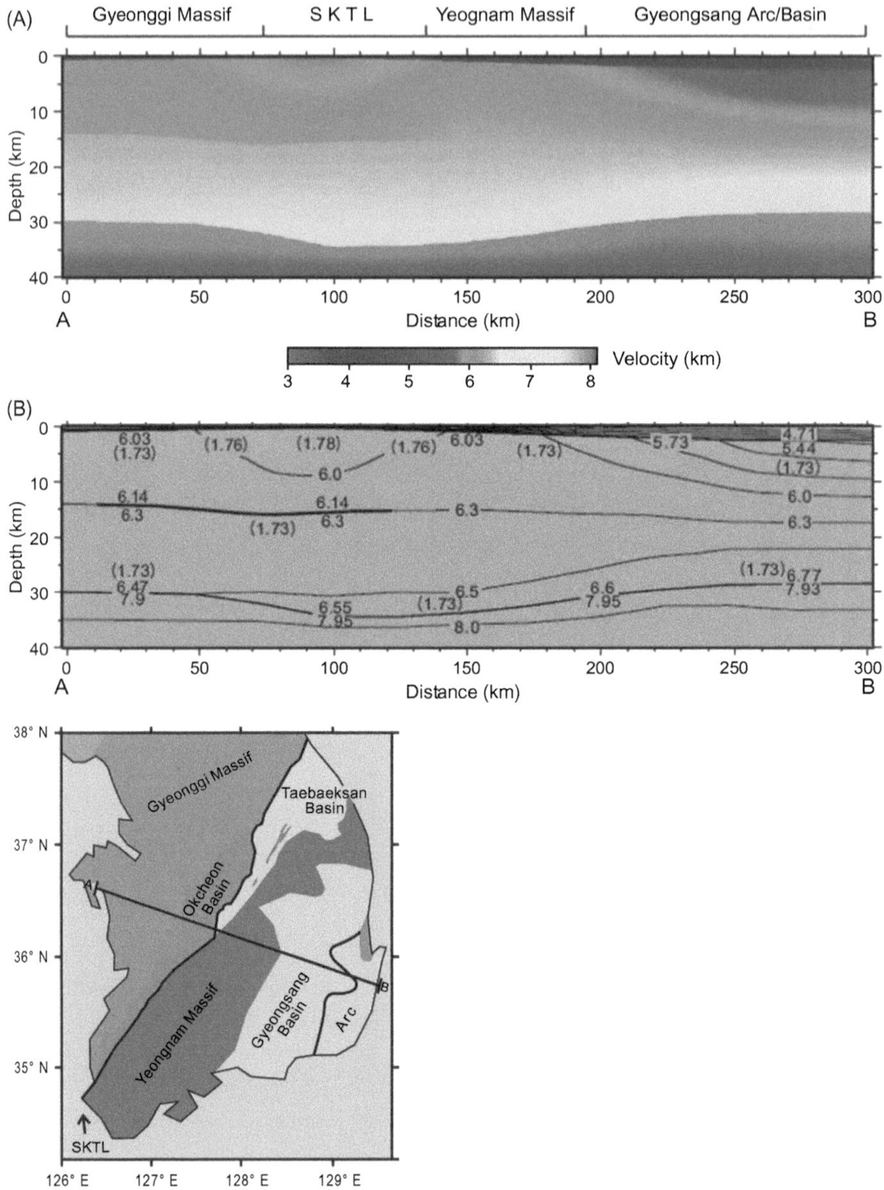

Figure 8.17 Vertical section (A) and interpretation (B) of P-wave velocity across the southern part of the Korean Peninsula. SKTL, South Korean Tectonic Line.
Source: Cho et al. (2006b) by permission of the American Geophysical Union.

8.3.1.7 Mungyeong Thrust

The thrust deformation along the SKTL also appears in the Mungyeong area where the Okcheon Group rocks thrust southeastward against the Joseon and Pyeongan supergroups (Figure 8.15). Here, the Precambrian massif (Gyeonggi Massif) consists of garnet-mica gneiss and mylonite, whereas the Okcheon Group comprises metasedimentary rocks (argillaceous and carbonaceous) with chert and interbedded limestone. The limestone and dolomite beds of the Joseon Supergroup have undergone intense ductile deformation, including recumbent isoclinal folds, sheath folds, and boudinage. Trilobite fossils (*Redlichia*) occur in relatively less deformed shale beds, which indicate an Early Cambrian age. However, the Pyeongan Supergroup rocks comprise strongly cleaved texture but little ductile deformation. Sandstone and shale (Jurassic Bansong Formation) also occur (Figure 8.15). Amphibolite bodies intruded into the Okcheon Group rocks, whereas granite plutons are largely dispersed over the entire area (Figure 8.15).

Although the definite field relationship between the Okcheon Group and Joseon Supergroup is hard to define, it is most likely that the former represents thrust over the latter along the faults that trend northeast to southwest (Figure 8.15). In places, the Pyeongan Supergroup rocks show high-angle dips and cleavages, which are in direct contact with the thrust blocks of the Okcheon Group. The piggyback basins of the Bansong Formation formed in the depressions by infilling of sediments eroded from the uplifted thrust blocks.

8.3.1.8 Nampo Group

The Nampo Group (Early to Middle Jurassic) of Chungnam Basin (about 30 km wide and 50 km long) within the Gyeonggi Massif consists of a nonmarine succession (1 km thick) of breccia, conglomerate, sandstone, and mudstone as well as coal deposits. It formed in alluvial fans and deltas and the associated braidplains, which prograded into lacustrine environments (Choi, 1987; Egawa and Lee, 2009). The Nampo Group provides evidence for intrablock formation and deformation of a nonmarine basin within the Gyeonggi Massif during compression caused by subduction of the Pacific Plate under the Eurasian continent. Lim and Cho (2012) made a detailed analysis of the deformation styles of the Ocheon thrust and the associated features in the Nampo Group (Figure 8.16). The majority of the succession and a basement slab were folded and displaced toward the west–northwest along the thrust during the first phase of deformation in the Middle to early Late Jurassic; the second phase (Late Jurassic) was dominated by reverse faults (east–west trending) and folds to form a dome-shaped structure, which was caused by a north–south shortening (Lim and Cho, 2012).

8.4 Crustal Structure

A crustal-scale seismic refraction survey across the southern part of the Korean Peninsula has revealed P-wave velocity structure of the crust (Cho et al., 2006b).

The crust under the Gyeonggi and Yeongnam massifs is 30 km thick (Figure 8.17). In the shear zone of the SKTL, however, it is about 10% thicker (34 km thick) than that of the stable craton, suggesting that the deformation had a significant effect on the isostatic adjustment of the crust. However, the arc-backarc region of Gyeongsang Province is rather thin (28 km thick), accompanied by a low-velocity layer near the surface (Figure 8.17).

9 Gyeongsang Arc System

9.1 Introduction

The eastern margin of the Asian continent experienced left-lateral wrench tectonics during the Cretaceous, induced by oblique subduction of the Izanagi Plate under the Asian continent (Engebretson et al., 1985; Kim et al., 1997; Klimetz, 1983; Watson et al., 1987). Major deformation occurred in the backarc region between the subduction zone and the Tan-Lu Fault in China, forming a number of sedimentary basins (Figure 9.1). The Korean Peninsula was also subject to both left-lateral wrench tectonics and continental arc volcanism (Chough and Sohn, 2010). The Gyeongsang Arc System comprises a volcanic arc (Gyeongsang Volcanic Arc) and the adjacent backarc basin (Gyeongsang Basin) behind the arc (Chough and Sohn, 2010) (Figure 9.2). The arc system formed a southwest extension of the Japanese Arc.

9.2 Gyeongsang Volcanic Arc

The arc comprises volcanic and plutonic rocks (Figure 9.3). The volcanics include andesitic and rhyolitic rocks, comprising andesitic lavas, agglomerates or tuff breccias, dacitic to rhyolitic lavas, and welded tuffs (Hwang and Kim, 2006; Jeong and Jwa, 2000; Kim et al., 1998; Park et al., 2006b; Yun, 1998). The majority of the volcanic rocks occurs in calderas and volcanotectonic depressions (Cha and Yun, 1988; Hwang and Cheong, 1998; Hwang and Kim, 1990, 1992; Hwang et al., 2002; Reedman et al., 1987). Sedimentary successions comprise conglomerates, sandstones, mudstones, and marlstones in patchy areas with common intercalations of lava flows and tuffs (Kim and Paik, 1997; Paik et al., 1996, 1997, 2006; Son et al., 2009), representing intraarc depressions and basins. According to the trace element compositions and rare earth element patterns of the volcanic rocks, these rocks belong to the typical calc-alkaline volcanic rocks related to the subduction (Kim et al., 1991b; Park et al., 2006b; Yun et al., 2000). The voluminous calc-alkaline volcanism was immediately followed by granitic plutonism (Figure 9.3). The granitic rocks comprise diorite/granodiorite with porphyritic texture and I-type characteristics, indicative of shallow-depth emplacement, either in the lower crust or in the upper mantle (Cho and Kwon, 1994; Hong, 1987; Jin, 1980; Jwa et al., 1990; Lee, 1992a). These events largely occurred posterior to about 80 million years ago.

Geology and Sedimentology of the Korean Peninsula. DOI: http://dx.doi.org/10.1016/B978-0-12-405518-6.00009-7

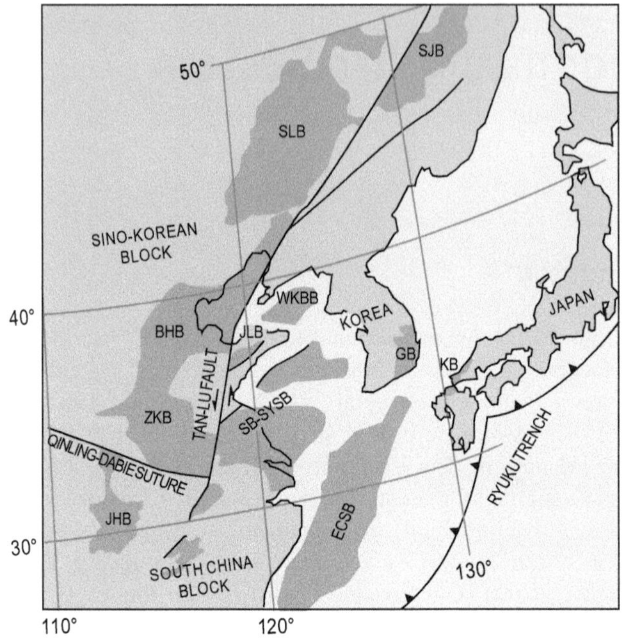

Figure 9.1 (A) Simplified tectonic map of northeast Asia, showing wide distribution of Cretaceous–Tertiary sedimentary basins (modified from Allen et al. (1997), Chen and Qin (1989) and Ren et al. (2002)). BHB, Bohai Basin; ECSB, East China Sea Basin; GB, Gyeongsang Basin; JHB, Jianghan Basin; JLB, Jiaolai Basin; KB, Kanmon Basin; SB, Subei Basin; SJB, Sanjiang Basin; SLB, Songliao Basin; SYSB, South Yellow Sea Basin; WKBB, West Korean Bay Basin; ZKB, Zhoukou Basin.

The arc was an extension of the Japanese Arc, which ran approximately parallel to the paleo-subduction zone of the Izanagi Plate (Figure 9.4). Together, the arc system was hundreds of kilometers wide and thousands of kilometers long.

9.3 Gyeongsang Basin

A large nonmarine basin, Gyeongsang Basin, formed in the southeastern part of the peninsula; a number of small-scale oblique-slip (sinistral) basins developed in the southwestern part (Figure 9.1). Gyeongsang Basin was filled with a thick succession, about 9 km thick, of siliciclastic rocks and volcanic deposits as well as carbonates and evaporites. The spores and pollen are suggestive of semi-arid conditions during this time. Only terrestrial fossils occur, including plant stems and leaves, fresh-water shells and algae, ostracodes, insects, dinosaur bones, and tracks (Choi, 1990; Houck and Lockley, 2006; Kim et al., 2008; Paik and Kim, 2006; Paik et al., 1988, 2007; Yang, 1976, 1982; Yi et al., 1993).

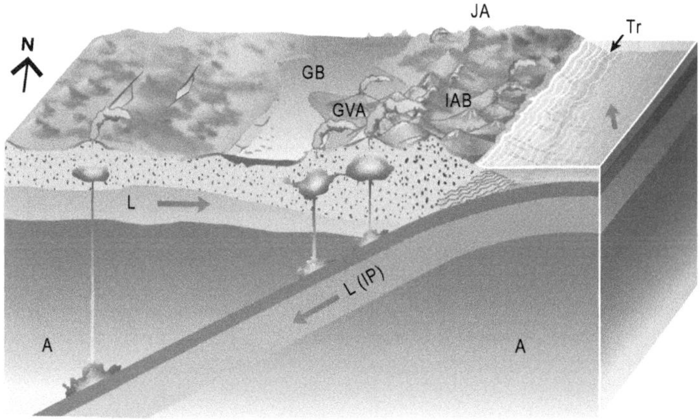

Figure 9.2 Tectonic framework of the northeast Asian margin during the Cretaceous, showing the distribution of sedimentary basins and volcanic arc. The Izanagi Plate subducted obliquely northward under the Asian continent causing left-lateral wrench deformation and continental arc volcanism in the southern part of the Korean Peninsula. GB, Gyeongsang Basin; GVA, Gyeongsang Volcanic Arc; JA, Japanese Arc; IAB, Intraarc Basin; L, Lithosphere; A, Asthenosphere; IP, Izanagi Plate; Tr, Trench.
Source: Chough and Sohn (2010) by permission of the Elsevier.

The Gyeongsang Basin comprises three subbasins: the Yeongyang subbasin, the Uiseong subbasin, and the Jinju subbasin (Chough and Sohn, 2010) (Figure 9.3). The Yeongyang subbasin is bounded by the Uiseong subbasin in the south by the Andong Fault and Cheongsong Ridge. The Uiseong subbasin is demarcated from the Jinju subbasin by a narrow gap near Hyeonpung village, an intrabasinal ridge of volcanics (Figure 9.3).

9.3.1 Jinju Subbasin

The sedimentary strata (sandstones and mudrocks) generally dip southeastward at low angles (Figure 9.5). The strata can be divided into several lithologic units (Table 9.1; Figure 9.6). They formed in alluvial, fluvial, and lacustrine environments (Chang, 1975, 1977; Cheong and Kim, 1996; Choi, 1986; Houck and Lockley, 2006; Paik et al., 2001, 2007). Sedimentation was initiated in a linear depression along the western margin with a progressive eastward down-fault of the basin floor. Sediments were largely supplied from the drainage basins of the Precambrian gneiss and schist as well as granite on the west. A number of depositional systems are inferred to have developed in the trough, including alluvial fans in the trough margins, braided rivers along the axis of the depression, and swamps and lakes in the lowlands (Cheong and Kim, 1996; Paik and Lee, 1994, 1998). The basin progressively expanded and received volcanic materials from the arc on the

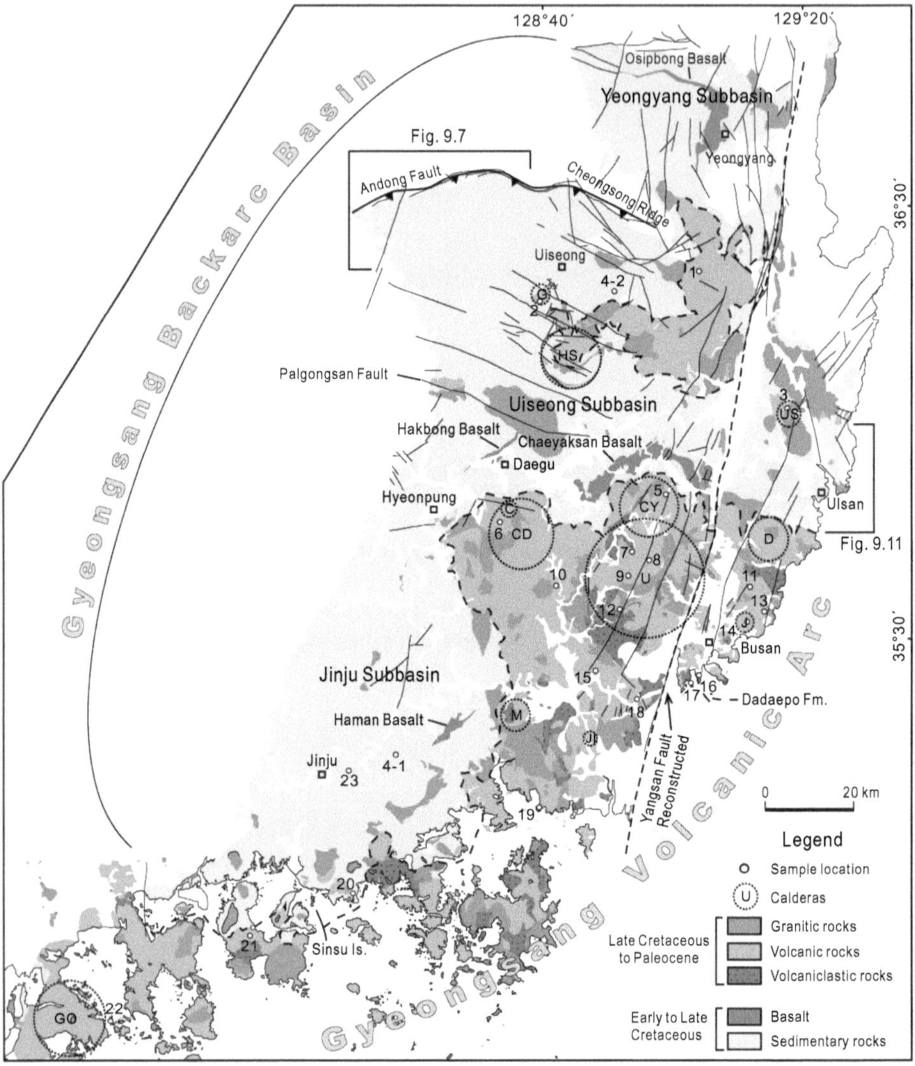

Figure 9.3 Simplified geological map of the Gyeongsang Arc System (Gyeongsang Volcanic Arc and Gyeongsang Backarc Basin). The sedimentary rocks and intercalating basalts are regarded as filling a backarc basin, whereas the volcanic/volcaniclastic rocks and the granitic rocks are regarded as constituting a volcanic arc platform. Note the subdivision of the backarc basin into three subbasins. Dotted circles with characters at center represent approximate boundaries of volcanotectonic depressions or calderas (Cha and Yun (1988)): Goseong (Go), Geumseongsan (G), Hwasan (HS), Unmunsa (U), Choijeongsan (C), Cheongdo (CD), Chaeyaksan (CY), Ulsan (US), Daeunsan (D), Jangsan (J), Jinrye (JI), and Mageumsan (M). Only the Hwasan caldera has almost perfectly circular shape.
Source: Chough and Sohn (2010) by permission of the Elsevier.

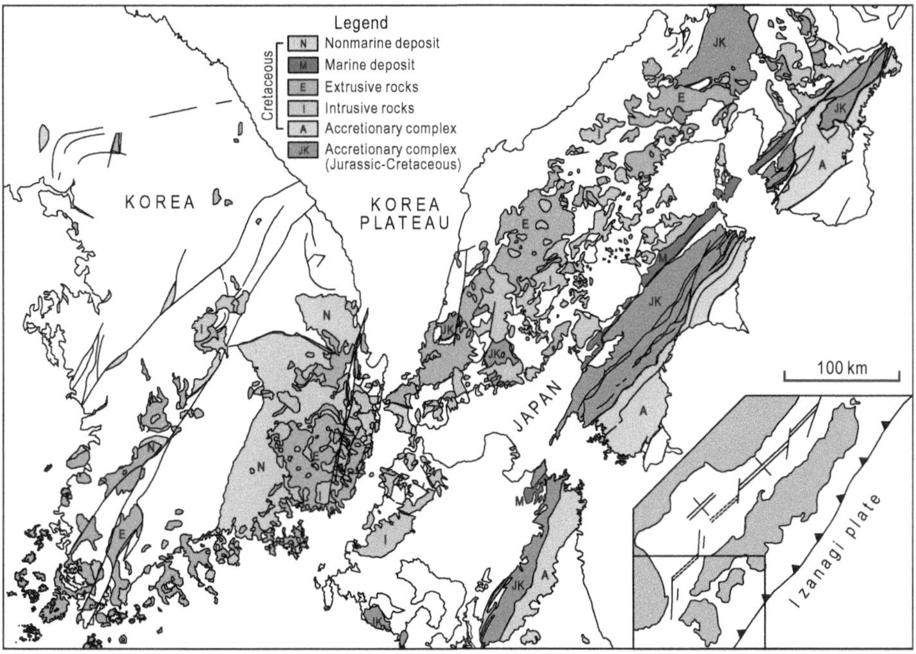

Figure 9.4 A paleogeographic plate reconstruction of the Japanese Arc relative to the Korean Peninsula during the Cretaceous. The distribution of extrusive volcanic rocks represents an occurrence of an arcuate volcanic arc associated with subduction of the Izanagi Plate. Geology of Japan after Teraoka and Okumura (2003). Inset after Yoon and Chough (1995).
Source: Chough and Sohn (2010) by permission of the Elsevier.

east and south. In particular, there was a huge explosive eruption of silicic magma which resulted in a basin-wide deposition of the Gusandong tuff (Jeon and Sohn, 2003) (Table 9.1). The tuff bed provides useful criteria for the stratigraphic correlation. Volcanic activity increased in the arc region with time.

9.3.2 Uiseong Subbasin

A detailed analysis of sedimentary facies is presently available only in the northwestern part of the subbasin (Jo, 2003a,b; Jo and Chough, 2001; Jo et al., 1997; Rhee et al., 1998). It reveals that the alluvial deposits of conglomerate, gravelly sandstone, sandstone, and mudstone can be grouped into five allomembers bounded by stratigraphic discontinuities (Figure 9.7). The discontinuities trend northwest to southeast, marked by distinct facies transitions, abrupt emplacement of conglomerate, and persistent mudstone beds (Figure 9.8). Sedimentary facies and architectural analyses reveal that each allomember formed a depositional system of fluvial channel networks draining toward the southeast with alluvial fans on the northern

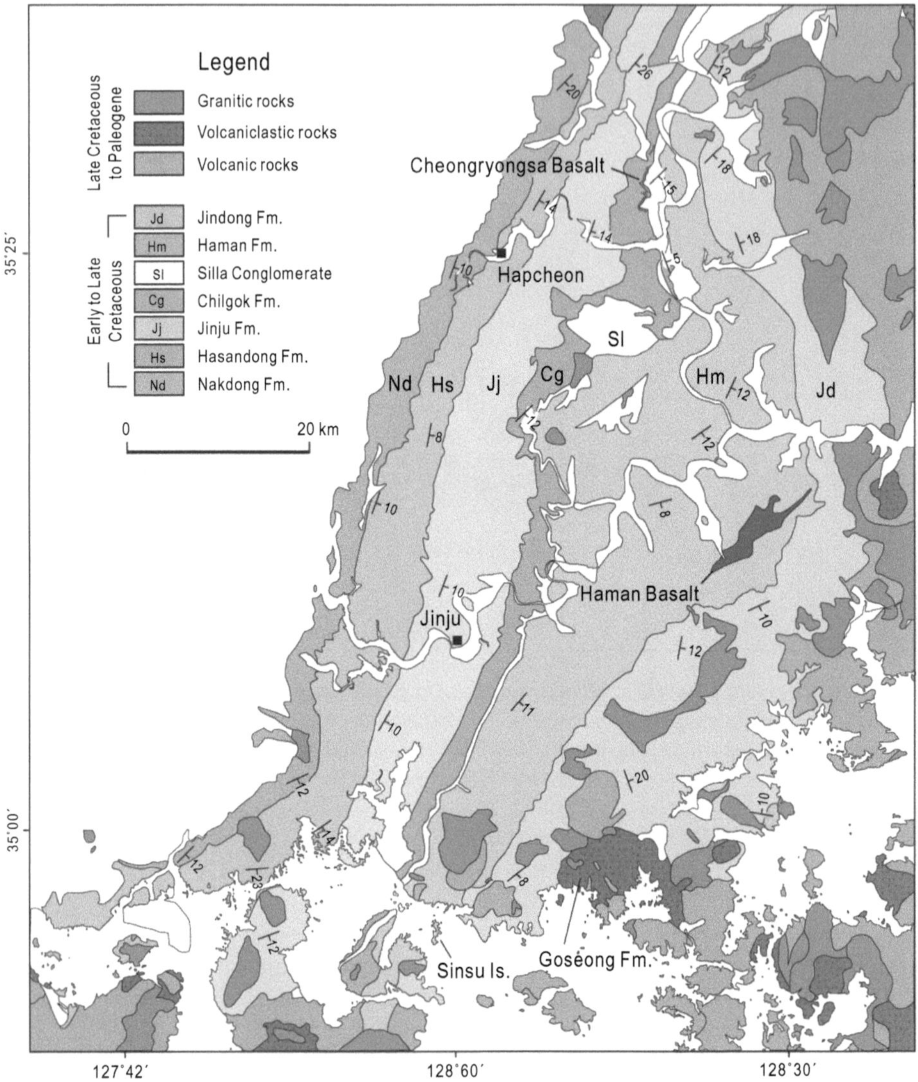

Figure 9.5 Geologic map of Jinju subbasin (Kim et al. (1998)) wherein the strata are generally undeformed, but tilted gently toward the east or southeast. The volcanic/volcaniclastic rocks, formerly designated as the Yucheon Group, are poorly differentiated. *Source*: Chough and Sohn (2010) by permission of the Elsevier.

margin (Figure 9.9). Each allomember can be characterized by distinctive architecture of channel fills, clast composition of conglomerate, and sandstone/mudstone ratio (Figure 9.10). Successive units show an eastward shift in the locus of deposition, suggesting basin-ward relocations of alluvial systems.

Table 9.1 Refined Lithostratigraghic Classification of the Gyeongsang Basin Fill (modified after Chang (1975, 1977, 1988), Chang et al. (1990, 2003), and Choi (1986)) and New Definition of the Gyeongsang Arc System

			Gyeongsang Arc System						
			Gyeongsang Volcanic Arc*		Gyeongsang Backarc Basin				
			Arc Platform	Intraarc basin & volcaniclastic apron	Jinju Subbasin	Uiseong Subbasin: Daegu area	Uiseong Subbasin: Uiseong area	Uiseong Subbasin: Ulsan area	Yeongyang Subbasin
Paleocene	58.7	Thanetian	Granitic rocks: Gaderi						
	61.7	Selandian							
	65.5	Danian	Rhyolitic rocks: Unmunsa / Samrangjin caldera						
Upper Cretaceous	70.6	Maastrichtian							
		Campanian	Andesitic rocks: Jangsan / Gyeongsan / Miryang / Changwon/ Palyongsan / Jusasan	Jeonggaksan Fm. / Jangmokri Fm. / Jangpyeongri Fm. / Dadaepo Fm.			Geumseongsan and Hwasan calderas		
				Goseong Fm.					
	83.5	Santonian				Geoncheonri Fm.	Sinyangdong Fm.		Sinyangdong Fm.
	85.8	Coniacian			Jindong Fm.	Chaeyaksan Bt.	Chunsan Fm.		Gisadong Fm.
	89.3	Turonian	Granodiorites: Chisulryeong			Songnaedong Fm.			
						Banyaweol Fm.		Joilri Fm.	
	93.5	Cenomanian	Gusandong Tuff (spanning)						Dogyedong Fm.
Lower Cretaceous	99.6		Sinsudo Diamictite (spanning)						
		Albian			Haman Fm.	Haman Fm.	Sagok Fm.	Sinhwari Fm.	Osipbong Fm.
					Silla Conglo.	Hakbong Bt.	Jeomgok Fm.		Cheongryangsan Fm.
									Gasongdong Fm.
	112.0	Aptian			Cheongryongsa Bt.				Dongwhachi Fm.
					Chilgok Fm.	Chilgok Fm.	Gugyedong Fm. Gumidong Fm. Iljik Fm. (A Jotap)	Banyeonri Fm.	Ullyeonsan Fm.
					Jinju Fm.	Jinju Fm.	Jinju Fm. (A Sinpyeong - Anpyeong)	Guyeongri Fm.	Gyeongjeongdong Fm.
	125.0	Barremian			Hasandong Fm.	Hasandong Fm.	Hasandong Fm. (A 2, 3)	Cheongokri Fm.	
	130.0 (Ma)	Hauterivian			Nakdong Fm.	Nakdong Fm.	Nakdong Fm. (A 1)		

* Stratigraphy poorly constrained. A = allomembers (Rhee et al., 1998; Jo and Chough, 2001), Conglo. = conglomerate, Bt. = basalt, Fm. = Formation

Source: Chough and Sohn (2010) by permission of the Elsevier.

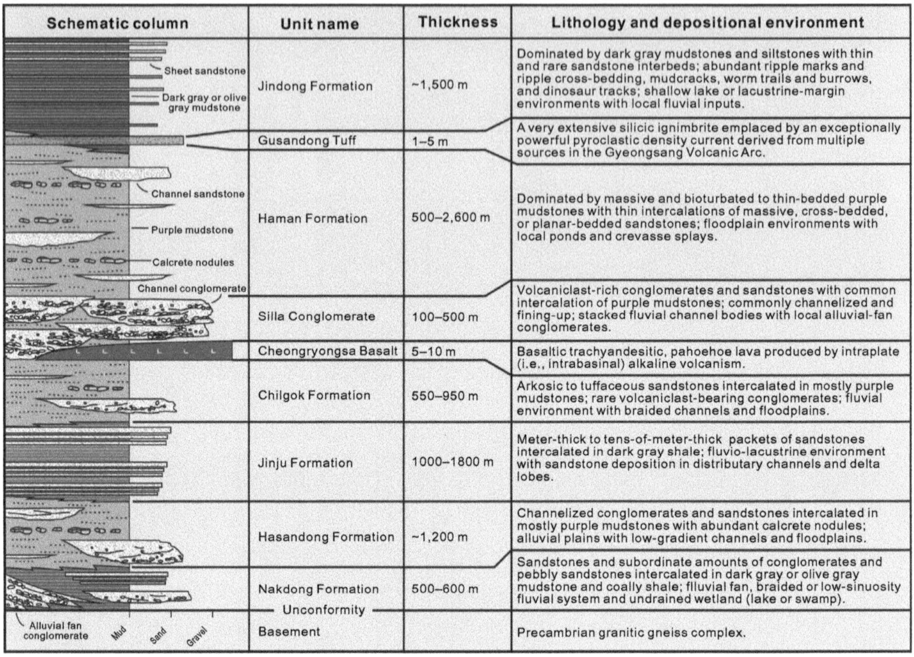

Figure 9.6 Schematic stratigraphic column of the basin fill in the Jinju subbasin with brief descriptions and interpretations of the constituent lithostratigraphic units.
Source: Chough and Sohn (2010) by permission of the Elsevier.

The deposits of each unit commonly comprise three major components: thick sandstone, thin sandstone, and mudstone-dominated bodies (Table 9.2). The thick sandstone bodies consist of stacked element groups bounded by laterally persistent erosional surfaces, probably representing superposition of channel belts (Figure 9.9A). Within individual element groups, elements are laterally superposed, indicating growth and superposition of bars and movement of associated channels. The thickness of bar deposits and dip angles of laterally inclined strata suggest that channels were a few to 10 m in depth and 9 m to several tens of meters in width for the Sinpyeong-Anpyeong unit. Paleocurrent data indicate southward- or southeastward-draining channel systems (Figure 9.7). The thin sandstone bodies (generally less than 2 m thick) are characterized by planar bed geometry, representing sand sheets and splays on floodplains. Laminated limestones and stromatolites are commonly associated with gray mudstone beds. The mudstone-dominated bodies represent well-drained floodplains in the proximal part and poorly drained floodplains with shallow lakes in the distal part (Figure 9.9).

The thick sandstone bodies formed in major channels and channel belts, which are characterized by lateral superposition of numerous bar and channel deposits, limited extent of individual lateral-accretion sets, and the presence of mid-channel bar deposits. These are suggestive of deposition in braided rivers (Figure 9.9).

Table 9.2 Description and Interpretation of Sedimentary Facies, Gyeongsang Basin

Facies	Descriptions	Interpretation
Clast-supported conglomerate (Cc)	Cc1: Pebble to boulder; poorly sorted; disorganized; graded/coarse-tail inverse grading; 0.15–2 m thick; erosional base; distinct/diffuse top; commonly form a basal part of hollow fills	Flood deposits (Allen, 1981; Maizels, 1989, 1993; Todd, 1989)
	Cc2: Boulder-dominated; tightly packed with coarse sand matrix (< 5%); elongate clasts are subparallel to bedding plane; up to 10 m thick (amalgamated); frequently grades into massive/stratified sandstone; contain mudstone chips; inversely graded; interbedded with mudstone/fine sandstone wedges; relatively flat and slightly erosional base; diffuse top	Clast-rich debris flow (Shultz, 1984)
Matrix-supported conglomerate (Cms)	Granule to cobble-grade clasts; poorly sorted; very coarse to coarse sand matrix; elongate clasts are subparallel to bedding plane; 0.3–1 m thick; disorganized; clasts are dispersed or form aggregates; commonly overlain by laminated fine standstone (facies Si) or purple mudstone (facies Mp); flat base and top	Cohesionless sandy debris flow (Bull, 1977; Hampton, 1975; Larsen and Steel, 1978)
Horizontally stratified gravelly sandstone (GSh)	Granule to cobble-grade clasts; poorly sorted; elongate clasts are parallel to bedding plane; occasional imbrication; 0.3–4 m thick; alternation of discontinuous gravel (or gravel-rich) layers decimeter thick and sandstone layers (a few cm); underlain by facies Cc; grade into facies Sm, Sh or Mp; gravels form bands, stringers or patches	Sheetfloods (Blair, 1987; Blair and McPherson, 1994); bed-load pulsation (Reid and Frostick, 1984)

(Continued)

Table 9.2 (Continued)

Facies	Descriptions	Interpretation
Planar cross-stratified gravelly sandstone (GSp)	Coarse sand, granule to pebble; ca. 1 m thick; discontinuous gravel layers of a few gravels thick are alternated with sandstone layers (<0.1 m thick) to form couplets or sets; slightly tangential cross-stratification; variable dip angle (up to ca. 20°); bounded by flat base and convex-up/flat top; sets are bounded by reactivation surfaces	Bar-slipface deposition with bedload on bar surfaces (Baker, 1973; Bluck, 1980; Harms et al., 1975; Steel and Thomson, 1983)
Massive sandstone (Sm)	Fine to very coarse sand; well sorted to very poorly sorted; 0.05–several meter thick; massive; occur as lenticles in mudstone (facies Mp and Md); relatively diffuse top and distinct base (erosive); graded in the topmost part; gravels (granule to pebble-grade) and very coarse sand grains are included; some units are graded	Rapid suspension fallout from highly concentrated turbulent flows (Smith, 1986)
Horizontally stratified sandstone (Sh)	Medium to fine sand; well sorted; up to 3–4 m thick (amalgamated); repetition of crudely stratified medium to fine sand layers which are laterally discontinuous and gradational; variable base and diffuse top; grade into facies Sm/Mp	Upper flow regime stratification (Picard and High, 1973); very flat dune migration (Jopling, 1967; McBride et al., 1975; Smith, 1971)
Planar cross-stratified sandstone (Sp)	Coarse to medium sand; moderately to well sorted; 0.2–0.5 m (set) thick; up to 2 m (coset) thick; form solitary set or locally tabular sets; convex-up (or inclined), indistinct base in contact with disorganized gravel layers; flat top	Avalanching along bar/dune margins (Harms and Fahnestok, 1965; Harms et al., 1975)

(Continued)

Table 9.2 (Continued)

Facies	Descriptions	Interpretation
Trough cross-stratified sandstone (St)	Medium to coarse sand; moderately to well sorted; <2 m thick; lenticular shape sets resting on erosional, concave-up base; cosets are bounded by gravel-bearing layers; asymmetric trough cross-strata	Migration of dunes (Harms et al., 1975) or scour fills (Harms and Fahnestock, 1965)
Laminated sandstone (Si)	Medium to very fine sand; well sorted; 10 cm thick; well laminated (ca. 1 mm thick); fissile due to intercalation of mica flakes; underlain by massive sandstone (facies Sm)	Suspension settling and traction; bursting of turbulent boundary layers (Bridge, 1978; Bridge and Best, 1988; Cheel and Middleton, 1986; Paola et al., 1989)
Purple mudstone (Mp)	Silt; a few centimeter to several meter thick (amalgamated); homogeneous; contain dispersed sand grains; fine gravels or calcareous nodules are included; some beds show block ped structures and slickensided texture	Rapid fallout from overbank flood; oxidation after deposition (Besley and Turner, 1983; van Houten, 1968; Walker, 1967)
Dark mudstone (Md)	Silt; 10 cm thick (10 m: amalgamated); homogeneous, dark greenish (5GY 4/1) or black; include plant fragments; locally alternate with coals; calcareous	Suspension sedimentation in swamps or flood basins

Source: Rhee et al. (1998) by permission of the Wiley-Blackwell.

Detailed reconstruction of bars and channels reveals a range of processes including growth, coalescence, and erosion of bars and migration, switching, and filling of channels. Lateral accretion of bars and the associated channel migration were the most important processes.

An extensive conglomerate deposit (allomember 3) with interbedded sandstone and purple siltstone, referred to as the Sanghori deposit, is a spectacular example of a streamflow-dominated alluvial fan system (Figure 9.9B). The deposit is about 150 m thick and characterized by stacked sheet bodies of conglomerate and sandstone (facies assemblages 1 and 2) alternating with interbedded sandstone and purple siltstone (facies assemblage 3) (Jo et al., 1997). Facies assemblage 1 represents

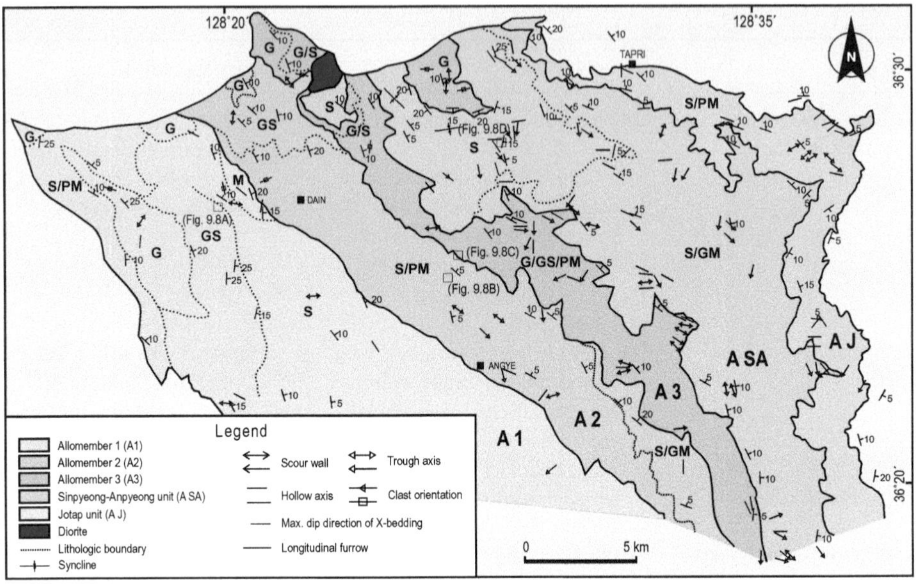

Figure 9.7 Geologic map of the northwestern part of the Uiseong subbasin with paleocurrent data. For location, see Figure 9.3. Allomembers (A) 1, 2, 3, SA (Sinpyeong–Anpyeong), and J (Jotap) correspond to Nakdong (A 1), Hasandong (A 2 and 3), Jinju (A SA), and Iljik (A J) formations. G, conglomerate; G/S, conglomerate/sandstone; GS, gravelly sandstone; G/GS/PM, conglomerate/gravelly sandstone/purple mudstone; S, sandstone; S/PM, sandstone/purple mudstone; M, dark gray mudstone.
Source: Rhee et al. (1998) and Jo and Chough (2001).

gravel-bed streams dominated by low-relief sheet bars. Facies assemblage of interbedded purple siltstone and tabular, stratified sandstone with pedogenic structures and burrows formed in floodplains. The streams were hundreds of meters wide and a few meters deep during flash floods.

In the northeastern and eastern parts of the basin (Ulsan area), the strata dip generally westward (Figure 9.11). The sandstones include radiolarian chert fragments that are identical to those of the Permian–Jurassic accretionary complex of southwest Japan (Mitsugi et al., 2001). It implies that sediments were largely derived from the uplifted accretionary complex and volcanic arc on the east (Figure 9.3).

9.3.3 Yeongyang Subbasin

The sedimentary strata (about 4500 m thick) consist largely of conglomerates, sandstones, and purple mudstones (Figure 9.12). Sediments were deposited in alluvial to fluvial environments with subordinate amounts of olive gray mudstone, chert, and siliceous mudstone formed in lacustrine environments. The topmost

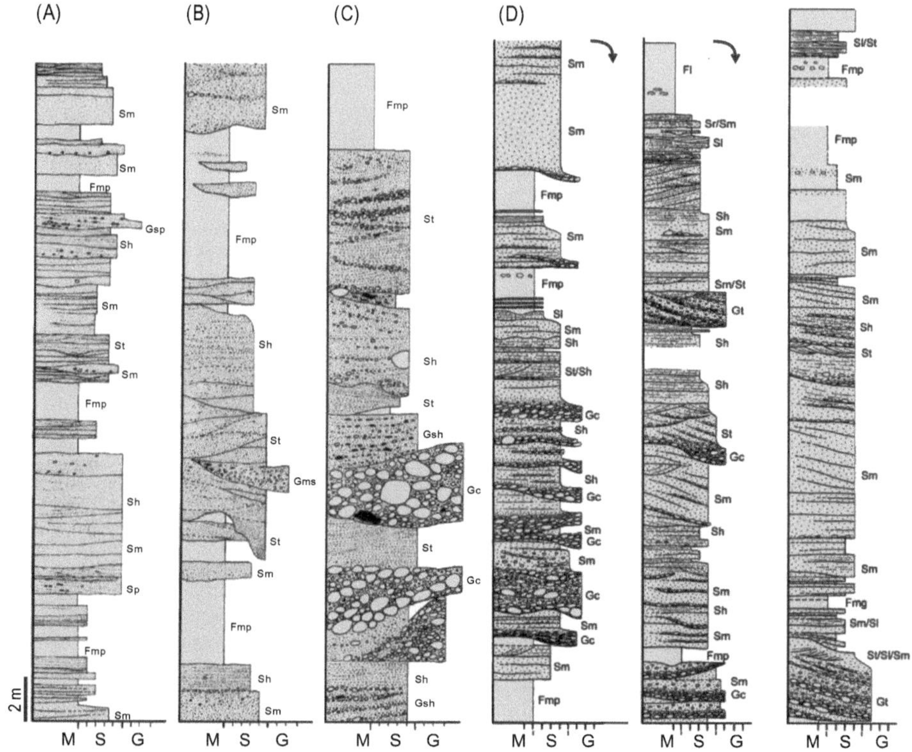

Figure 9.8 Representative sedimentary log of allomembers in the northwestern part of Uiseong subbasin. (A) allomember 1 (gravelly sandstone and sandstone), (B) allomember 2 (conglomerate, gravelly sandstone, and purple mudstone), (C) allomember 3 (conglomerate, conglomerate/sandstone, conglomerate/gravelly sandstone/purple mudstone), (D) allomember Sinpyeong-Anpyeong (sandstone and gray mudstone with subordinate conglomerate, gravelly sandstone, and purple siltstone). M, mudstone; S, sandstone; G, conglomerate; Gms, matrix-supported conglomerate; Gc, clast-supported conglomerate; Gt, trough cross-stratified conglomerate; Gsp, planar cross-stratified gravelly sandstone; Gsh, horizontally stratified gravelly sandstone; St, trough cross-stratified sandstone; Sp, planar cross-stratified sandstone; Sh, horizontally stratified sandstone; Sl, laminated sandstone; Sr, ripple cross-laminated sandstone; Sm, massive sandstone; Fmp, purple massive fines (Rhee et al. (1998) and Jo and Chough (2001)). For locations, see Figure 9.7.

stratigraphic unit (about 200 m thick) is composed of dark gray to black shale with minor sandstone and conglomerate. A detailed analysis of sedimentary facies is only available for the Pyeonghae area (Rhee and Chough, 1993a; Rhee et al., 1993). A provenance study suggests that sandstone beds contain chert fragments with radiolarian faunas of the Permian–Middle Jurassic that were derived from the Japanese arc (Chang et al., 1990, 2003; Kamata et al., 2000; Mitsugi et al., 2001).

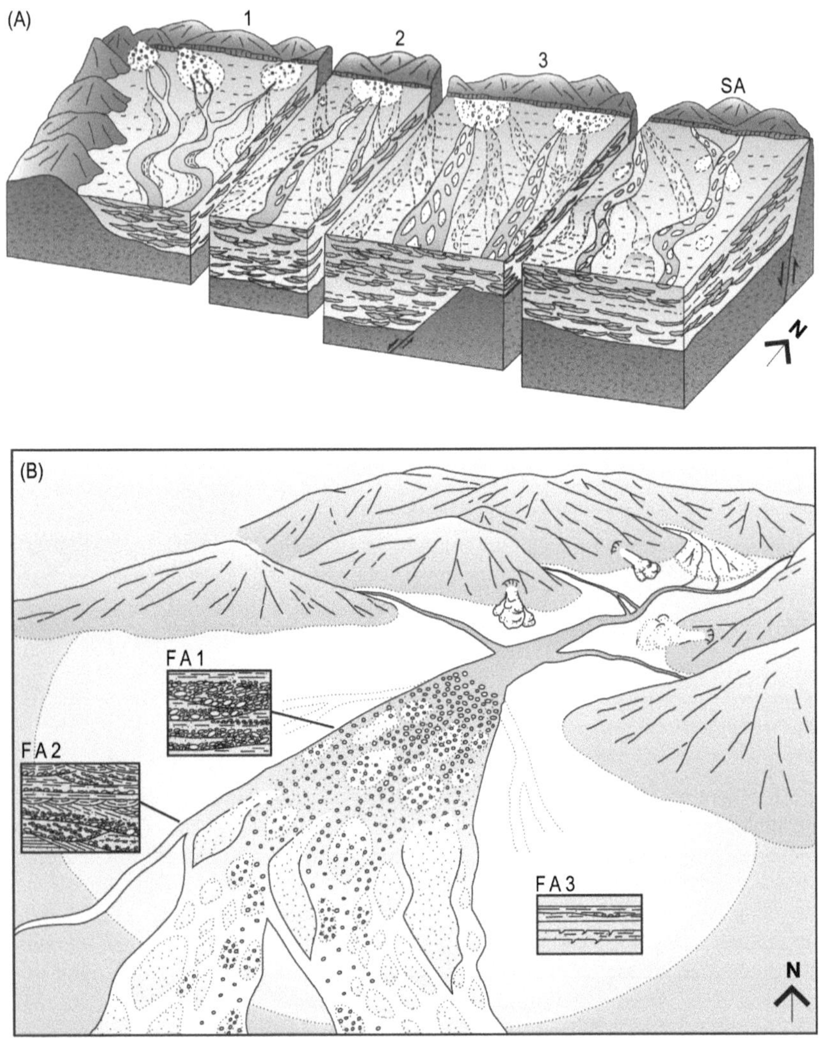

Figure 9.9 (A) Depositional model of allomembers 1, 2, 3, and SA. The northern margin is fault-bounded whereas the western boundary is represented by unconformable onlap onto Precambrian basement. In each allomember, channel types and stacking patterns vary successively eastward due to the development of dip-slip motion associated with strike-slip deformation. Modified after Rhee et al. (1998) and Jo and Chough (2001). (B) Depositional model of the Sanghori conglomeratic deposit on the northern margin of the Gyeongsang Basin. FAl, gravel-bed streams dominated by sheet bars; FA2, sandy braided streams with longitudinal/transverse bars and dunes; FA3, floodplain.
Source: (A) Chough et al. (2010) by permission of the Elsevier. (B) Jo et al. (1997) by permission of the Elsevier.

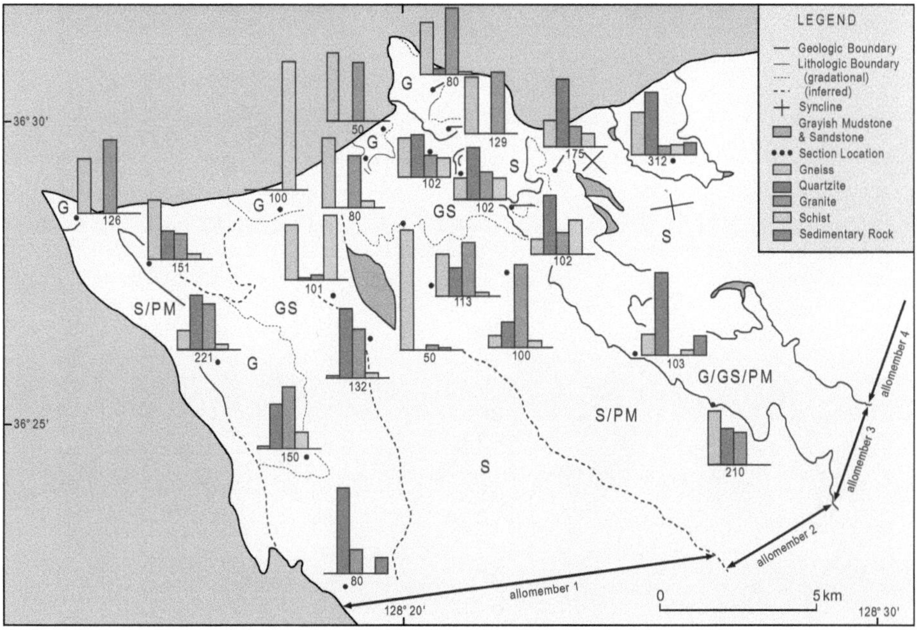

Figure 9.10 Histograms of clast composition. Each allomember shows differing composition, suggesting a local source. The number of counts is shown below abscissa. *Source*: Rhee et al. (1998) by permission of the Wiley-Blackwell.

9.4 Arc Volcanism

In the early stage of basin development, volcanogenic sediments were frequently intercalated in the sedimentary successions. For example, in the Nakdong Formation, volcanogenic mudstone contains glass shards and diagenetic alteration products (Cheong and Kim, 1996). Arkosic sandstones rich in subhedral, zoned plagioclase grains are also of volcanogenic origin. The sandstones of the Jinju Formation are especially rich in plagioclase grains of volcanic origin (Noh and Park, 1990). The Chilgok Formation contains abundant layers of tuffaceous sandstones and conglomerates with basaltic volcaniclasts (Chang, 1975; Kim et al., 2000). A pahoehoe lava flow unit (5–6 m thick), the Cheongryongsa basalt, occurs in the Jinju subbasin, which can be traced for about 12 km (Jeon and Sohn, 2008). It contains plagioclase phenocrysts, characteristic of intraplate alkaline magma (Kim et al., 2000).

9.5 Intraarc Basins

Small-scale isolated basins occur in the arc platform. The sedimentary successions in these basins comprise conglomerates, sandstones, dark gray and purple siltstones

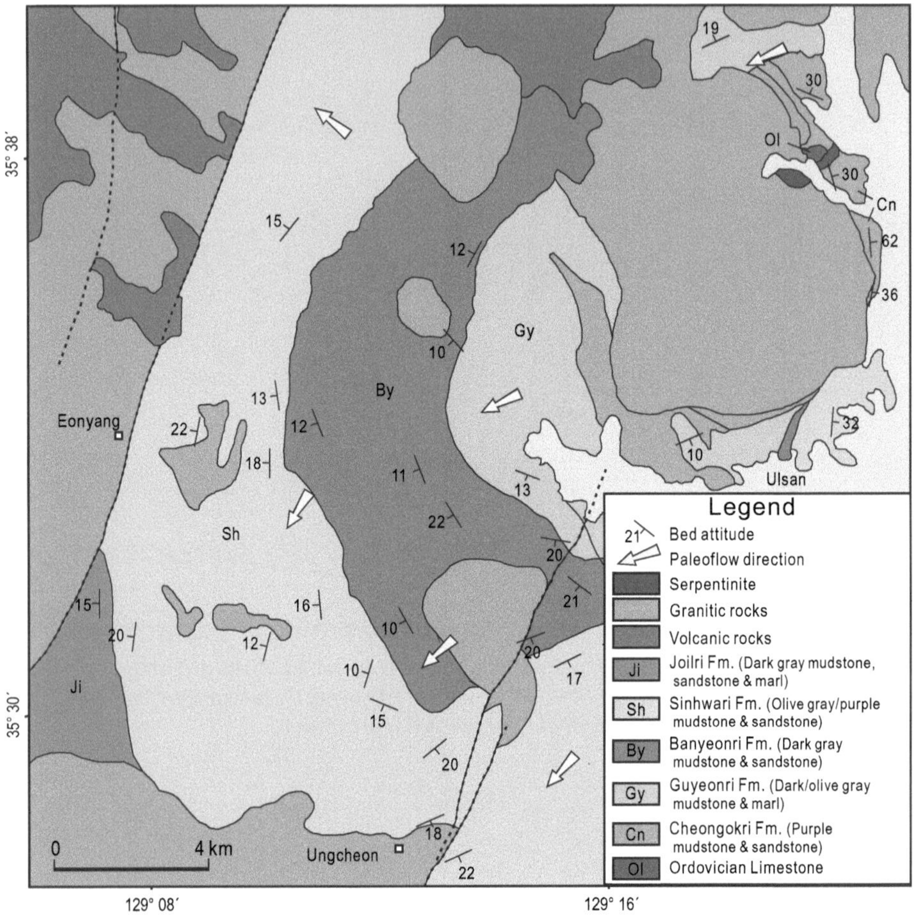

Figure 9.11 Geologic map of the eastern part (Ulsan area) of the Uiseong subbasin (modified after Chang, 1988). The paleo-flow pattern is generally southwestward. For location, see Figure 9.3.
Source: Chough and Sohn (2010) by permission of the Elsevier.

and mudrocks, and cherts as well as volcanogenic deposits (Figure 9.13) (Cho et al., 2011; Kim and Paik, 1997; Paik et al., 1996, 1997; Shinn et al., in prep.). Conglomerates commonly contain volcanic rock fragments such as basalt, andesite, and tuff as well as quartzite and chert. Quartzite clasts were derived from the basement, whereas volcaniclasts originated from the arc platform. Chert originated from chemical precipitation of silica in a silica-rich alkaline lake when pH level decreased during fresh water input (Chough et al., 1996).

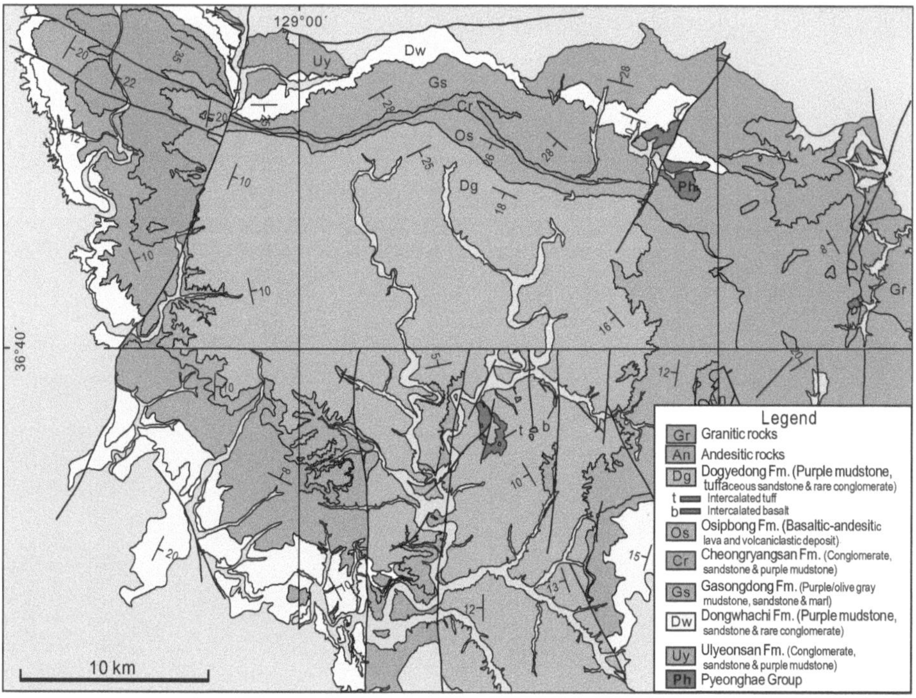

Figure 9.12 Geologic map of the Yeongyang subbasin (compiled after Kim and Lee (1970), Kim and Park (1970), Kim et al. (1963a,b), and Lee and Lee (1963)).
Source: Chough and Sohn (2010) by permission of the Elsevier.

9.6 Evolution of Gyeongsang Arc System

A continental arc system (the Gyeongsang Arc) formed in the southeastern part of the Korean Peninsula during the Cretaceous (Figure 9.2). The arc system was initiated in the Early Cretaceous by oblique subduction of the Izanagi plate under the Asian continent. The arc platform (Gyeongsang Volcanic Arc) consists of calk-alkaline volcanic rocks and granite plutons that are commonly associated with cauldron structures (Figure 9.14). A large continental basin (Gyeongsang Basin) formed in the backarc region. The basin was subdivided into three subbasins, demarcated by the basement highs and the arc platform, whose development history was somewhat independent of each other.

Both the Jinju and Uiseong subbasins were initiated as continental depressions along the basin margins (Figure 9.15). Deposition occurred in stream-dominated

Figure 9.13 Outcrop photographs of the Dadaepo Formation. (A) Panoramic view of the lower Dadaepo Formation, characterized by thick and massive purple mudstones and conglomerate/sandstone interbeds. (B) An exposure showing the sharp contacts between the massive purple mudstone of the lower Dadaepo Formation, a few meter thick and massive acidic tuff layer in the middle, and the well-bedded sandstone/gray mudstone succession of the upper Dadaepo Formation. (C) Graded and cross-stratified conglomerates of the lower Dadaepo Formation intercalated within purple mudstones with erosional lower surfaces. (D) Hummocky cross-stratified arkosic sandstones of the upper Dadaepo Formation. The coin in D is 2.3 cm in diameter.

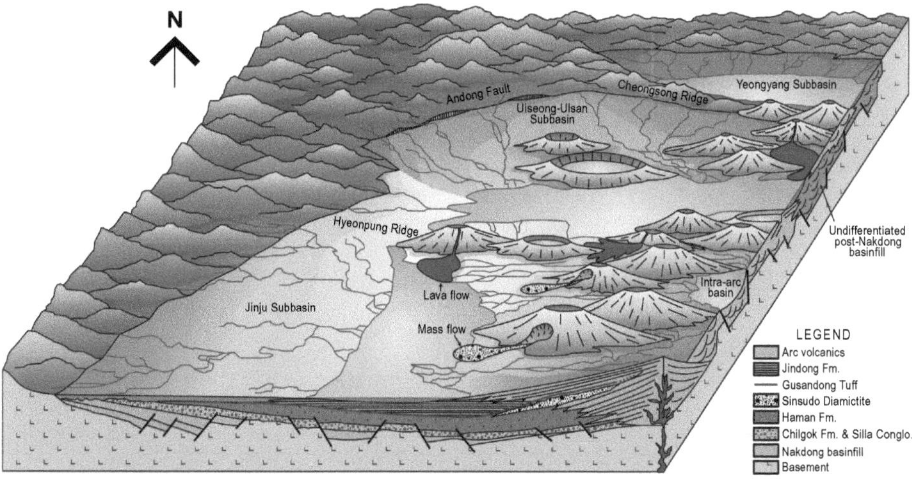

Figure 9.14 Three-dimensional block diagram of the Gyeongsang Arc System during its late development stage. Sediments were delivered into the backarc basin through a number of fluvial systems sourced from the western and northern basin margins and the eastern to southern arc platform. The arc platform was dotted with numerous volcanic centers and comprised small intraarc basins. Lava flows and volcaniclastic materials derived from the arc platform are both interlayered with and overlying the backarc strata.
Source: Chough and Sohn (2010) by permission of the Elsevier.

alluvial fans in the north and west and in braided streams along the basin axis with the local development of floodplains and lakes. Sediments were derived from the western and northwestern margins as well as from the east and south, draining the volcanic arc (Figure 9.16). These subbasins were flooded with shallow lake water despite the input of large amounts of volcanogenic sediments from the arc (Figure 9.17). The Yeongyang subbasin received sediments from the fault-bounded northern margin and the arc platform in the northeast. Conglomerates contain abundant radiolarian-bearing chert clasts, which originated from the accretionary prism of the Japanese Arc.

Explosive to effusive and felsic to mafic volcanism was common during the deposition of the basin fills. Extrusion of basaltic lavas poured into the lowland basin, forming numerous interlayers as well as water- and airborne volcaniclastic materials from the arc. The arc was emerged as a highland about 100 million years ago. Volcanic aprons and intraarc basins formed along the margins and interior of the arc platform, respectively.

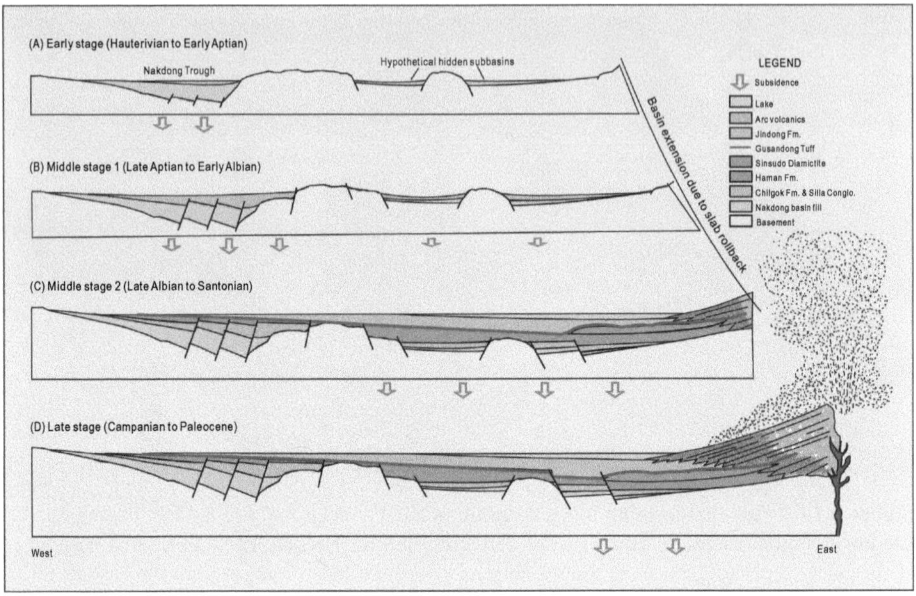

Figure 9.15 Schematic two-dimensional evolution of the stratigraphy in the Gyeongsang Backarc Basin. (A) Formation of the Nakdong Trough and other hypothetical subbasins during the early stage of basin evolution. (B) Expansion (extension) of the early trough and deposition of the Chilgok and Silla Conglomerate formations. (C) Formation of a broad fluvio-lacustrine basin during the deposition of the Haman and Jindong formations. (D) Impingement of volcaniclastic aprons from the arc platform during the late stage of basin evolution. Indirect evidence for arc volcanism is found from the earliest strata (the Nakdong basin fill). The locus of subsidence shifted progressive eastward during the basin evolution probably because of slab rollback. The basin fill ages of the early to middle stages are still controversial (Choi, 1985, 1987; Choi and Park, 1987; Lee et al., 2010b; Park, 1986; Seo, 1985).
Source: Chough and Sohn (2010) by permission of the Elsevier.

The Gyeongsang Arc System formed in a convergent margin setting where the paleo-Pacific Plate subducted under the Asian continent. The arc represents an extensional system, similar to the Central America Arc, suggesting weak coupling of the subducting and overriding plates. It extended for thousands of kilometers, accompanied by numerous extensional basins, including the Gyeongsang and Kanmon basins. Apart from the volcanic activities centered in the arc, the basin area also experienced extrusive volcanism, volcano-tectonic extension, and expansion with accelerated subsidence.

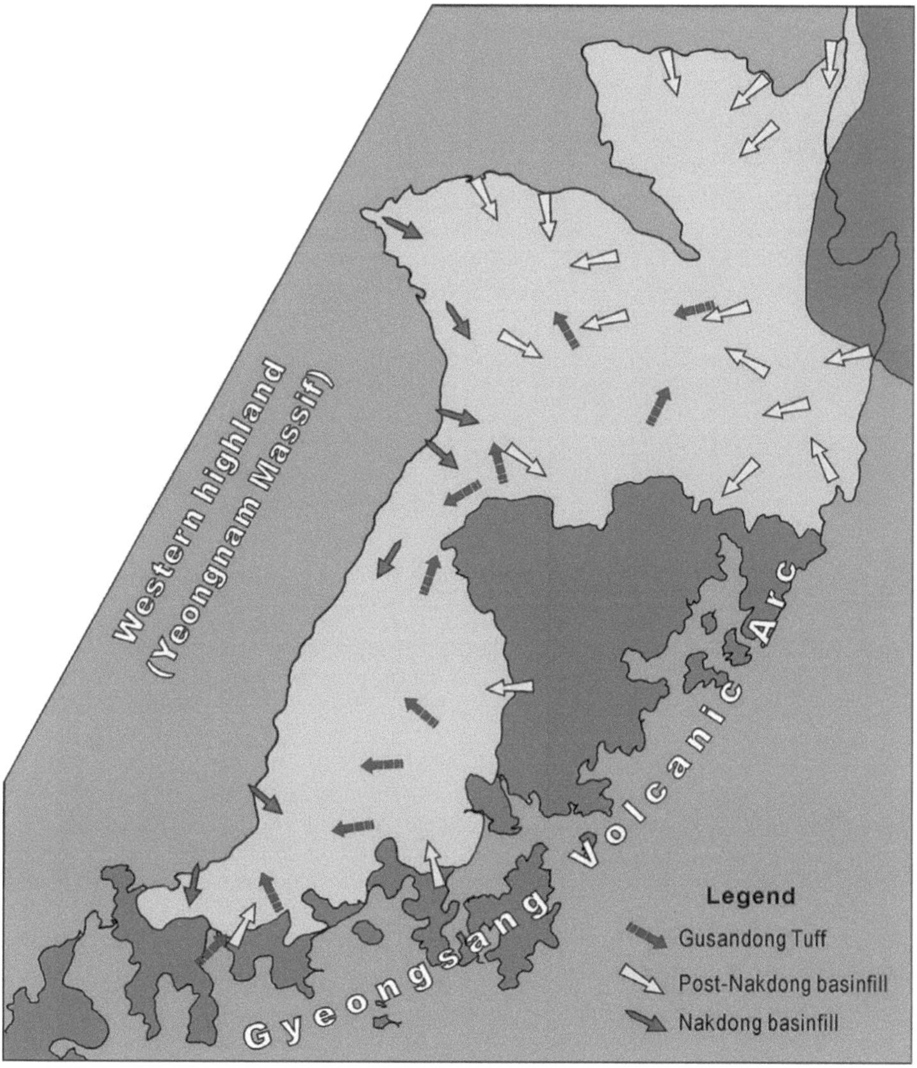

Figure 9.16 Paleoflow pattern within the Gyeongsang Basin (compiled from Chang (1988), Chang et al. (1990), Cheong and Kim, (1996), Jo et al. (1997), Jo and Chough (2001), Kim (1994), Mitsugi et al. (2001), Rhee et al. (1998), Sohn et al. (2009), and new observations). Paleocurrent data are explained in the text.
Source: Chough and Sohn (2010) by permission of the Elsevier.

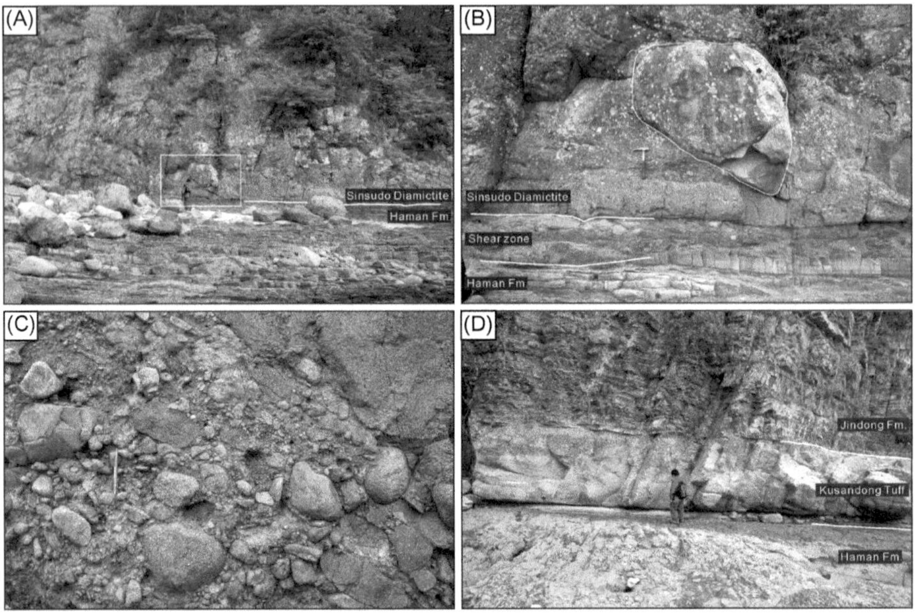

Figure 9.17 Outcrop photographs of the Sinsudo Diamictite and adjacent strata in Sinsu Island. (A) The Haman Formation in Sinsu Island consists of dark gray shale and thin sandstone interbeds with common ripple marks. Beyond the island, the formation is dominated by purple mudrocks. The formation is overlain by the Sinsudo Diamictite, which comprises a single disorganized megabed, 30–40 m thick. (B) Close-up of the basal part of the Sinsudo Diamictite. A meter-long sandstone block within the diamictite is outlined. A decimeter-thick shear zone is present between the diamictite and the underlying Haman Formation. (C) Close-up of a large conglomerate block within the diamictite, which consists of clast-supported, subrounded, weakly imbricated, andesitic to basaltic clasts that are pebble to boulder in size. (D) An exposure of the Gusandong Tuff over- and underlain by fluviolacustrine sedimentary strata of the Haman and Jindong formations with sharp and planar contacts. The hammer in (B) is 30 cm long; the pencil in (C) is 14.5 cm long.

10 Other Cretaceous Basins

10.1 Introduction

In the southwestern Korean Peninsula, a series of strike-slip faults, the Gongju and Gwangju fault systems, occur trending northeast to southwest (Figure 10.1). They are characterized by normal and *en echelon* patterns associated with high-angle transtensional faults. These major faults were activated by sinistral strike-slip movements in the Late Jurassic to the Cretaceous, forming small-scale rhomboidal basins (Chun and Chough, 1992, 1995). These elongated basins were filled with alluvial to lacustrine sediments during the Early to Late Cretaceous, controlled by left-lateral strike-slip faults (Figure 10.1).

10.2 Sihwa Basin

10.2.1 Sedimentary Facies Analysis

The Sihwa Basin is located in the Namyang and Sihwa areas, Gyeonggi Province (Figure 10.1). The succession rests on the basement of gneiss and schist (Precambrian) and granite (Jurassic) (Figure 10.2). The Sihwa Formation comprises alluvial deposits of conglomerate, gravelly sandstone, sandstone, and reddish brown siltstone/mudrock. Clasts are composed mainly of granite gneiss, banded gneiss, green or blue schist, and quartzite with subordinate amounts of limestone, granite, and volcanic rocks. Based on lithology, bed geometry, and sedimentary structures, 11 sedimentary facies have been identified (Table 10.1; Figure 10.3). The deposits can be grouped into three architectural elements based on constituent facies, bed geometry, stacking pattern, and bounding surface characteristics (Kim et al., 2009).

10.2.1.1 *Alluvial Fan (Element I)*

It consists of monotonous stacking of tabular or crudely stratified conglomeratic units (Figure 10.4). Clasts are mostly of pebble to cobble grade with subordinate amounts of boulders (up to 70 cm long) and generally angular to subangular. Element I deposits formed by either debris flows (facies association IA) or stream-flows (facies association IB).

Facies association IA consists entirely of tabular conglomerate beds (25–150 m in thickness) (Figure 10.4). Clasts are predominantly angular and inversely or

Geology and Sedimentology of the Korean Peninsula. DOI: http://dx.doi.org/10.1016/B978-0-12-405518-6.00010-3

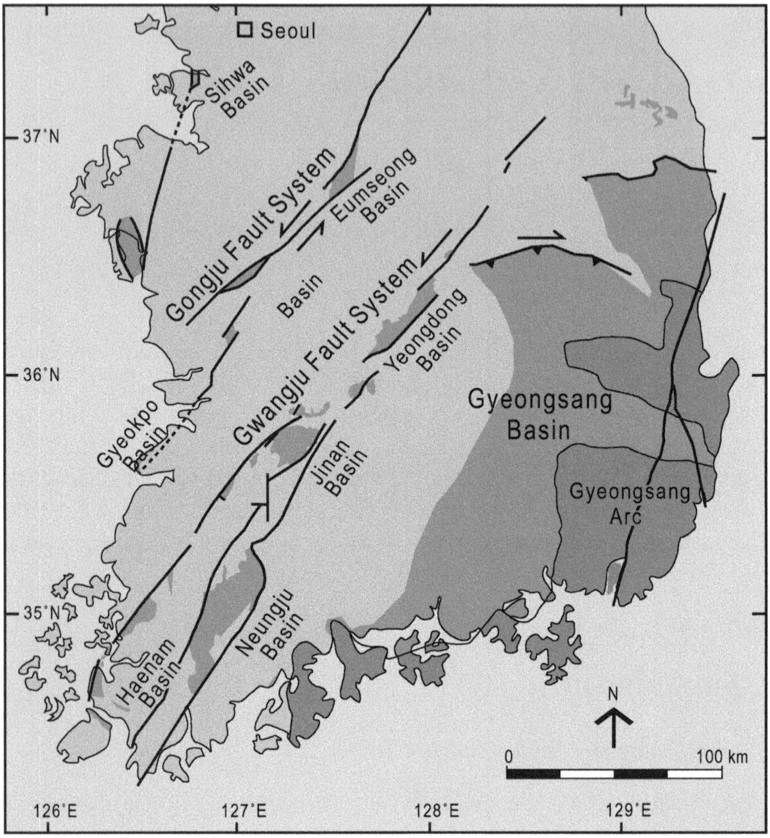

Figure 10.1 Distribution of Cretaceous sedimentary basins in the Korean Peninsula.

inverse-to-normally graded. Matrix material comprises poorly sorted, coarse-grained sandstone and siltstone. The common inverse grading of gravel clasts with a framework support suggests cohesionless debris flows.

Facies association IB occurs in the northern basin sector (Figure 10.3). Each facies unit is mostly planar bedded, although the bases are very irregular with common downtruding clasts and local scoop-shaped scours. The overall characteristics of the crudely stratified conglomerate/sandstone couplets are similar to those of sheetflood deposits. The planar couplets of pebble–cobble conglomerate and pebbly sandstone can be explained by successive bedload and suspension fallout.

10.2.1.2 Ephemeral Braided Stream (Element II)

Element II is characterized by multi-storey, sheet-like or upward-widening conglomeratic channel fills. It occurs mainly in the northwestern part of the basin. The

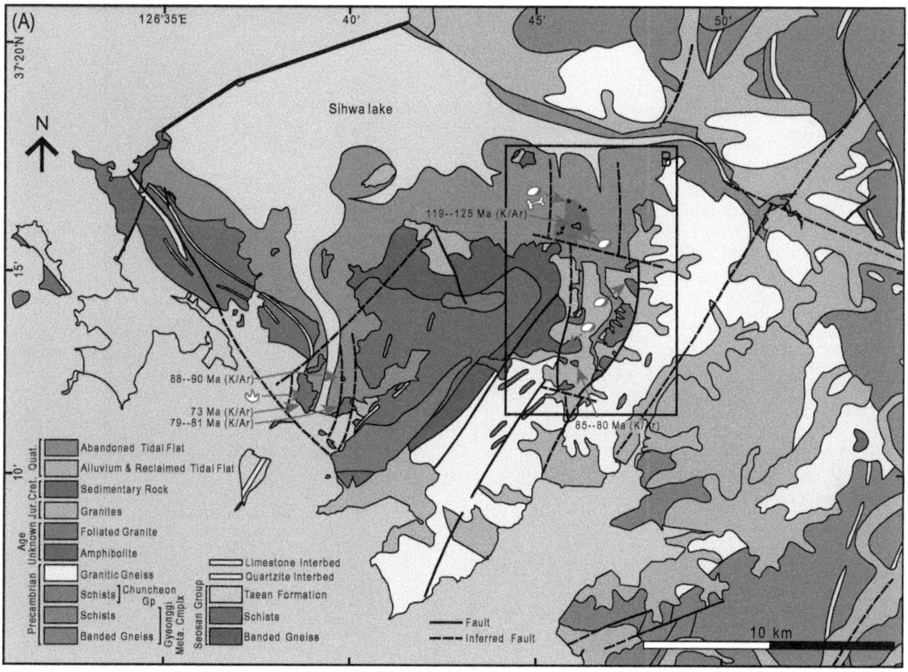

Figure 10.2 (A) Regional geologic map of the Sihwa area. Cretaceous sedimentary rocks are unconformably underlain by basement of Precambrian metamorphic and Mesozoic plutonic rocks. Locations of dinosaur remains and dated volcanic samples are also depicted. (B) Geologic map of the Sihwa Basin (Hwasung City (2005), Park (2000), and Park and Kim (1972)). Locations of measured sections and dinosaur remains are denoted. (C) Detailed map of the northwestern part (Hanyom area) of the basin (Hwasung City, 2005).
Source: (A) Kim et al. (2009) by permission of the Elsevier.

erosional base commonly involves numerous scours of more than 1 m wide and several decimeters deep (Figure 10.5). Siltstone/mudrock-plugged scours are locally present along the upper boundary.

Facies association IIA forms hollow-fill units with sharp erosional scoop-shaped bases. Hollow walls are mostly asymmetrical with the deepest axis near the steep flanks. Individual hollow fills are mostly composed of single sets of trough cross-stratified gravelly sandstones or conglomerates. The openwork lenticular conglomerate is interpreted as channel-lag deposits.

Facies association IIB comprises composite deposits of clast-supported lenticular conglomerate and inclined or horizontally stratified gravelly sandstone. The clast-supported lenticular conglomerate is characterized by planar-convex, concave-planar, or biconvex geometries (Figure 10.5). The clast-supported lenticular conglomerate is interpreted as low-relief longitudinal gravel-bar deposits.

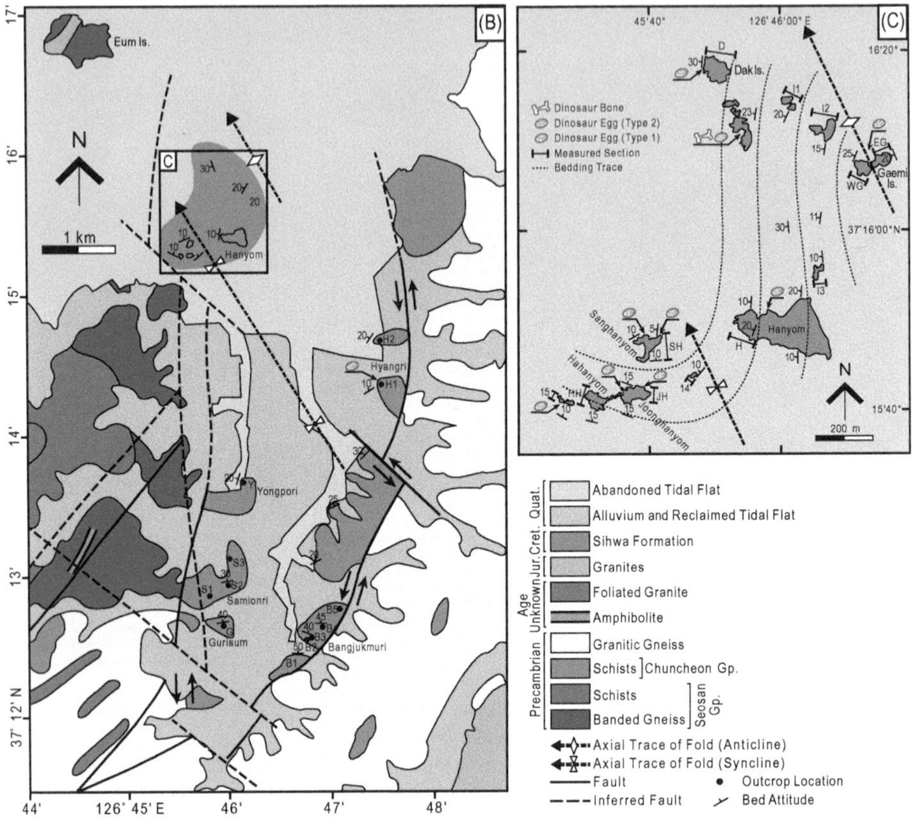

Figure 10.2 (Continued)

10.2.1.3 Channel Margin to Floodplain (Element III)

This element is characterized by fining-upward stacks (up to 30 m thick) of (i) single- or multi-storey channel fills with limited lateral extent (FA IIIA), (ii) poorly sorted, graded conglomerate/siltstone couplets with diffuse gravel sheets or shallow hollow fills (FA IIIB), and (iii) poorly sorted, homogeneous or diffusely graded, reddish brown fine-grained deposit with gravel pockets or lenses (FA IIIC) (Figure 10.6).

Facies association IIIA (crevasse channels) forms single- or multi-storey small-scale channel fills. The channel fills are generally more than 3 m wide with a thickness of less than 1.5 m, showing an overall bowl-shaped geometry (Figure 10.6). These channel fills consist mainly of horizontally stratified gravelly sandstones with minor amounts of openwork lenticular conglomerate (Figure 10.6). Each channel-fill unit is isolated.

Table 10.1 Description and Interpretation of Sedimentary Facies in the Sihwa Formation

Facies	Description	Interpretation
Tabular conglomerate (facies Gt)	0.25–1 m thick; laterally continuous (>10 m); planar sharp lower boundary and slightly irregular upper boundary with common protruding clasts; inversely or inverse-to-normally graded; matrix- or clast-supported; common parallel oriented clasts; pebble to cobble-size clast common with occasional outsized boulders; poorly sorted, coarse-grained sandstone matrix; common lensoidal or pocket-shaped siltstone intercalations with pebbles; internal, laterally overlapping wedges; occasional armored gravel concentrations along upper boundary	Cohesionless debris flow (Nemec and Steel, 1984; Kim et al., 1995b; Sohn et al., 1999)
Gravel sheet (facies Gs)	A few clast to decimeters thick, laterally continuous (>10 m); tabular or wedged with flat or shallowly concave-up base and planar or slightly convex-up upper boundary; densely packed pebble-cobble clasts with a paucity of sandstone matrix in the central part; less densely packed, pebble-dominant wedged margins with greater amounts of matrix material of silty sandstone; laterally transitional into discontinuous pebble-cobble trains; stacked repeatedly or interlayered with facies Fh units	Bedload deposition as gravel sheets or splays by highmagni-tude flood flows (Bristow et al., 1999; Rhee and Chough, 1993a,b;Stear, 1983)
Gravel pocket (facies Gp)	One to a few clast thick; less than a few meters wide; lenticular or pocket-shaped, clast-supported granule-pebble deposit, or loosly packed, discontinuous pebble train; slightly contorted in places; encased within facies Fh units	Bedload deposition as gravel streams or lineaments by weak distal flood flows (Jo, 2003a)
Open-work lenticular conglomerate (facies Glo)	Resting above a scour hollow (decimeters to meters in width and decimeters in depth), completely filling-up or forming a thin basal veneer, one to several-clast-thick; mainly cobble-grade clasts with occasional pebbles and boulders; tightly interlocked with open framework or small amounts of fine- to coarse-grained interstitial sandstone	Channel lag or gully-fill deposit (Blair, 1999, 2000; Nemec and Postma, 1993)

(Continued)

Table 10.1 (Continued)

Facies	Description	Interpretation
Clast-supported lenticular conglomerate (facies Glc)	10–50 cm thick; lenticular, a few meters wide (max. 20 m), either planar-convex, concave-planar or biconvex; sharp lower boundary and relatively indistinct upper boundary; cobble-boulder core and pebbly fringes; laterally or vertically transitional into facies GySix or GySh units fringes: thinner than core; common clast imbrication; normally or more commonly inversely graded; coarse-grained sandstone matrix	Low-relief longitudinal gravelbar (Boothroyd and Ashley, 1975; Mall, 1985; Jo et al., 1997; Karpeta,1993)
Trough cross-stratified conglomerate to gravelly sandstone (facies Gtx)	Resting above a scour hollow (decimeters to meters in width and decimeters in depth), completely filling-up or above basal veneer of facies Glo; alternations of tightly packed, pebble-cobble conglomerate (one- to several-clast-thick) layer and massive or trough cross-stratified coarse sandstone (up to 15 cm thick) layer; each layer gently dipping ($< 15°$), generally conforming the topography of the basal scour hollow; single sets predominant, cosets rare; each set commonly fining- and thinning-upward	Infills of minor channels (Jo and Chough, 2001; Khadkikar, 1999; Miall, 1977, 1985)
Crudely stratified conglomerate/ sandstone couplets (facies GSs)	30–80 cm thick; laterally continuous (> 10 m); generally abruptly normally graded with lower conglomerate and upper sandstone divisions Conglomerate division: laterally discontinuous a few meters apart; a-few-clast thick; sharp planar, locally scooped erosional base; tightly interlocked clasts with a paucity of sandstone matrix Sandstone division: usually thicker than lower division, moderately sorted, coarse-grained sandstone, massive or normally graded with well-laminated tops	Sheetflood (Blair, 1987, 1999, 2000)

(Continued)

Table 10.1 (Continued)

Facies	Description	Interpretation
Graded conglomerate/ siltstone couplet or conglomerate/ sandstone/ siltstone triplet (facies GZg)	Decimeters to >1 m thick; laterally continuous a few decameters; tabular bedded; diffusely or indistinctly bounded; distribution-type graded with pebble-cobble silty conglomerate at the base and normally graded siltstone at top, intervened or not by an interval of silty sandstone; stacked repeatedly showing an overall fining- and thinning-upward trend. Successive bedload and suspension sedimentation from unconfined, rapidly waning conglomerate division: loosely packed clasts with matrix- to clast-supported fabric; partly, diffusely and discontinuously flood flows stratified with local clusters; deformed in places; common clast imbrication; sandy siltstone matrix sandstone division: ungraded or graded; unstratified; moderately sorted; dominantly coarse-grained; silty interstitial material siltstone division; poorly sorted, distribution-type or coarse-tail graded with sandy base; nonfissile; red-brown in color	Successive bedload and suspension sedimentation from unconfined, rapidly waning flood flows
Horizontally stratified gravelly sandstone (facies GySh)	A few decimeters thick; a few meters to decameters wide; alternating layers (1–10 cm thick) of pebbly coarse-grained sandstone and medium- to fine-grained sandstone; each layer (sub-)parallel, slightly undulatory or low-amplitude wavy; common lateral layer splitting or convergence; individual layers either ungraded, normally or inversely graded or locally interstratified; pebbles either flat-lying or imbricated; occasional outsized cobbles and boulders; laterally juxtaposed with or transitionally overlying facies Glc or GySh units	Low-relief or plane-bedded bar (Allen, 1983) or upper-plane bedded shoals or interbar plains (Allen, 1973; Harms et al., 1982)

(Continued)

Table 10.1 (Continued)

Facies	Description	Interpretation
Low-angle inclined stratified gravelly sandston (facies GySix)	A few decimeters thick; a few meters wide; alternating layers (1–10 cm thick) of pebbly coarse-grained sandstone and medium to fine-grained sandstone; each layer subparallel and low-angle inclined cross-stratified, showing a divergent fanning or off lapping pattern; strata generally conforming the lower topographic relief with tangentially or less commonly angularly downlap terminations; pebbles either flat-lying or imbricated; laterally juxtaposed with facies Glc or GySh units	Lateral or downstream accretion in bar fringes
Red-brown homogeneous or graded fine-grained deposit (facies Fh)	50 cm to more than 10 m thick; laterally continuous more than decameters; red to brown in color; distribution-type graded with poorly sorted sandy siltstone bases and progressively well-sorted and weakly fissile siltstone tops; occasional outsized pebbles and cobbles; common calcareous nodules and vertical burrows; transitionally or abruptly overlying facies Gs or GZg units or stacked repeatedly by itself; occasionally including facies Gp units	Retarded suspension settling from weak floodflows (Jo, 2003a; Turner, 1980)

Source: Kim et al. (2009) by permission of the Elsevier.

The scour incisions with common overhanging and/or stepped walls suggest crevasse channel incision. The single-storey units indicate rapid channel filling by ephemeral flows during episodic events. The predominance of horizontally stratified gravelly sandstones indicates bedload deposition under upper flow regime.

Facies association IIIB (crevasse splays) commonly consists of alternation of discontinuous gravel sheets and fine-grained deposit as well as graded conglomerate/siltstone. The gravel sheet is tabular or wedged. Clasts range in size from pebble to cobble grade and are predominantly oriented parallel to bedding plane. Clasts are loosely packed and commonly imbricated.

The close association of this facies association with the crevasse channel deposit of FA IIIA and the incorporation of large amounts of gravel clasts suggest voluminous and repeated bedload supply from the main trunk channel, interpreted as crevasse splays. This gravel facies unit is analogous to the sand splays or sheets of channel margins or floodplains.

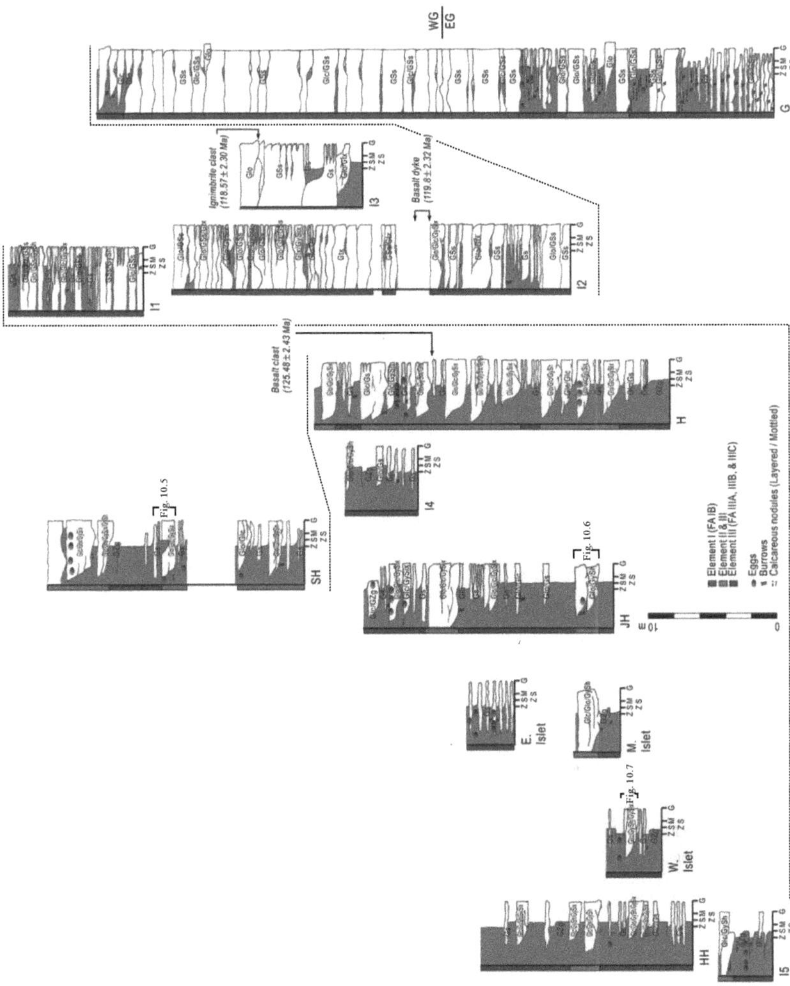

Figure 10.3 Summary of columnar sections in the northern part of the Sihwa Basin. See Figure 10.2C for the location. Stratigraphic correlation is based on both the bedding and the location of dinosaur eggs. Columns are classified into three element groups, i.e., elements I, II, and III, and III. Z, SZ, M, and C indicate siltstone, sandy siltstone, medium sandstone, and conglomerate, respectively. Dinosaur egg clutches generally occur in the uppermost part of the channel deposits or the overlying floodplain deposits, whereas solitary eggs occur on the bottom of channels. For facies codes, see Table 10.1.
Source: Kim et al. (2009) by permission of the Elsevier.

Facies association IIIC (floodplain fines) is represented by reddish brown fine-grained deposit, transitionally overlying the crevasse splay deposit (Figure 10.7). Each unit is commonly more than 50 cm thick and is laterally continuous over the entire outcrop with uniform thickness. Discontinuous granule–pebble lenses, trains,

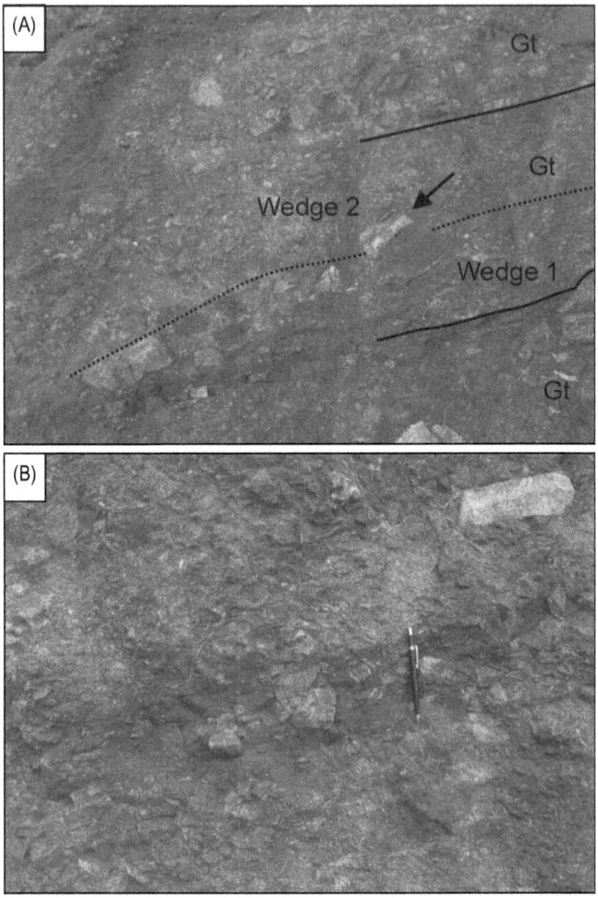

Figure 10.4 Photographs of FA IA deposits of debris-flow-dominated fan. (A) FA IA consists of stacked tabular conglomerates (facies Gt) with thin discontinuous red siltstone intercalations (from section B4). Note that a conglomerate bed consists of two overlapping wedges. Hammer (30 cm long) for scale. A large clast (arrowed) is not a floating clast in the middle part of the bed but a protruding clast above the lower wedge (wedge 1). (B) The intercalated siltstone lens. Note the irregular and diffuse lower boundary and the incorporation of pebbles. Pencil (14.5 cm long) for scale. (For interpretation of the references to color in this figure legend, the reader is referred to the online version of this book.)

and pockets are common. Outsized pebbles and cobbles are occasionally incorporated. Calcareous nodules or mottles commonly occur. Vertical burrows (less than 5 cm in diameter and up to 20 cm long) are also common and usually filled with sediment penetrated from the overlying bed. This facies succession shows an overall fining- and thinning-upward trend.

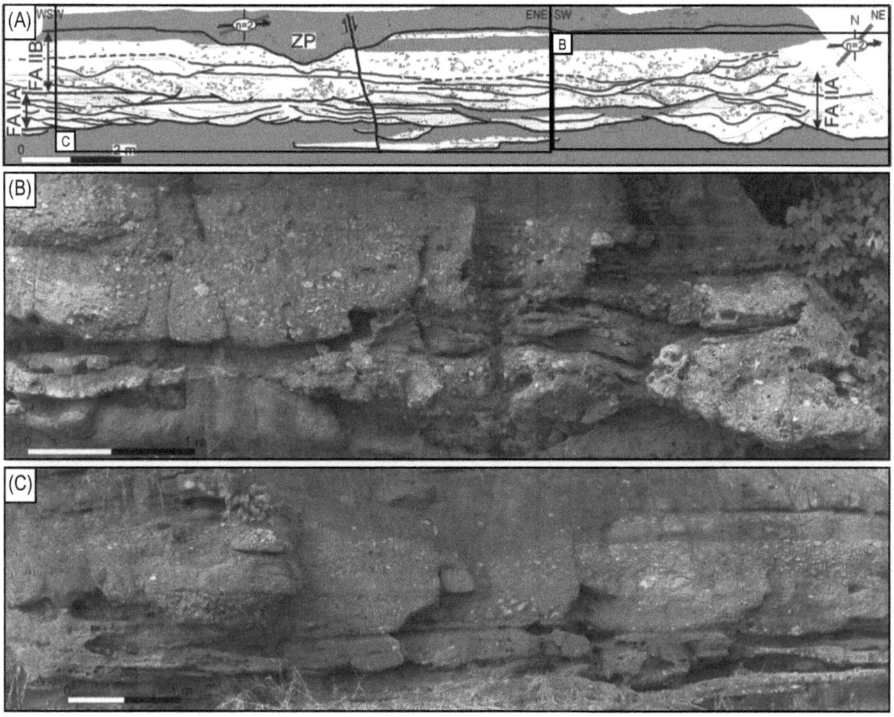

Figure 10.5 (A) Line drawing of a sheet-type channel body (element II) at the lowermost part of Section SH. The channel body consists of cut-and-fill units (FA IIA) in the lower part and low-relief bar deposits of FA IIB in the upper part. Note a siltstone-plugged (ZP) channel fill at the upper boundary. (B, C) Photographs showing the parts of (A). For facies codes, see Table 10.1.
Source: Kim et al. (2009) by permission of the Elsevier.

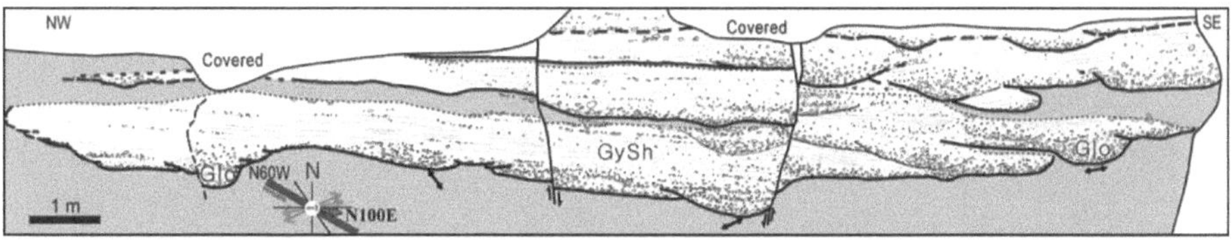

Figure 10.6 Sketch of vertically stacked, small-scale, multi-storey channel bodies of FA IIIA (from section JH). Each channel fill consists mainly of hollow fills of facies Glc and Glo in the lower part and transitionally overlying horizontal units of facies GySh. Bidirectional arrows indicate the axial directions of the scours at the channel base. For facies codes, see Table 10.1.
Source: Kim et al. (2009) by permission of the Elsevier.

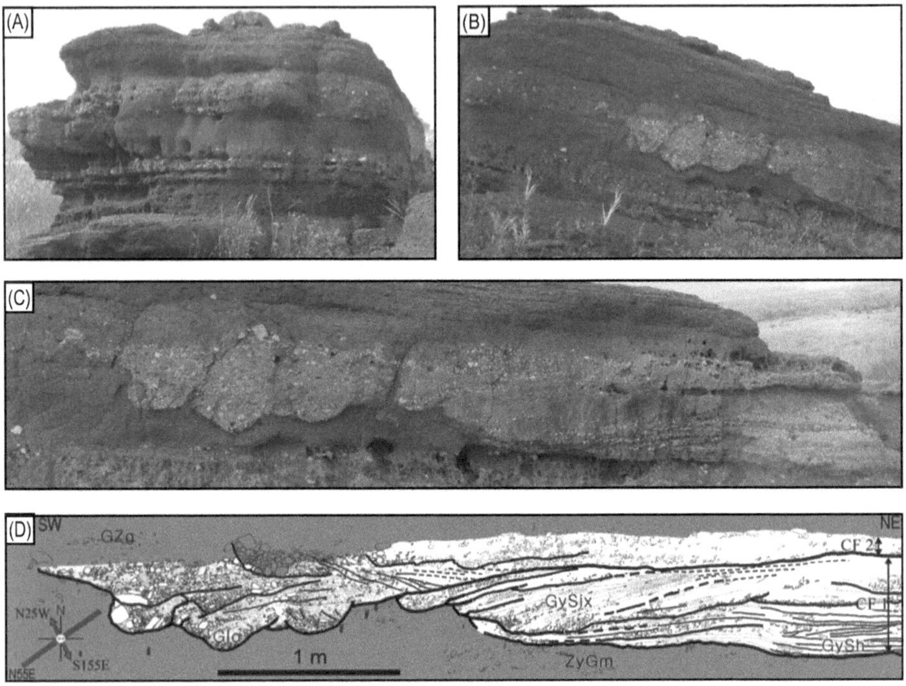

Figure 10.7 Photograph (A–C) and detailed sketch (D) of a small-scale but multi-storey channel fill of FA IIIA embedded in FA IIIC deposits (from an islet between sections JH and HH). For facies codes, see Table 10.1.
Source: Kim et al. (2009) by permission of the Elsevier.

This facies association is interpreted as a floodplain deposit based on the predominance of fine-grained deposits. The associated calcareous nodules and burrows indicate post-depositional modification by pedogenic processes and bioturbation. The reddish brown color reveals subaerial exposure and subsequent sediment reddening under oxidizing conditions. Red siltstone formed in well-drained proximal floodplains.

10.2.2 Depositional Processes

Figure 10.8 summarizes the paleoflow pattern and the distribution of each architectural element as well as the constituent facies associations. Alluvial-fan deposits (Element I) occur mainly along the eastern basin margin, forming three discrete depositional bodies. These alluvial fans are regarded as basin marginal or basin transverse depositional systems issued from the eastern basin margin. The braided-stream channel fills (Element II) occur in the northwestern part of the basin. The channel fills display a southward paleoflow. The scour axes of crevasse channels indicate a diverse paleoflow, roughly perpendicular to those of the main channel deposits of Element II.

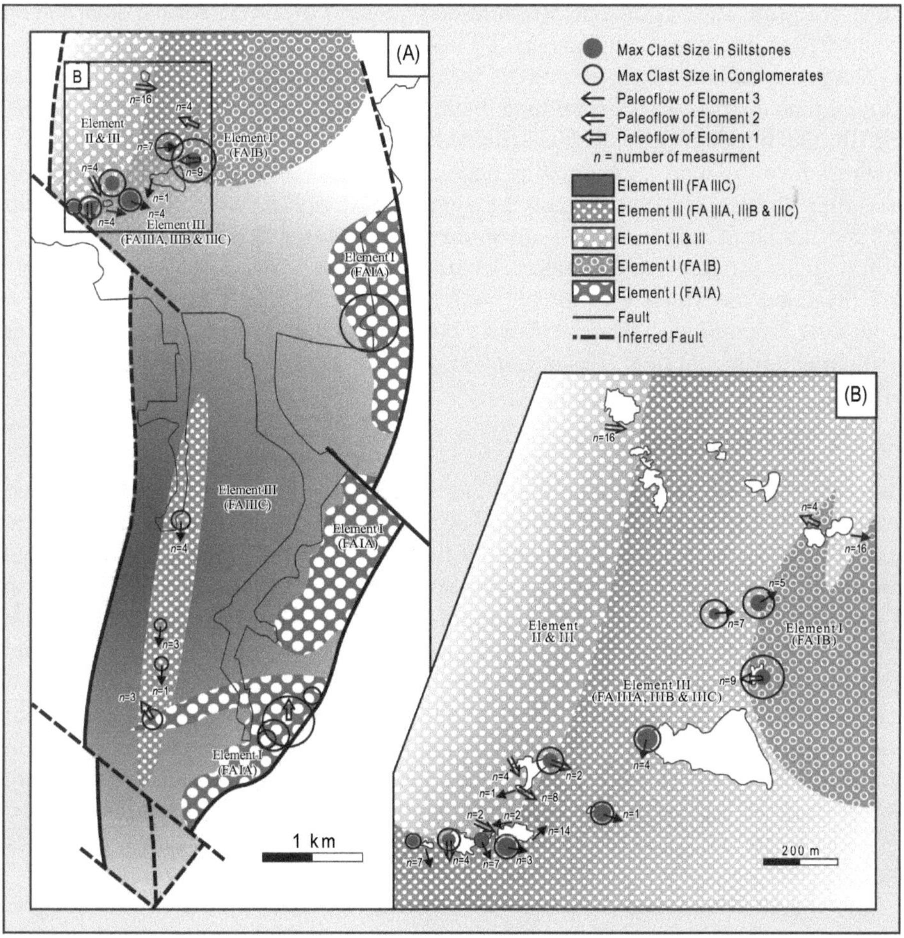

Figure 10.8 Schematic distribution map of each element and facies association in the entire Sihwa Basin (A) and in the Hanyom area (B). Palaeoflows (measured from channel axis orientation and/or clast imbrications) and the maximum clast size (average often largest clasts) are also depicted by the arrows and the circles. Circle diameter is proportional to the clast size. Open circles represent the clast size in gravelly deposits, whereas the closed circles indicate the size of scattered clasts in silty deposits (facies GZg and Fh).
Source: Kim et al. (2009) by permission of the Elsevier.

Further downstream, floodplain fines (FA IIIC) prevail with sparse crevasse channel fills and gravel sheets, covering most of the western and central parts of the basin. The downstream architectural variations indicate progressively less confined and diminishing braided streams with more common spillover but lesser crevasse incision in the distal part. The prevalence of reddish brown floodplain fines in the basin center indicates a well-drained condition and poor development of lakes or pools. Such an indistinct termination of channel network is suggestive of a terminal-fan environment (e.g., Miall, 1996; Rhee and Chough, 1993b).

The juxtaposition of the basin-axial terminal-fan system with the basin-marginal alluvial fans is also corroborated by a distinct internal architecture and stacking pattern of the braided channel fills (Element II). On alluvial-fan surfaces, braided channels form by lobe sedimentation and subsequent flow divergence at the termini of a trunk channel or chute beyond the intersection point between the channel floor and the fan surface. The flow is invariably in the lower flow regime, resulting in preferential upstream and/or lateral accretion of lobes. The low flow depth/grain size ratio and the absence of gravel bedforms with avalanche faces are indicative of less or unconfined, shallow flows, but do not necessarily indicate a steep-gradient environment (e.g., Heinz and Aigner, 2003).

10.2.3 Depositional Model

The sedimentary fill of the Sihwa Basin can be divided into two depositional successions: the lower terminal-fan succession distributed widely in the western and central parts of the basin and the sharply overlying alluvial-fan succession forming discrete bodies along the eastern margin. A two-stage evolution of the Sihwa Basin is suggested (Figure 10.9). The terminal-fan succession is characterized by an aggradational stacking pattern, i.e., each channel fill of the proximal braided stream is vertically superposed, not being interconnected but demarcated by the intervening channel-marginal to floodplain fine-grained deposits. The basin-marginal alluvial-fan successions are characterized by a progradational stacking pattern preferentially in a northward direction despite the partially radial dispersal pattern. The overall basin-fill architecture and the variations in stacking pattern of each succession can be understood in terms of the interplay of basin subsidence (tectonics) and sediment supply.

The aggradational stacking pattern of the lower terminal-fan succession suggests relatively rapid subsidence of the basin floor at a rate exceeding or balanced with that of the sediment supply in the early stage of basin formation. The southward dispersal pattern and the basin-wide distribution of fine-grained deposits suggest an abundant sediment supply from the drainage basin with large catchment areas and high-order drainage networks beyond the northern basin end. Generally, a basin-axial system is fed by the largest drainage basin among others and thus forms the widest depositional unit (Gawthorpe and Colella, 1990). In contrast, the progradational pattern of the upper alluvial-fan succession indicates a retarded rate of basin subsidence below the sediment supply rate in the late stage of basin evolution. The overall asymmetrical cross-basin distribution of depositional systems with gravelly alluvial fans in the eastern basin margin and a floodplain in the western margin reflects a half-graben structure of the basin with steep-gradient fault-bounded eastern margin (footwall block) and gently sloped flexural western margin (hanging wall block).

The predominance of ephemeral braided streams along with the reddish brown floodplain deposits indicates an arid to semi-arid paleoclimate. This climatic condition is favorable for the development of ephemeral braided streams because the

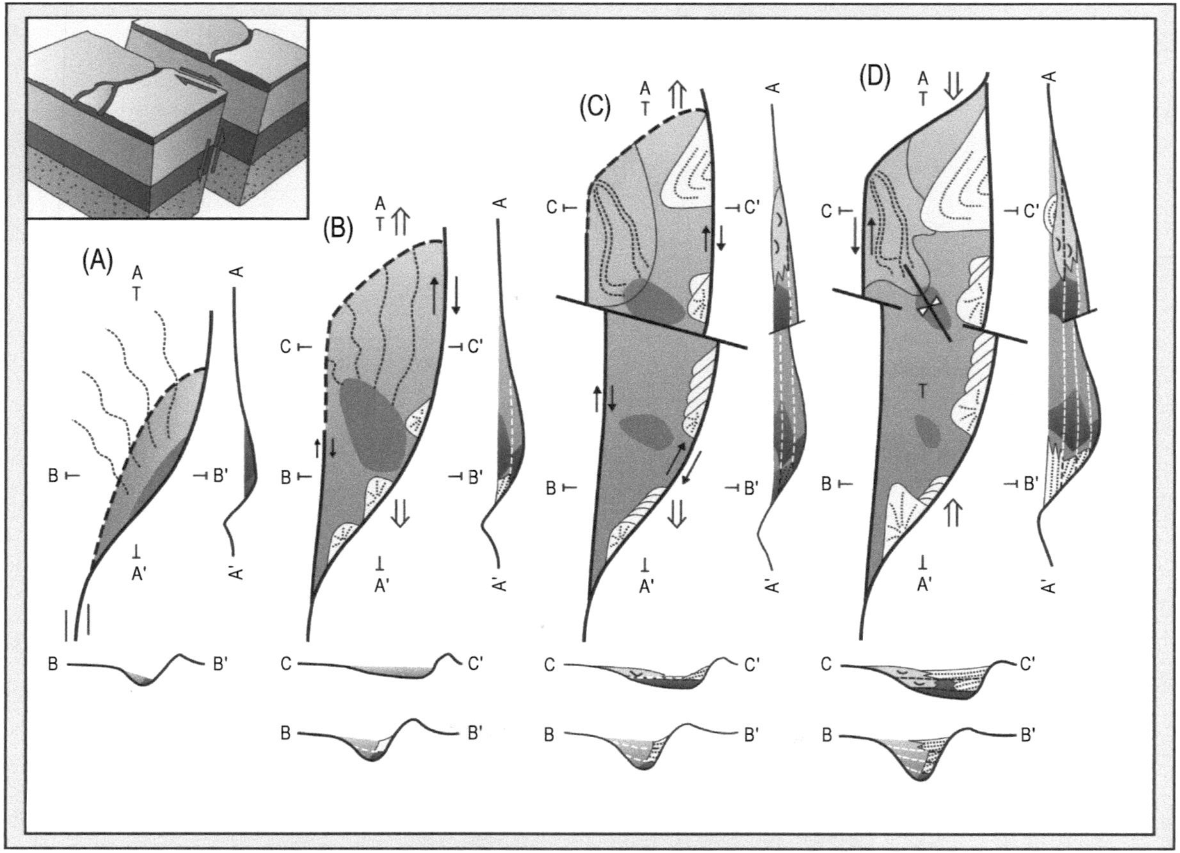

Figure 10.9 Reconstructed evolution model of the Sihwa Basin.
Source: Kim et al. (2009) by permission of the Elsevier.

seasonally concentrated precipitations, poor vegetation, and ample production of coarse-grained colluviums during the dry seasons are prone to triggering high-discharge flash floods. The development of calcareous nodules in the floodplain deposits also suggests an arid to semi-arid climate since the formation requires prolonged excessive evaporation. A warm and arid to semi-arid climate in the Korean Peninsula during the Cretaceous has been widely recognized by the occurrence of distinctive palynological assemblage (Choi, 1985), pedogenic characteristics (Paik and Chun, 1993; Paik and Kim, 2003; Paik and Lee, 1998), terminal-fan facies assemblages (Rhee and Chough, 1993b), and distinctive sedimentary deposits, such as radial calcite oolites embedded in lacustrine stromatolites (Woo et al., 1991).

10.2.4 Dinosaur Eggs and Nests

Type 1 eggs are of large size (11–15 cm in longest diameter) and have a thin shell (0.85–1.23 mm), whereas type 2 eggs (dark gray to black) are of small size (9.5–10.3 cm in longest diameter) and have a thick shell (3.4–5.0 mm) (Figure 10.10). Both types are characterized by spherical to subspherical shapes, an absence of surface protuberance or ornaments, and an abundance of open pore canals (ca. 50% of the surface area). The pore canal structure and inner shell surface texture, however, show different characteristics. Type 1 eggs are characterized by depressed circular openings (0.34–0.45 mm in diameter) on the outer surface and interconnected beneath, forming a labyrinth-type network with no linear vertical columns. The inner shell surface is smooth and unpatterned. Type 2 eggs comprise longitudinal, wavy, and commonly bifurcating furrows or striations (submillimeter wide and averagely 0.8 mm apart) on the outer surface. They also show wedge-shaped, outward fanning, and bifurcating columnar canals with faint vertical growth lines. The common multicanaliculate pore system of both types of eggs suggests that the eggs were laid and buried under moist soils (Lee et al., 2000b).

More than 140 dinosaur eggs, either isolated or clustered, have been identified from several stratigraphical horizons (Figure 10.10). They mainly occur in the cut-and-fill conglomerates of main channel (FA IIA) or crevasse channel (FA IIIA), and gravelly siltstones of the crevasse splay (FA IIIB) deposits. The eggs commonly form concentrated horizons of two to six eggs in vertical section and are locally arranged in a circular pattern. These are indicative of nesting habits in floodplains and abandoned channels. The eggs are, however, completely stuffed with the substrate of gravels and silt, either breached or holed, although the original oval shape is retained. The layering and burying of eggs in pits are indicative of sauropod nesting habits. Moisture is required for multicanaliculate type eggs in semi-arid to arid climatic conditions. Preferential nesting habitat is near-channel or abandoned channel areas.

10.2.5 Tando Subbasin

The sedimentary succession of the Tando subbasin occurs along the coastline of Tando islet. It consists of conglomerate, sandstone, and reddish brown siltstone as

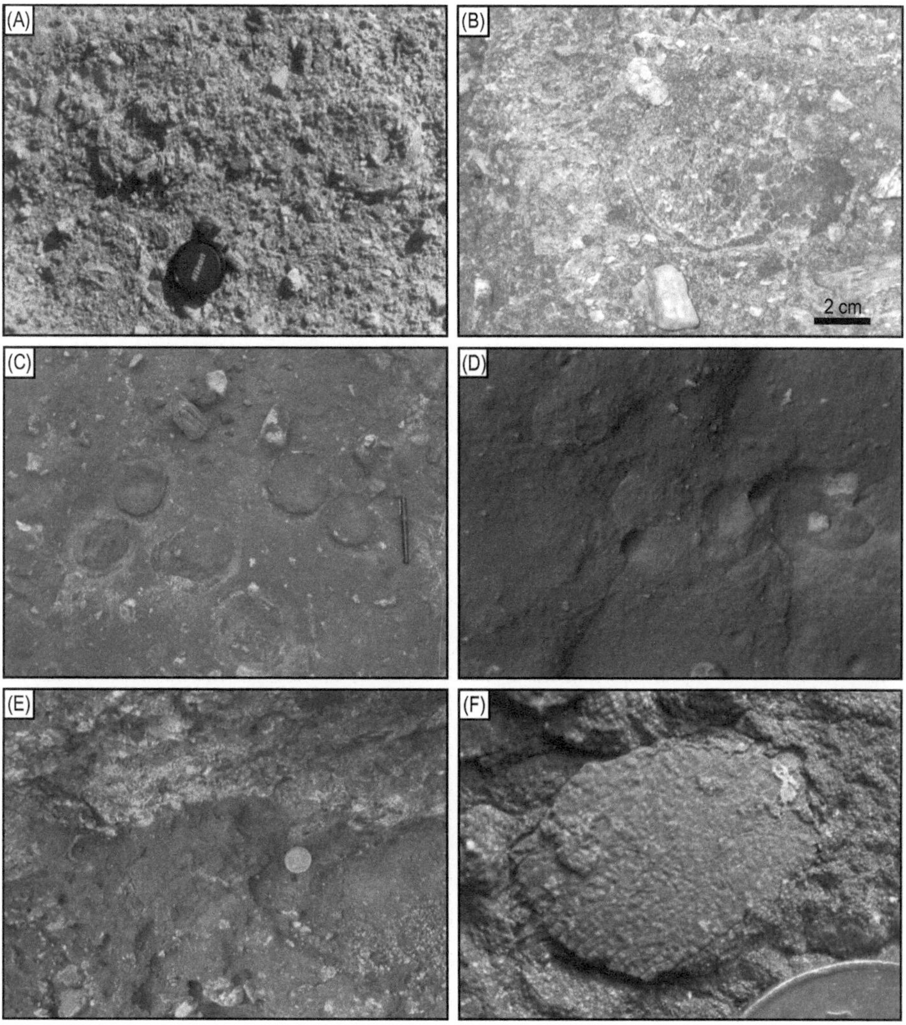

Figure 10.10 Intact oval-shaped dinosaur eggs filled with substrates, which occur mainly together up to several units in the same horizon and repeatedly in different stratigraphic horizons. Lens cap (5.5 cm in diameter) for scale. (A) Topless eggs in trough cross-stratified conglomerate at Sanghanyom. (B) Topless eggs in tabular conglomerates at Sanghanyom. (C) Concentrated dinosaur eggs in gravelly siltstone at Joonghanyom section. Pen (13 cm in length) for scale. (D) Dinosaur eggs in slightly gravelly siltstone. Lens cap (5.5 cm in diameter) for scale. (E) Type 2 dinosaur eggs in gravelly siltstone at Gaemisum section. Coin (2.3 cm in diameter) for scale. (F) Enlarged view of the shell surface of Type 2 eggs. Coin (2.3 cm in diameter) for scale.

Figure 10.11 Photograph of graded conglomerate, stratified sandstone, and interlayered chert and cherty mudstone, Tando islet.

Figure 10.12 Volcanic debris flow doposit (lahar) showing the margins of scoured sandstone and chert layers (Sihwa Formation), Tando islet.

well as chert and cherty sandstone and mudstone (Figure 10.11). The conglomerate and sandstone beds are commonly graded and sharply under- and overlain by cherty mudstone and chert with a significant lateral continuity. These interlayers are overlain and partly cut by volcanic mass flow deposits (lahars); the basal part maintains crude stratification without erosion (Figure 10.12).

The basin most likely formed by oblique-slip movement in which alluvial fans and the associated floodplains formed in down-dropped block. The lateral continuity of individual beds along the shoreline reveals, however, that the floodplain was subsequently submerged with a rise in water level and leveled by incoming turbidity currents under subaqueous conditions. With abundant supply of rhyolitic volcanic ash, cherty sandstone, and mudstone, as well as chert formed on the lake floor, triggered by the episodic inflows of river water into a silica-rich alkaline lake. The fresh water input led to an abrupt decrease in pH, which induced precipitation of inorganic silica. The interlayered graded and massive sandstone beds are suggestive of prevalent density undercurrents. The overlying lahar deposit is indicative of volcanic activity in the late stage of basin evolution.

10.3 Gyeokpo Basin

10.3.1 Geological Setting

The Gyeokpo Basin is located at the tip of a strike-slip fault, the Gongju fault (Figure 10.13). It rests on Precambrian gneiss and quartz monzonite and comprises volcaniclast-rich siliciclastic deposits (Gyeokpori Formation). The Gyeokpori Formation (about 290 m thick) is underlain by a volcanic complex of basaltic-andesitic flows and tuffs and overlain by a silicic volcanic succession of rhyolite and tuff (Figure 10.13). The quartz monzonite was dated at 128.0 ± 3.0 Ma, whereas the rhyolitic flow was dated at 83.6 ± 2.4 Ma using whole rock K–Ar methods (Lambiase and Bosworth, 1995). It suggests that the basin developed during the mid-Cretaceous (Barremian–Santonian) (Kim et al., 2003). The Gyeokpo Basin formed by left-lateral strike-slip faults, as a pull-apart or stepover basin, bounded by northeast-southwest-trending master faults (Figure 10.13) (Kim et al., 2003).

10.3.2 Sedimentary Facies and Depositional Processes

The deposits can be grouped into eight facies associations, based on the constituent facies, interbed relationships, and bed geometry. Table 10.2 outlines description of individual sedimentary facies and interpretation of depositional processes.

10.3.2.1 Subaqueous Talus in the Jukmak Area (Facies Association I)

It consists of crude tabular-bedded breccias with scarce interbeds of sandstone. Each bed is mostly disorganized or slightly inversely graded. Clasts are up to 2 m in long axis, whereas the matrix consists of angular coarse sands and fine pebbles. Both are composed of fragments of quartz monzonite. This facies association exclusively occurs in the Jukmak area, overlying the basement of quartz monzonite. It most likely represents a subaqueous talus at the foot of an escarpment, dominated by cohesionless debris flows.

10.3.2.2 Delta Plain (Facies Association II)

Facies association II consists of an amalgamated succession of lobate or hollow-fill, lenticular conglomerates. Clasts are angular to round (up to 2 m long) and disorganized with interlocked cobbles and boulders. Interstices are void or filled with

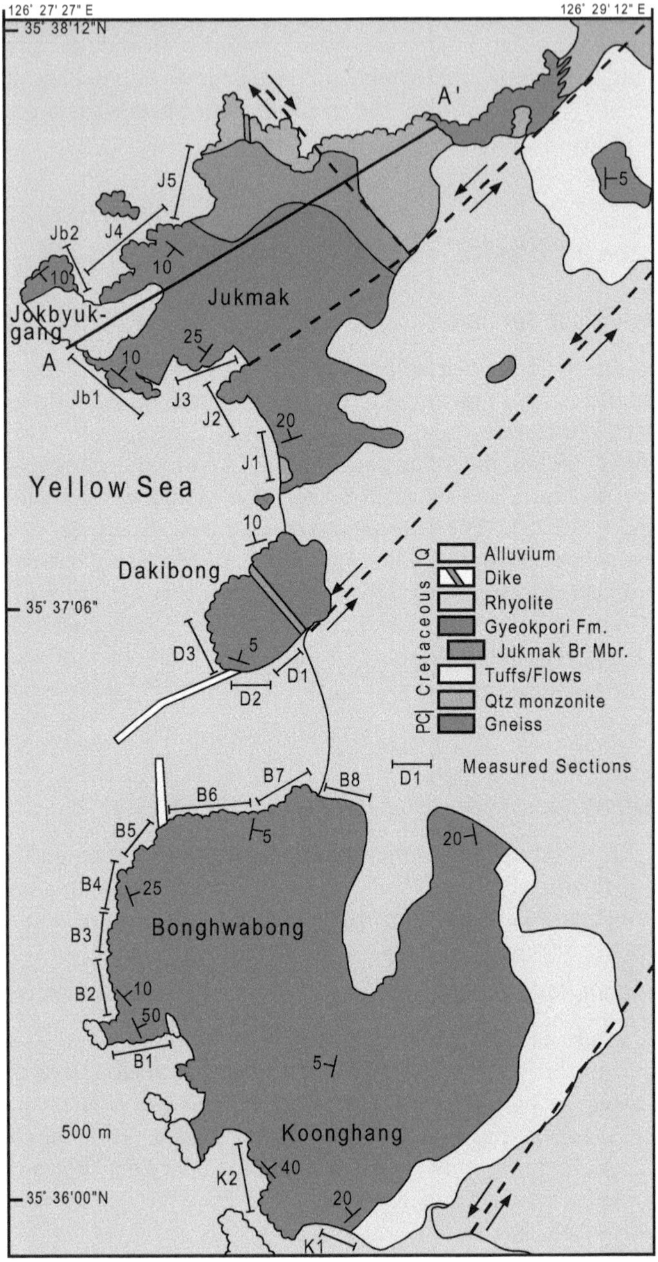

Figure 10.13 Geological map of the Gyeokpori area. Dashed lines indicate inferred faults (modified from Lee et al. (1999b)). Bars represent measured sections. B, Bonhwabong; D, Dakibong; J, Jukmak; Jb, Jokbyukgang; Fm, Formation; Br Mbr, breccia member; Qtz, quartz.

Source: Kim et al. (2003) by permission of the Wiley-Blackwell.

Table 10.2 Description and Interpretation of Sedimentary Facies in the Gyeokpori Formation

Facies	Description	Interpretation
Tuff (Facies T)	Mainly composed of fine-grained ash; greenish gray or leucocratic in color; mostly homogeneous or slightly graded; but locally subtly banded or weakly fissile; includes accretionary lapilli and carbonate nodules; a few decimeter to meter thick; interlayered with black shales; laterally persistent; some beds correlated for a few kilometer	Sub-lacustrine ash fall
Crudely bedded breccia (Facies Bb)	Thick to very thick; mostly disorganized; locally inversely graded in the basal part; subangular to angular clasts; coarse sand- to boulder-grade particles (up to 2 m long); almost monomictite of quartz monzonite; direct contact with basement (quartz monzonite)	Sub-lacustrine ash fall avalanche; cohesionless debiris flows and rock falls (Nemec and Steel, 1984)
	Tabular units: clast-supported with minor coarse sand to pebble matrix. Hummocky units: low-relief convex-up geometry; tightly interlocky clasts with minimal matrix material	Tabular units: cohesionless debris flows hummocky units: rock falls
Tabular conglomerate to gravelly sandstone (Facies Gt)	Generally medium- to thick-bedded sometimes thin- or very thick-bedded; laterally continuous or wedged in longitudinal section; tabular or lobate in transverse section; generally nonerosional base; irregularly bedded with common projected clasts; mostly clast-supported, locally matrix-supported; common boulder-sized clasts (up to > 2 m long); medium sand to granular matrices;	Cohesionless debris flows or density-modified grain flows (Kim et al., 1995b; Lowe, 1982; Nemec and Steel, 1984; Surlyk, 1984)

(Continued)

Table 10.2 (Continued)

Facies	Description	Interpretation
	disorganized, inversely graded, normally graded or crudely stratified; common rapid lateral changes in grading pattern; commonly associated with beds of Facies Gs, GSb, Gsg and Sl	
Lenticular conglomerate to gravelly sandstone (Facies Gl)	Variable in dimension, a few decimeters to tens of meters in width and a few decimeters to meters in height; subangular to rounded clasts; common large boulders (> 2 m long).	High-magnitude flood flows (Allen, 1981; Gloppen and Steel, 1981)
	Lobate units: tightly interlocked large clasts; openwork or fine sand to silt matrix; openwork in the core; tabular or trough cross-stratified pebbly downflow margins; shingled or vertically stacked	High-magnitude flood flows (Allen, 1981; Jo et al., 1997)
	Hollow-fill units: present on top of or beside lobate units; tightly packed clasts at the bottom; massive or crudely stratified pebbly sandstone in the upper part	Hollow-fill type: channel deposits (Friend, 1983)
Gravel sheet, lens, and pocket (Facies Gs)	Thick to very thick; alternation of tightly packed, gravel-rich layer and sand-rich layer with some floating clasts; common parallel orientation or imbrication of clasts; often steeply inclined relative to underlying beds	Debris falls of individual grains or grain assemblages (Nemec, 1990; Kim et al., 1995b; Sohn et al., 1997)
Bedded conglomerate/ sandstone couplets (Facies GSb)	Abrupt decrease in clast size and abundance across the lower inversely graded conglomerate to the upper normally graded and stratified sandstone divisions; common protruding clasts above the conglomerate division;	Bipartite flows consisting of lower cohesionless debris flows and upper turbidity currents (Falk and Dorsey, 1998; Hampton, 1972; Surlyk, 1984)

(*Continued*)

Table 10.2 (Continued)

Facies	Description	Interpretation
	gradual transition of matrix population between the two divisions; common convergence of sandstone strata onto conglomerate division.	
Graded and stratified conglomerate/ sandstone couplets (Facies GSg)	Gradual transition from lower graded or ungraded conglomerate to upper graded and/or stratified (laminated) sandstone divisions; pebble-to cobble-graded clasts, either flat lying or imbricated; protruding elongate boulder clusters; commonly associated with beds of Facies Gt and Sl	Bipartite high-concentration turbidity currents with lower traction carpets and upper turbulent flows (Postma et al., 1988; Sohn et al., 1997)
Medium- to thick-bedded sandstone (Facies Sb)	Up to >5 m thick; massive, normally graded, inversely graded, laminated or stratified layers; common outsized clasts (max. 1–6 m long), scattered or clustered, several times larger than the layer thickness, oriented either parallel or transverse to the palaeocurrent	Sandstone layers: high-concentration turbidity currents (Kim et al., 1995b; Lowe, 1982; Surlyk, 1984); outsized clasts: debris falls (Kim et al., 1995b; Nemec, 1990)
Laminated to thin-bedded sandstone (Facies Sb)	Well-sorted fine to very coarse sandstone with common outsized clasts, up to boulder grade; common mudstone chips; sharp and flat bases; massive, normally graded, inversely graded or laminated	High-concentration turbidity currents concomitant with debris falls (Nemec, 1990; Kim et al., 1995b)
Thickly bedded, graded sandstone/mudstone couplets (Facies SMg)	Overall distributional grading from sandstone to mudstone divisions; normally graded, inverse to normally graded or crudely stratified lower sandstone division; normally graded or weakly laminated upper mudstone division; common mudstone chips and plant debris; sharp bases;	Low-concentration turbidity currents (Bouma, 1962; Chun and Chough, 1995)

(*Continued*)

Table 10.2 (Continued)

Facies	Description	Interpretation
	a few tens of centimeter to 2–10 m thick	
Laminated to thin-bedded sandstone/mudstone couplets (Facies SMl)	Graded or ungraded lower sandstone and the overlying finely laminated mudstone with sharp boundaries; common mudstone chips and plant debris; laterally persistent; individual layers laminated to thickly bedded; common syndepositional deformation; occasional outsized clasts	Low-concentration turbidity currents (Bouma, 1962; Chun and Chough, 1995)
Homogenous mudstone (Facies Mh)	Structureless; dark gray to black in color; noncalcareous, abundant plant debris; occasional trace fossils	Suspension settling and bioturbation
Finely laminated mudstone (Facies Ml)	Finely laminated; light to dark gray in color; occasional ripple marks; noncalcareous; sporadic carbonate nodules, abundant plant debris, outsized pebbles in some places; commonly deformed, folded or convoluted	Fine-grained turbidity currents (Bouma, 1962; Chun and Chough, 1995)

Source: Kim et al. (2003) by permission of the Wiley-Blackwell.

poorly sorted matrix of sandstone and siltstone. Lobate units display disorganized fabric in the center and cross-stratification in the margins. Hollow-fill units are commonly graded and packed with pebble–boulder clasts.

Bouldery lobate units most likely represent gravel bars formed by high-magnitude flood flows, whereas sediment segregation from the boulder cores to pebbly margins suggests sorting by weakened flows. These processes of bar formation and hollow incision occur in braided streams of delta-plain environments.

10.3.2.3 Delta Slope (Facies Association III)

It is characterized by steeply inclined tabular conglomerate and sandstone. The foreset is 10–25 m in height. Each bed is commonly clast supported and disorganized or inverse-to-normally graded, most likely formed by cohesionless debris flow or debris fall on a steep slope.

10.3.2.4 Base of Slope to Prodelta (Facies Association IV)

It consists of gently inclined sandstone facies with minor conglomerates. Each bed is commonly normally graded and sharply based, generally persistent for tens of meters. However, conglomerate beds are mostly lenticular. This association is prevalent in the southern part (Koonghang and Bonghwabong sections).

10.3.2.5 Nested Gilbert-Type Delta (Facies Association V)

This facies association is characterized by a small-scale tripartite geometry of topset, foreset, and bottomset strata within the slope deposit.

10.3.2.6 Delta Lobe (Facies Association VI)

This association consists of mound-shaped or lobate bodies of sandy conglomerate. Each bed is up to 9 m high and 30 m wide (Figure 10.14). They are conformably under- and overlain by sandy black shale deposits, forming an isolated or coalesced package. Coarsening-upward trends are prominent. Conglomeratic units are mostly inversely graded and contain outsized floating or protruding clasts. Each bed either thickens downdip, forming a blunt terminus, or thins, being transitional to concentrated gravel clasts at the terminus.

The mound-shaped external form and the coarsening-upward stacking pattern suggest progradational delta lobes (Kim and Chough, 2000). The development of inverse grading, incorporation of large outsized clasts, and downdip coarsening with clast-concentrated termini in the conglomeratic units suggest cohesionless debris flows or modified grain flows (Nemec and Steel, 1984). However, the normally graded sandy units indicate sandy turbidity currents.

10.3.2.7 Subaqueous Fan (Facies Association VII)

This association is characterized by an overall fining-upward succession which comprises channelized units of tabular conglomerate and graded and crudely stratified conglomerate/sandstone couplets in the lower part, commonly graded, medium- to thick-bedded sandstones in the middle part, and laminated to thin-bedded sandstones and mudstone in the upper part (Figure 10.15). In the middle part, deposits become tuffaceous and show a monotonous stacking pattern of tabular bedded units with occasional fining-upward package.

The lower conglomeratic deposits suggest cohesionless debris flows and gravelly high-concentration turbidity currents (Lowe, 1982), whereas the middle sandstone-dominated and the upper mudstone-dominated successions indicate sandy, high-concentration and fine-grained, low-concentration turbidity currents, respectively. The channelized conglomerates in the lower part may indicate feeder channels, the fining-upward sandstone packages in the middle part may reflect suprafan lobes, and the upper mudstone-dominated succession can be interpreted as outer-fan deposits (Normark et al., 1993).

Figure 10.14 (A) The Gyeokpori Formation, showing a wedge-shaped sandstone body (facies association VI), marked by bidirectional arrow, prograding northward. Chaeseokgang, looking north from the seawall causeway. (B) Enlarged view of part (A). Note the outward dipping of individual beds. For location of the section (D3), see Figure 10.13.

10.3.2.8 Basin Plain (Facies Association VIII)

This is characterized by a planar bedded, monotonous succession of laminated sandstone, graded (or laminated) sandstone/mudstone couplets, very finely laminated mudstone, and homogeneous mudstone with interbedded tuffs. Individual beds can be correlated over a few kilometers. Syndepositional deformation structures, such as folds, convolute bedding, flame structures, and rip-down clasts, are common. This facies association occurs in the upper parts of the Bongwhabong, Dakibong, and Jukmak sections (Figure 10.15).

The monotonous stacking of tabular, laterally continuous sandy shale units, either graded or laminated, suggests deposition from expansive turbulent flows or pelagic sedimentation (Chun and Chough, 1995). The sandstone facies are

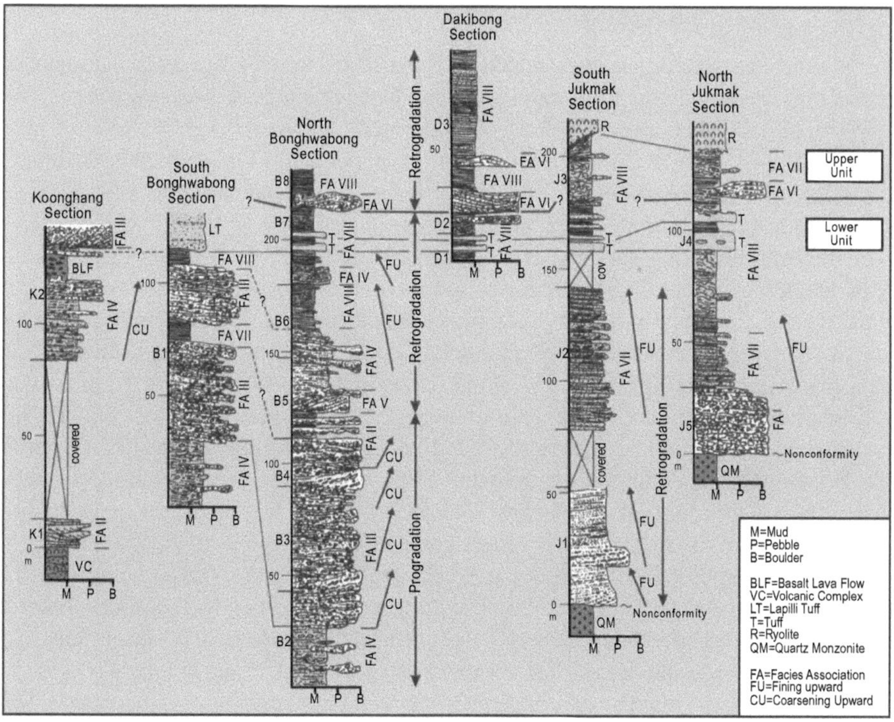

Figure 10.15 Composite sedimentary logs and stratigraphic correlation of the Gyeokpori Formation along the Gyeokpori coast. For section locations, see Figure 10.13. Correlation is made using key beds (e.g., green tuff and deformed shale). The formation comprises two succession units. The lower unit shows either a prograding to retrograding, coarsening-to fining-upward trend (at north Jukmak section, about 115 m thick). The upper retrograding fining-upward unit is more than 75 m thick (at Dakibong section). For facies association code, see Table 10.2. (For interpretation of the references to color in this figure legend, the reader is referred to the online version of this book.)
Source: Kim et al. (2003) by permission of the Wiley-Blackwell.

indicative of high- and low-concentration turbidity currents, whereas the mudstone facies suggest distal, fine-grained turbidity currents. Some thickly bedded, homogeneous mudstones reflect prolonged suspension settling from riverine effluents. The associated tuff beds indicate spasmodic volcanic activity in the hinterland.

10.3.3 Basin-Fill Processes

The succession can be grouped into two large-scale units, i.e., the lower succession and the upper succession. The lower unit is well exposed in the Koonghang, Bonghwabong, and Jukmak sections, whereas the upper unit appears in outcrops of the upper Bonghwabong, Dakibong, and Jukmak sections (Figure 10.15).

10.3.3.1 Lower Succession

The Koonghang section comprises channelized cross-stratified gravelly sandstones, coarsening-upward deposits of tabular-bedded conglomerates and sandstones, and steeply inclined, very thick-bedded conglomerates (Figure 10.15). A basaltic lava-flow unit (~ 13 m thick) is interlayered in the upper part. The section represents a prograding delta succession: proximal delta plain or distal alluvial origin, distal prodelta, and prograding delta slope or foreset.

In the Bonghwabong section, deposits are less rich in volcaniclasts than those in the Koonghang section. The section shows an overall coarsening-to-fining-upward trend, whereas the upper 105 m is a fining-upward package composed of nested, small-scale conglomeratic Gilbert-type delta, sandy prodelta, and sandy black shale basin-plain deposits (Figure 10.15). The coarsening-upward lower part represents a progradational, steep-sloped, coarse-grained delta system, whereas the fining-upward upper part indicates a retrogradation with drowning of the delta system. The Bonghwabong delta system formed slope aprons, each of which prograded northeastward and shifted northward. The delta slopes are up to 25 m in height, which may approximate the paleowater depths at the basin margin.

In the Jukmak sections, two separate packages of coarse-grained deposits are identified: one is in the north Jukmak section and the other in the south Jukmak section (Figure 10.15). The former is unconformably underlain by basement rock (quartz monzonite) and shows an overall fining-upward trend (Figure 10.15). Its lower part consists mainly of conglomeratic deposits (proximal subaqueous fan) that are either channelized or thickly bedded and contain large amounts of quartz monzonite fragments, whereas in the middle part, deposits become tuffaceous and are composed of alternations of tabular bedded sandstones and mudstones (distal subaqueous fan). The upper part comprises basin-plain black shale deposits interlayered with tuff beds. The latter (north Jukmak section) is also unconformably underlain by the quartz monzonite basement and is dominated by monomictic breccia of quartz monzonite in the basal part. This section also shows a fining-upward trend but with an abrupt change in grain size. The lower breccia is gradationally overlain by tabular conglomerates, which are, in turn, abruptly overlain by sandy black shale (Figure 10.15). The breccia and conglomerate deposits in the lower section represent subaqueous talus at the base of a steep escarpment, whereas the upper black shale deposit indicates a basin plain.

10.3.3.2 Upper Succession

The upper succession mainly occurs in the upper parts of the Bonghwabong, Dakibong, and Jukmak sections (Figure 10.15). It starts with deposition of lobate or mound-shaped bodies of gravelly deposits (small-scale delta lobes) above basin-plain shales and tuffs of the lower succession (Figure 10.14). Steeply inclined, angular downlap onto the underlying black shale (Figure 10.14B) indicates that the delta lobes were mainly constructed by mass flows, which terminated abruptly at the base of the delta front. The lack of alluvial delta-plain facies indicates that the delta lobes grew mainly under subaqueous conditions.

Based on their locations, the delta-lobe deposits can be grouped into two packages: one is the Dakibong delta-lobe assemblage, north of the Bonghwabong gravelly delta system, and the other is the Jokbyukgang delta-lobe assemblage, south-west of the north Jukmak subaqueous talus system (Figure 10.15). The former prograded north-northeastward, whereas the latter prograded south-southwestward. These delta-lobe assemblages suggest rejuvenated sediment progradation from the basin margins.

10.3.4 Evolution of Gyeokpo Basin

On the northern margin, a small-scale, steep-sloped, aggradational talus system formed directly adjacent to the basin plain, whereas large-scale, less steep-sloped, progradational delta systems formed on the southern margin. This structural asymmetry is typical of a half-graben basin. Furthermore, progradational stacking of deposits on the southern basin margin is indicative of greater sediment supply than that of the northern footwall margin.

The first stage represents deposition of coarse-grained sediments at the base of the lower succession (Figure 10.16). Four distinct depositional systems developed at this stage: the Bonghwabong gravelly delta system and the Koonghang volcaniclastic delta system on the southern basin margin, the south Jukmak sandy subaqueous fan system north of a basement high and the north Jukmak subaqueous breccia talus system on the northern basin margin. Progradational and aggradational stacking patterns indicate a slow dip-slip motion of the basin-bounding faults in the early stage of basin formation.

The second stage started with deposition of nested small-scale Gilbert-type deltas above the drowned, large-scale delta system (Bonghwabong section) (Figure 10.17). This indicates an onset of transgression and thereby local backstepping of depositional systems. Further rise in base level can be inferred from the retrogradational stack of prodelta and basin-plain deposits above the nested deltas. This suggests basinwide transgression probably caused by rapid subsidence at the basin margins due to large dip-slip displacement along the bounding faults with a small component of strike-slip offset (e.g., Blair and Bilodeau, 1988).

The third stage commenced with emplacement of delta-lobe assemblages above the topmost sandy black shale deposit of the lower succession (Figure 10.17). These delta-lobe deposits are smaller in dimension, finer-grained, and located more basin-ward than the deposits of marginal systems of the lower succession. This lake-ward shift of depocenter indicates loss of accommodation in the basin margins, probably related to inversion of the basin-bounding faults (Lambiase and Bosworth, 1995). The repeated retrogradational stacking of delta-lobe deposits and the basin-plain deposits resulted from autocyclic processes, i.e., lateral shift of delta lobes as a result of migration of feeder channels.

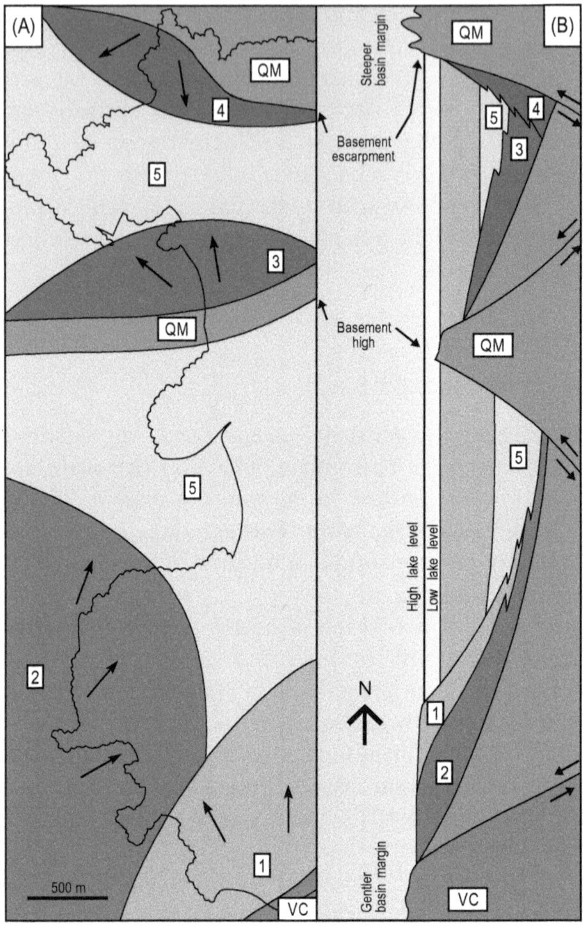

Figure 10.16 (A) Schematic map reconstruction of depositional systems in the lower succession unit of the Gyeokpori Formation. Arrows denote general progradation direction of each system. 1, Koonghang volcaniclastic delta system; 2, Bonghwabong gravelly delta system; 3, south Jukmak sandy subaqueous fan system; 4, north Jukmak subaqueous breccia talus system; 5, fine-grained deposits in the basin plain. QM, quartz monzonite; VC, volcanic complex. (B) Cross-sectional reconstruction of palaeoenvironments. The southern margin of the Gyokpo Basin was gentler in slope and shallower in water depth than the northern margin where a breccia talus system (4) developed at the foot of the basement escarpment. On the southern basin margins, a thick volcanic complex of lava flows and tuffs was emplaced and supplied large amounts of sediment basinward, forming coarse-grained delta systems with variable amounts of volcaniclastics. In the basin center, a basement high was present, and a subaqueous fan system (3) built out to the north on its northern flank. *Source*: Kim et al. (2003) by permission of the Wiley-Blackwell.

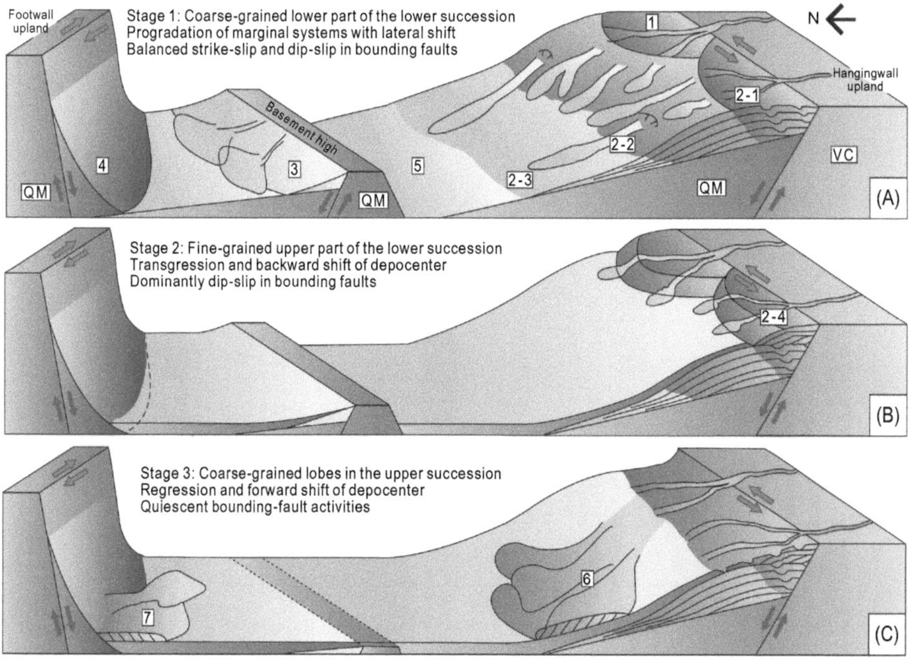

Figure 10.17 Basin-filling model for the development of sedimentary facies, distribution of facies association, and stratigraphic architecture of the Gyeokpori Formation. 1, Koonhang volcaniclastic delta system; 2, Bonhwabong gravelly delta system; 2-1, delta plain (FA II); 2-2, delta slope (FA III); 2-3, prodelta (FA IV); 2-4, nested Gibert-type delta (FA V); 3, south Jukmak sandy subaqueous fan system (FA VII); 4, north Jukmak subaqueous breccia talus system (FA I); 5, basin plain (FA VIII); 6, Dakibong delta-lobe assemblage (FA VI); 7, Jokbyukgang delta-lobe assembalge (FA VI). QM, quartz monzonite; VC, volcanic complex. Blue and red arrows denote the strike-slip and dip-slip directions of the bounding faults, respectively. (For interpretation of the references to color in this figure legend, the reader is referred to the online version of this book.)
Source: Kim et al. (2003) by permission of the Wiley-Blackwell.

10.4 Gunsan Basin

10.4.1 Seismic Units and Age Constraints

The Gunsan Basin is part of the Northern South Yellow Sea Basin (Figure 10.18). It displays a rhombic geometry and contains thick nonmarine deposits (>7000 m thick) of the Cretaceous to Tertiary. The basin comprises three fault-bounded sub-basins along the central uplift (Figure 10.19). The entire succession can be divided into three seismic units, bounded by two erosional surfaces (Figure 10.20).

Seismic unit 1 (SU-1) is confined to the fault-bound depression and truncated by an erosional surface, either ES1 or ES2 (Figure 10.20). Unit SU-1 consists of

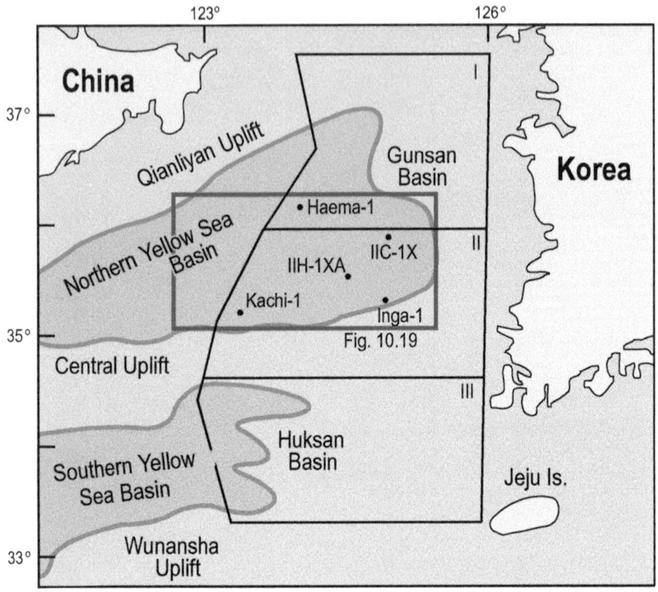

Figure 10.18 A simplified tectonic map of the Yellow Sea, showing the distribution of Cretaceous–Tertiary sedimentary basins.
Source: Allen et al. (1997), Chen and Qin (1989), and Ren et al. (2002).

poorly sorted sandstone and reddish dark gray mudstone, intercalated with limestone and volcanic rocks (Figure 10.21). Deposition was initiated in the Cretaceous, followed by tectonic inversion in the Oligocene. Unit SU-2 is characterized by high frequency and strong reflectors onlapping onto the erosional surface (ES1) (Figure 10.20). The unit unconformably overlies either unit SU-1 or acoustic basement, truncated by an erosional surface. It comprises interbedded sandstone and mudstone of the early Miocene (Figure 10.21). Sediments deposited in fluvial to alluvial or lacustrine environments (Ryu et al., 2000; Yi et al., 2003). The top of unit SU-2 represents the end of the early Miocene. Unit SU-3 comprises well-defined, continuous reflectors, generally concordant to the lower boundary (ES2) (Figure 10.20). It largely comprises sandstone, interbedded mudstone, conglomerate, and lignite of the middle Miocene (Figure 10.21).

10.4.1.1 Western Subbasin

The western subbasin formed as an asymmetric wedge-shaped half-graben bounded by the Kachi fault and extensional synthetic (or antithetic) normal faults (Figures 10.19 and 10.20). The wedge-shaped strata are offset by normal faults in the east of the Kachi fault, which is indicative of syntectonic growth. Kachi-1 well reveals that the half-graben fills began in the early Cretaceous. The hanging-wall block of the inverted Kachi fault was deformed by tight folds and reverse faults,

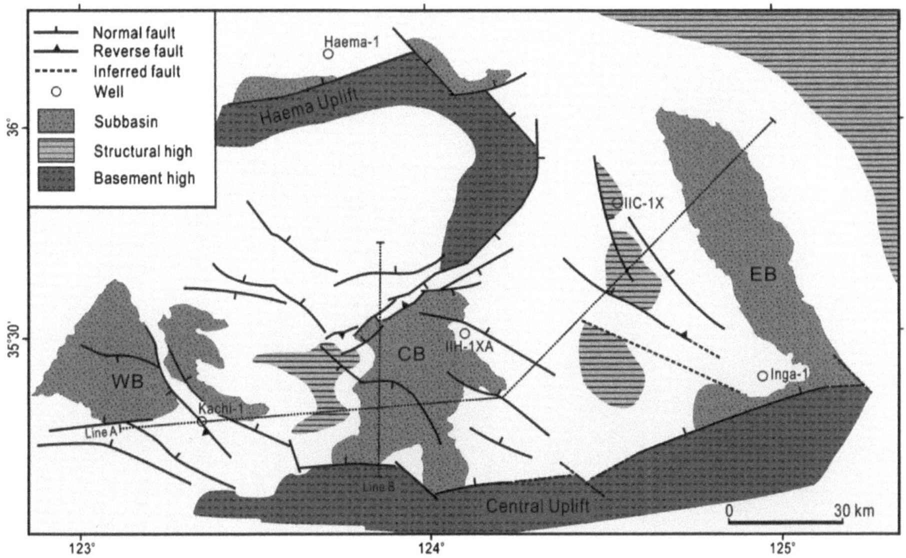

Figure 10.19 Geologic structures of Gunsan Basin, characterized by three fault families: E(NE)−W(SW), NW−SE, and NE−SE trending faults. WB, western subbasin; CB, central subbasin; EB, eastern subbasin.
Source: Shinn et al. (2010) by permission of the Elsevier.

forming a pop-up structure. The inversion of the half-graben was due to regional contraction in the Oligocene.

10.4.1.2 Central Subbasin

The central subbasin was controlled by south-bounding faults and oblique intrabasinal faults (Figures 10.19 and 10.20). Reverse reactivation of the intrabasinal normal fault took place in the Oligocene, evinced by the presence of anticlinal folds and the related faults in the hanging wall. Peneplanation of the inversion relief is indicative of coeval uplift and erosion, forming an erosional unconformity (ES1) (Figure 10.22). The inversion ceased with deposition in the Early Miocene. The wedge-shaped geometry and the divergent onlap pattern of the Early Miocene strata are suggestive of syntectonic sedimentation, controlled by reactivated, divergent strike-slip faults in the northern margin. The negative flower structures and *en echelon* arrangement of the associated faults are suggestive of a transtensional stress regime related to the right-lateral strike-slip movement.

10.4.1.3 Eastern Subbasin

The eastern subbasin is a wedge-shaped half-graben developed along the south-bounding fault (Figure 10.19). A synclinal fold with a northwest-southeast-trending axis formed as a result of post-depositional contraction (Figure 10.20). The

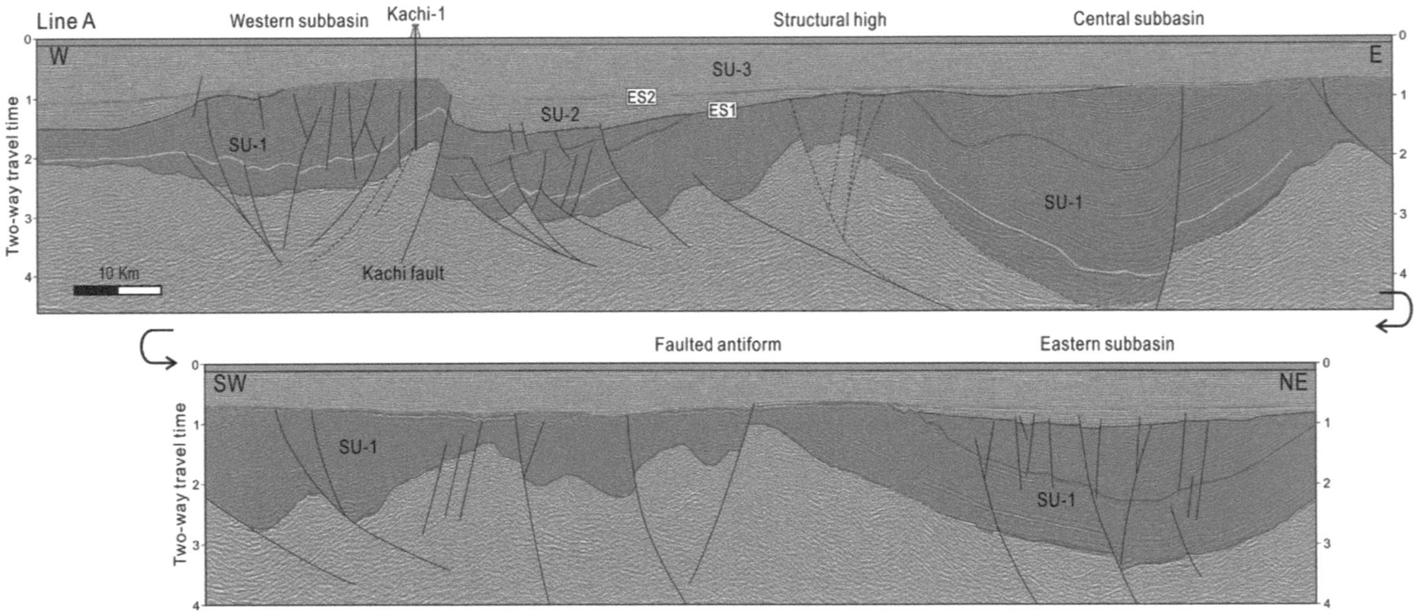

Figure 10.20 A regional seismic profile and its interpretation. For location of the seismic section, see Figure 10.19. Three seismic units (SU-1, SU-2, and SU-3 in ascending order) and two unconformity surfaces (ES1 and ES2) are identified. The section, subparallel to the south-bounding fault, displays inverted half-graben in the western subbasin and a synclinal fold in the central and eastern subbasins. These subbasins are separated by intervening structural highs.
Source: Shinn et al. (2010) by permission of the Elsevier.

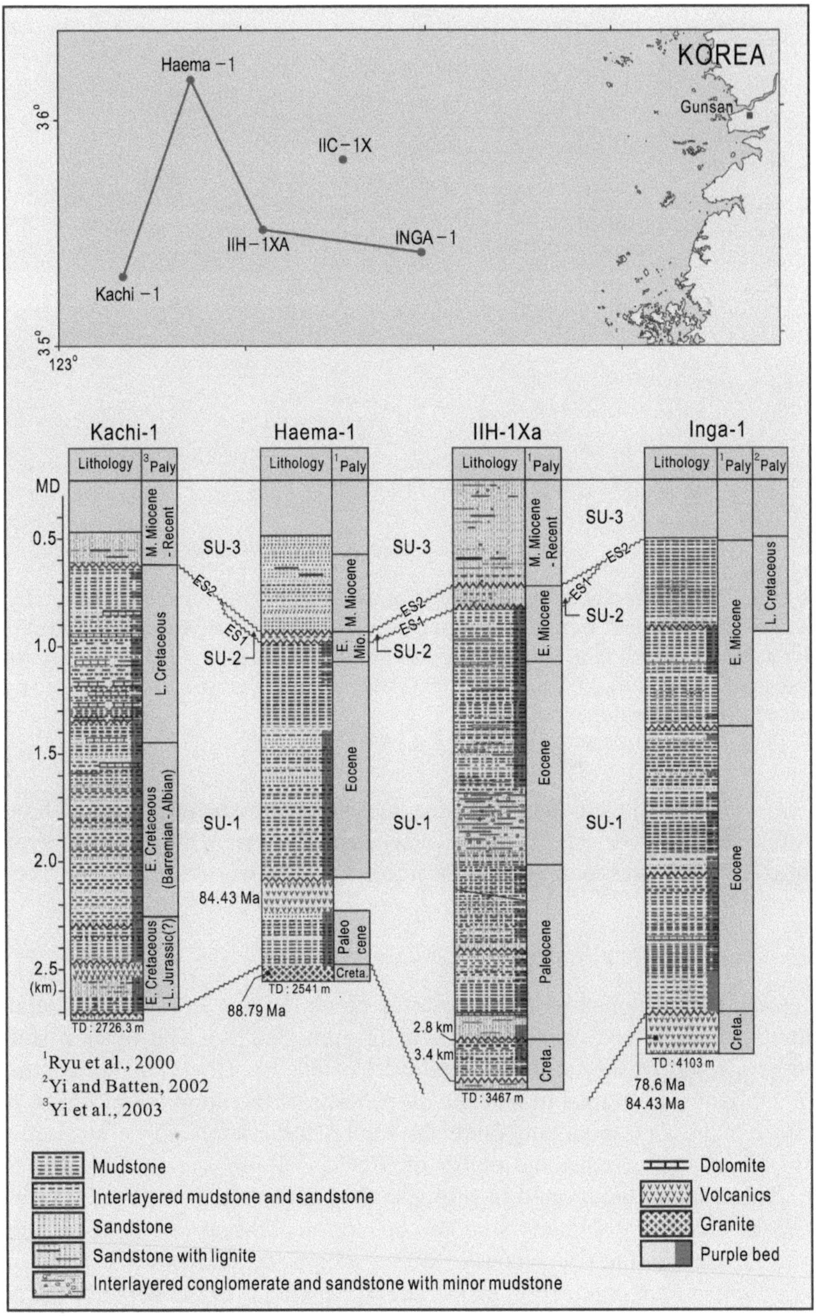

Figure 10.21 Lithology and biostratigraphy of exploratory wells (Ryu et al. (2000), Yi and Batten (2002), and Yi et al. (2003)) compared to seismic units and key horizons. Note the age determined by K/Ar dating method. For location of exploratory wells, see Figure 10.19. SU, seismic unit; MD, measured depth; TD, total depth; Paly, palynologic study.
Source: Shinn et al. (2010) by permission of the Elsevier.

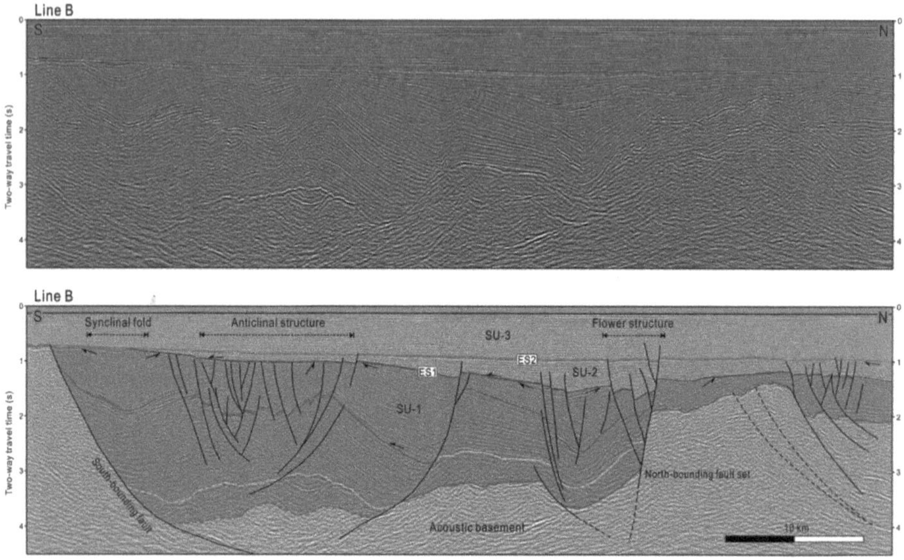

Figure 10.22 A representative seismic profile of the central subbasin and its interpretation. For location of the seismic section, see Figure 10.19. The central subbasin is bounded on the south by a curved normal fault and on the north by subvertical fault set. The south-bounding fault was inactive in the early Miocene, whereas the north-bounding fault propagated upward and formed flower structures.
Source: Shinn et al. (2010) by permission of the Elsevier.

anticline with a conjugate fault set was most likely associated with oblique- or strike-slip motion of the underlying reverse fault. The orientation of both synclinal and anticlinal structures is suggestive of northeast-southwest-directed contraction.

10.4.2 Evolutionary History

The Gunsan Basin formed by an extensional block faulting in the early Cretaceous with three subbasins and intervening highs (Figure 10.23A). The western subbasin formed as a result of the northeast-southwest extension. However, the central and eastern subbasins developed in *en echelon* patterns of the intrabasinal faults, which are oblique to the south-bounding fault. The fault patterns are largely compatible with those of the oblique extensional model of McClay (1995) and McClay and White (1995). These subbasins formed in response to regional northeast-southwest extension, accommodated by oblique- or strike-slip motion of the north-bounding fault.

In the Oligocene, the Cretaceous–Eocene succession underwent tectonic inversion, resulting from a contraction subparallel (or parallel) to the strike of the south-bounding fault (Figure 10.23B). It was accommodated by synclinal folds in the central and eastern subbasins as well as push-up of normal fault segments due to reverse reactivation. The north-bounding fault in the central subbasin was also inverted into left-lateral movement in response to the changes in regional stress regime. Positive flower structures and folds along the northern margin are

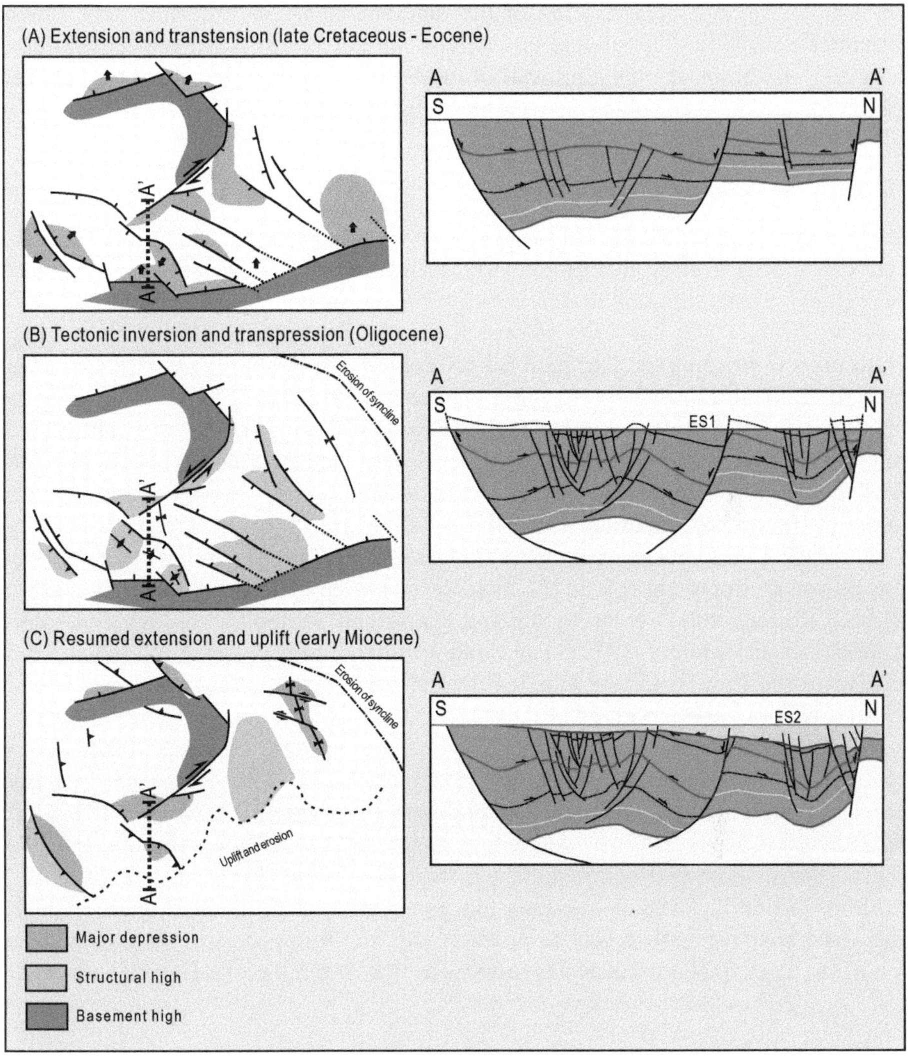

Figure 10.23 Evolution history of the Gunsan Basin. Simplified geologic cross-sections are reconstructed from interpretation of seismic section. (A) In the late Cretaceous to Eocene, the basin underwent oblique extension, indicated by oblique intrabasinal faults in *en echelon* arrays. (B) In the Oligocene, the basin was deformed by contraction, resulting in uplift and erosion of fault blocks and left-lateral transpression along the northern margin. (C) In the Early Miocene, the basin resumed extension and coeval uplift and erosion.
Source: Shinn et al. (2010) by permission of the Elsevier.

suggestive of transpressional oblique- or strike-slip reactivation of the preexisting transtensional (or normal) faults.

During the early Miocene, the basin resumed slow subsidence, accompanied with local uplift and erosion (Figure 10.23C). The northern margin of the central subbasin subsided by reactivation of the bounding fault, showing negative flower structures and widely distributed *en echelon* normal faults. However, the southern margin of the basin, the hanging wall of the Kachi fault, and the anticlines between the central and eastern subbasins were persistently uplifted.

10.4.3 Tectonic Implications

The Gunsan Basin was initiated in the Early Cretaceous and continued into the Late Cretaceous. During this time, the northeastern Asian continent was dominated by regional crustal thinning and large-scale mantle upwelling (Charvet et al., 1994; Lapierre et al., 1997; Li, 2000; Wu et al., 2005; Zhang et al., 1984). In the Late Cretaceous to Eocene, the Northern South Yellow Sea Basin progressively evolved through regional transtension, forming a large-scale rhombic geometry between the Qianliyan and the south-bounding faults (Figure 10.24A). The oblique intrabasinal faults in *en echelon* arrays are suggestive of transtension formed on an extensional overlap between the dextral strike-slip bounding faults. Dextral motion of the bounding faults was inherited from that of the Tan-Lu Fault, which most likely began in the Late Cretaceous. In the Paleogene, the Indian–Eurasian plate collision also played an important role in the tectonic evolution of East Asia. Far-field effect of the collision resulted in strike-slip and extensional faulting across northeast Asia (Tapponnier and Molnar, 1977). The rapid northward movement of the Indian Plate relative to the Eurasian Plate induced dextral transtension along the Tan-Lu Fault (Northrup et al., 1995; Ren et al., 2002). It suggests that dextral transtension persisted into the Eocene in part of the Tan-Lu Fault system. Pull-apart extension or transtension in the Korean Peninsula, however, ceased in the Paleogene. Gunsan Basin was then inverted by northeast-southwest contraction (Figure 10.24B). Since the Paleogene deposits are ill defined, it is not clear when the contraction began. The northeast-southwest contraction was triggered by an abrupt change from dextral transtension to sinistral transpression along the Tan-Lu Fault (Figure 10.24B). The sinistral transpression was associated with an abrupt increase in convergence rate of the Pacific–Eurasian plates relative to that of the Indian–Eurasian plates.

10.5 Eumseong Basin

10.5.1 Geologic Setting

The rhomb-shaped Eumseong Basin (about 7 km $\times$ 33 km in area) is a pull-apart basin bounded by two left-stepping sinistral master faults. It is in contact with cataclasite, microbreccia, and mylonite of the Precambrian gneiss in the basin margin (Figure 10.25) (Chun et al., 1994; Choi, 1996). Mylonitic foliation and minor fault orientations suggest that the major faults trend N13°E/82°W in the eastern part and N38°E/75°E in the western part of the basin (Cheong, 1987). The basin fill consists of seven lithofacies units: volcanics, conglomerate, conglomerate/purple mudstone, purple mudstone, green mudstone, greenish gray mudstone/gray sandstone, and

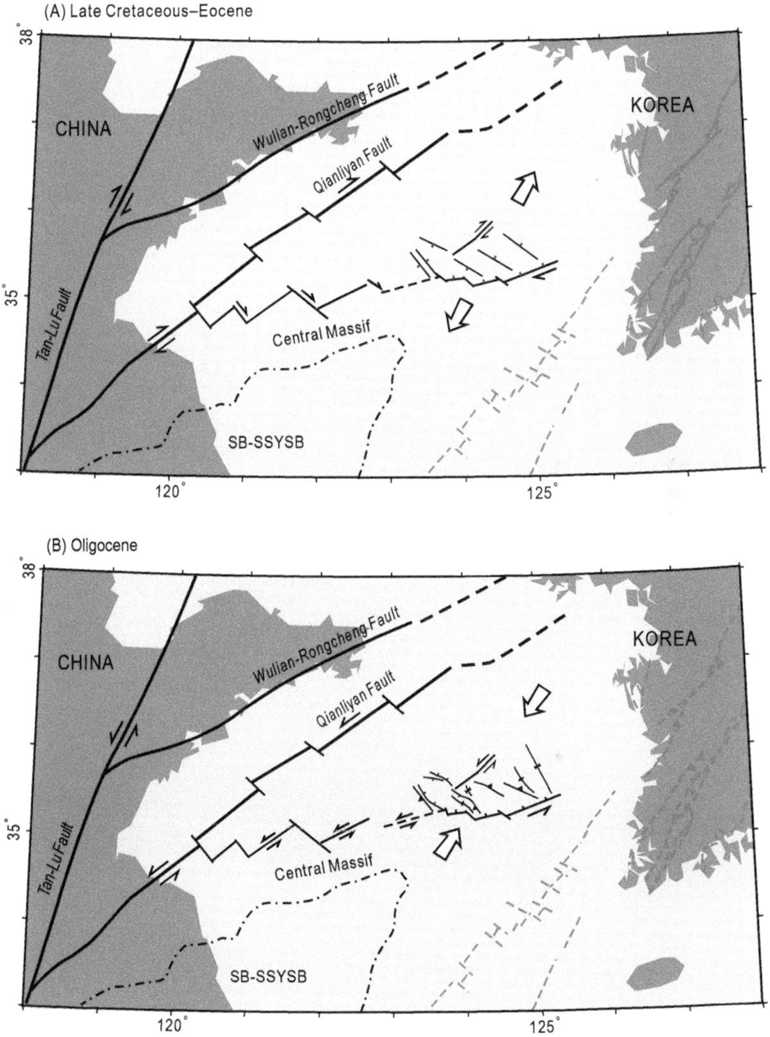

Figure 10.24 Evolution of regional stress regime and geologic structures. (A) Strike-slip opening of the central Yellow Sea due to regional transtension caused by dextral movement of the Tan-Lu Fault. Strike-slip faults in the Korean Peninsula were dormant in the Paleogene. (B) Contraction caused by abrupt change from dextral transtension to sinistral transpression of the Tan-Lu Fault.
Source: Shinn et al. (2010) by permission of the Elsevier.

dark gray mudstone (Table 10.3; Figure 10.26). Various fossil assemblages of plants (conifers and Ginkgoales), invertebrates, and microfossils (charophyta) are suggestive of temperate climate and lacustrine environments (Chun et al., 1994; Song et al., 1990). Charophyte in the green mudstone are of the Hauterivian–Aptian (Choi et al., 1995).

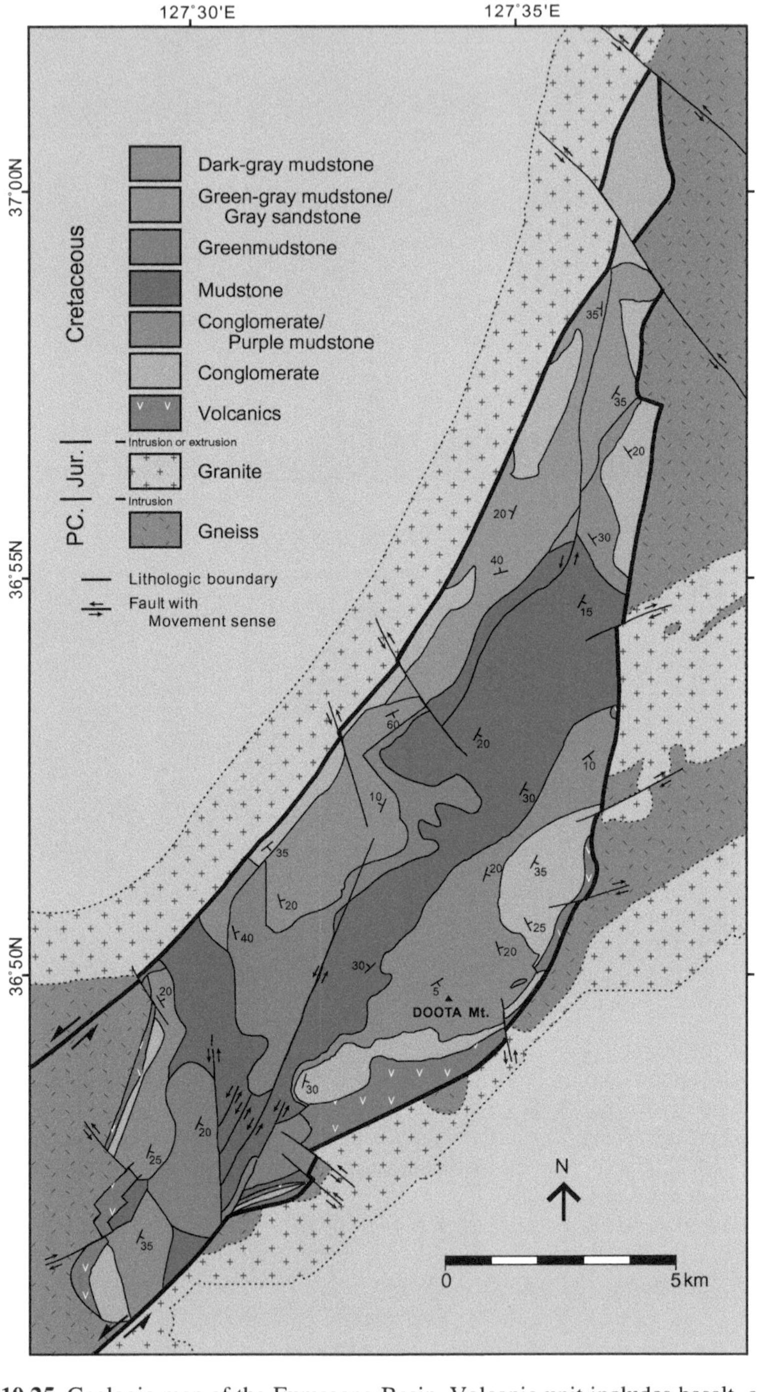

Figure 10.25 Geologic map of the Eumseong Basin. Volcanic unit includes basalt, andesite, and tuffaceous sediments.
Source: Ryang and Chough (1999) by permission of the Elsevier.

Table 10.3 Description of Sedimentary Facies and Inferred Depositional Processes of the Eumseong Basin

Facies Type	**Description**	**Interpretation**
Matrix-supported disorganized conglomerate (C1a)	Decimeter thick; poorly defined beds showing partly inverse or inverse-to normal grading; pebble- to cobble-grade and very angular to subrounded clasts; sand to mud matrix; nonerosional base; present mainly in the basin margin	Deposition from visco-plastic debris flows (Costa, 1988; Shultz, 1984)
Clast-supported disorganized conglomerate (C1b)	Discontinuous beds; decimeter thick; amalgamated with interbedded pebbly sandstone; partly matrix-rich; poorly sorted, very coarse-to-coarse sand matrix or occasionally purple (pebbly) siltstone matrix bearing pebble-size clasts; randomly oriented clasts; angular to subangular pebble-to cobble-size clasts	Deposition from high-concentration flow (Allen, 1981; Maizels, 1993; Morison and Hein, 1987)
Stratified conglomerate (C2)	Amalgamated; decimeter- to decameter-thick; indistinctly stratified or frequently low-angle cross-stratified; gravel patches with partly open-work fabric (up to 20 cm thick) are common; very coarse to coarse sand matrix; angular to subrounded, pebble- to cobble-size clasts; parallel-oriented elongate clasts [a(t)], occasionally imbricated [a(t), b(i)]	Bedload transport, traction of gravels and winnowed by stream flows (Harms et al., 1982)
Conglomerate encased in siltstone (CE)	Present in hollows of purple or green (pebbly) siltstone; generally thick (up to several meters); ribbon-shaped or sheetlike geometry; sharp, erosional base and relatively diffuse upper boundary; amalgamated with pebbly sandstone; disorganized, clast-supported in the lower part of	Water flood and subsequent stream flow; coarsegrained deposits—channel lag or entrapment in scoured hollows; fine grains, rapid settling from overflow (Harms et al., 1982; Morison and Hein, 1987; Siegenthaler and Huggenberger, 1993)

(Continued)

Table 10.3 (Continued)

Facies Type	Description	Interpretation
	hollows; crudely stratified or low-angle cross-stratified, matrix-rich in the upper part of hollows; very coarse to coarse sandstone matrix; occasionally inversely to normally graded; partly open-work fabric; pebble- to cobble-size clasts; parallel-oriented elongate clasts [a(t)]; partly imbricated [a(t),b(i)]	
Pebbly siltstone (PZ)	Variable in thickness (decimeters to several meters); poorly sorted; disorganized and partly stratified; randomly dispersed clasts; includes sandstone stringers, isolated pebbly sandstone lenticles; gravel patches and trains	Muddy debris flow; rapid fall-out deposition from high-concentration flow (Costa, 1988; Shultz, 1984)
Stratified pebbly sandstone encased in siltstone (PSE2)	Present in hollows of purple siltstone; sheetlike geometry; variable in thickness (up to 2 m); indistinctly stratified or cross-stratified pebbly sandstone; poorly to well sorted very coarse to medium sand matrix; conglomerate or pebbly sandstone layers commonly form discontinuous stringers or patches	Unconfined high-concentration flow and subsequent stream flow (Maizels, 1993)
Purple sandy siltstone (Zp)	Purple (5R 3/4); ubiquitous facies; generally thick (a few meters); poorly sorted; common inclusion of granules and sand grains; some calcareous nodules and sand-filled desiccation cracks	Rapid fall-out deposition of fine grains from tractive overflow on floodplain; oxidizing condition during and after deposition
Green sandy siltstone (Zg)	Green (10G 4/2); thin to very thick (decimeters to a few meters); common inclusion of dispersed granules and sand grains	Rapid settling of suspended grains below water level in pond or lake; reduction condition during and after deposition

(*Continued*)

Table 10.3 (Continued)

Facies Type	Description	Interpretation
Purple mudstone (Mp)	Purple (5R 3/4); major facies in the depocenter; generally thick (tens of meters); occasionally alternated with green mudstone facies; homogeneous; some calcareous nodules	Gradual settling of fine grains on lacustrine margin, controlled by fluctuating lake level; oxidizing condition during and after deposition
Green mudstone (Mg)	Green (10G 4/2); major facies in the depocentre; generally thick (tens of meters); occasionally alternated with purple and dark-gray mudstone facies; homogeneous; some calcareous nodules; some pyrite crystals	Gradual settling of fine grains below water level in pond or lake; reduction condition during and after deposition

Source: Ryang and Chough (1999) by permission of the Elsevier.

10.5.2 Sedimentary Facies

The basin is bound on the south by irregular marginal boundary and volcaniclastic sediments. The sinistral angular relationship of faulting suggests that the southeastern part of the basin was under compression, whereas the southwestern part was fractured in extensional regime (Choi, 1996). A high-resolution magnetotelluric survey suggests that the basin experienced pull-apart opening with rapid subsidence (Ryang et al., 1999). Coarse-grained alluvial systems in the northeastern part converge into an axial channel system in the northwestern part, whereas the axial channel system on the west prograded southward, interlayered with dark gray mudstone of lacustrine origin (Figure 10.27). Along the eastern margin, sediments were transported northward along the fault, forming an elongate thick wedge.

In the southeastern part of the basin, the conglomerate deposits can be organized into five facies associations, representing distinctive depositional environments: facies association I (debris-flow-dominated alluvial fan), facies association II (stream-flow-dominated alluvial fan), facies association III (alluvial fan fringe), facies association IV (alluvial plain), and facies association V (floodplain/lake). The entire succession can be divided into two stratigraphic units: the lower Dootasan unit and the overlying Berjae unit, based on the facies associations and the architecture of channel fills and floodplain deposits (Ryang and Chough, 1997). The Dootasan unit is characterized by a basinward change from matrix-supported conglomerate to ribbon-shaped channel fills and sheet-like beds encased in purple siltstone with a thinning-upward trend. The matrix-supported conglomerate facies is indicative of deposition by debris flows, whereas the hollow-fill and sheet-like deposits encased in purple siltstone are suggestive of channel or sheetflood deposition on an alluvial plain. The Berjae unit is transitional from superposed channel

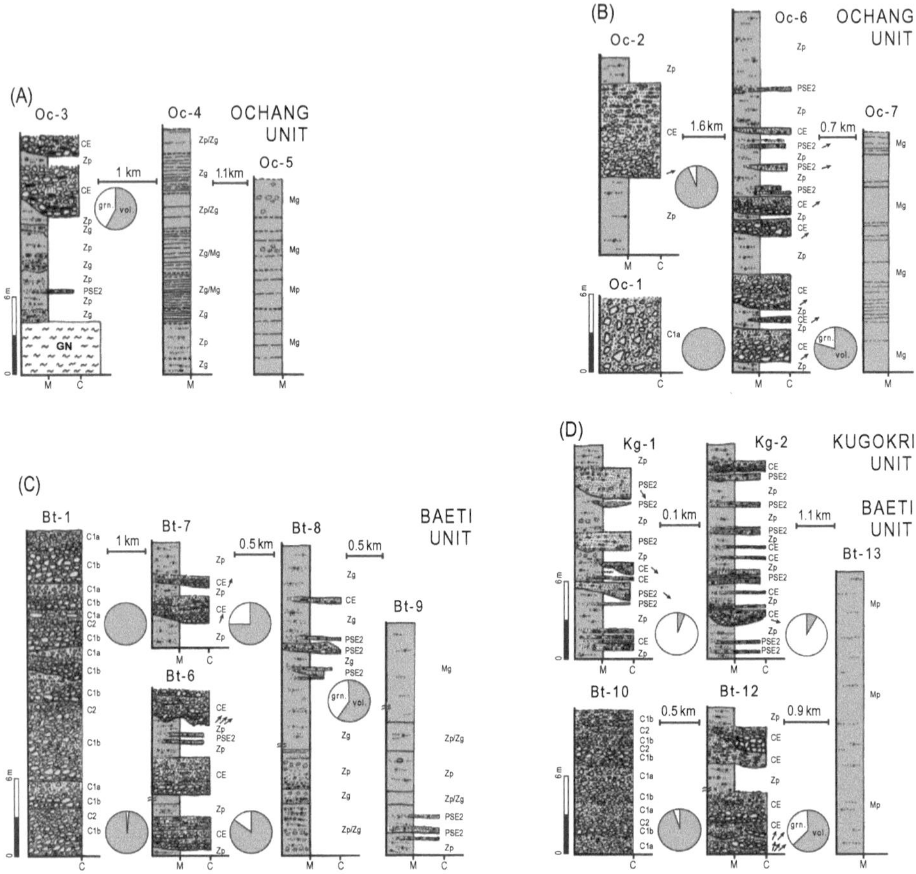

Figure 10.26 Measured stratigraphic sections arranged in downstream directions of paleoflow, representing alluvial-to-lacustrine systems from the basin margin to the depocenter (vol., volcaniclastics which includes basalt, andesite, and rhyolite; grn., granite and gneiss). For facies codes, see Table 10.3. Note that the clast compositions are relatively homogeneous at the basin margin whereas they become heterogeneous toward the depocenter. Distance between two adjacent sections is indicated. Arrows denote paleoflow directions and north is toward the top of the figure. (A) Ochang unit in the southernmost part of the basin. Note the lateral facies transition from Section Oc-3 to Sections Oc-4, 5 and the contact between the overlying green mudstone bed. (B) Lateral facies transition from Sections Oc-1, 2 to Sections Oc-6,7. (C) Baeti unit: lateral facies transition from Sections Bt-1, 6, 7 to Sections Bt-8, 9. (D) Baeti and Kugokri units: lateral facies transition from Section Bt-10 to Section Bt-12, 13 and from Section Kg-1 to Sections Kg-2, Bt-13. Note abrupt change in clast composition from dominant volcaniclastics in the Baeti unit to granite clasts in the Kugokri unit. (For interpretation of the references to color in this figure legend, the reader is referred to the online version of this book.)
Source: Ryang and Chough (1999) by permission of the Elsevier.

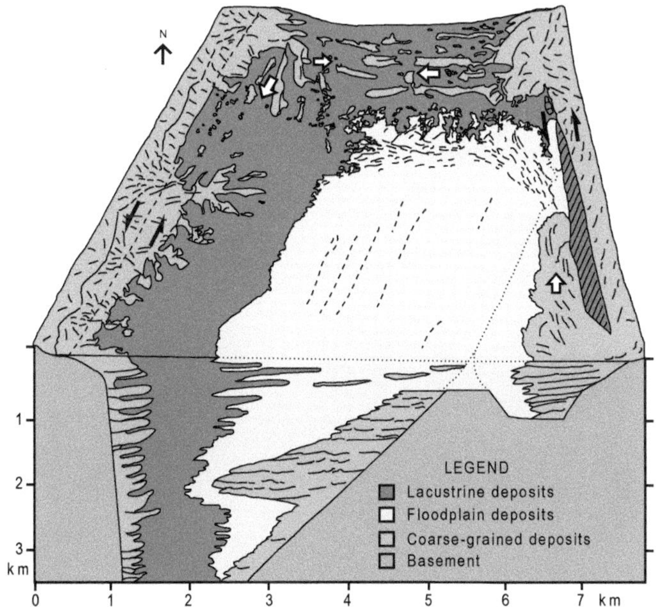

Figure 10.27 Reconstructed basinal form and filling model for Eumseong Basin. Coarse-grained alluvial systems in the northeast converge into an axial channel system in the northwestern part of the basin. The axial channel system on the west progressed southward, interlayered with dark-gray mudstone beds of lacustrine origin. Solid arrows indicate transtensional faulting with left-slip motion; open arrows show dominant directions of sediment transport.
Source: Ryang et al. (1999) by permission of the Elsevier.

fills/sheetflood deposits to conglomerates encased in siltstone. The stacked deposits are indicative of deposition in stream-flow-dominated alluvial fans. The Dootasan unit formed during a sinistral-fault movement with a relatively high rate of fault movement, whereas the Berjae unit formed in the late stage with a relatively low rate of fault slip (Ryang and Chough, 1997).

On the pull-apart margin in the southwestern part, the deposits can be divided into seven lithologic units: volcanics (andesite, basalt, and tuffaceous succession), conglomerate, conglomerate/purple siltstone, sandstone/green siltstone, purple mudstone, purple/green mudstone, and green mudstone. Clasts of the conglomerates are composed of granite, granite gneiss, and banded gneiss derived from the basement and calc-alkaline volcanic rocks along the basin margin. The conglomerate facies are transitional to pebbly sandstone, pebbly siltstone, and mudstone facies toward the basin center, showing a lateral facies change (Figure 10.26). Ripple marks and calcareous nodules are present in the dark gray mudstone and desiccation cracks. Calcareous nodules occur in the purple sandy siltstone. Based on the basinal structure and changes in clast composition and lithofacies, three alluvial–lacustrine depositional systems are recognized (Figure 10.28). Each system shows a

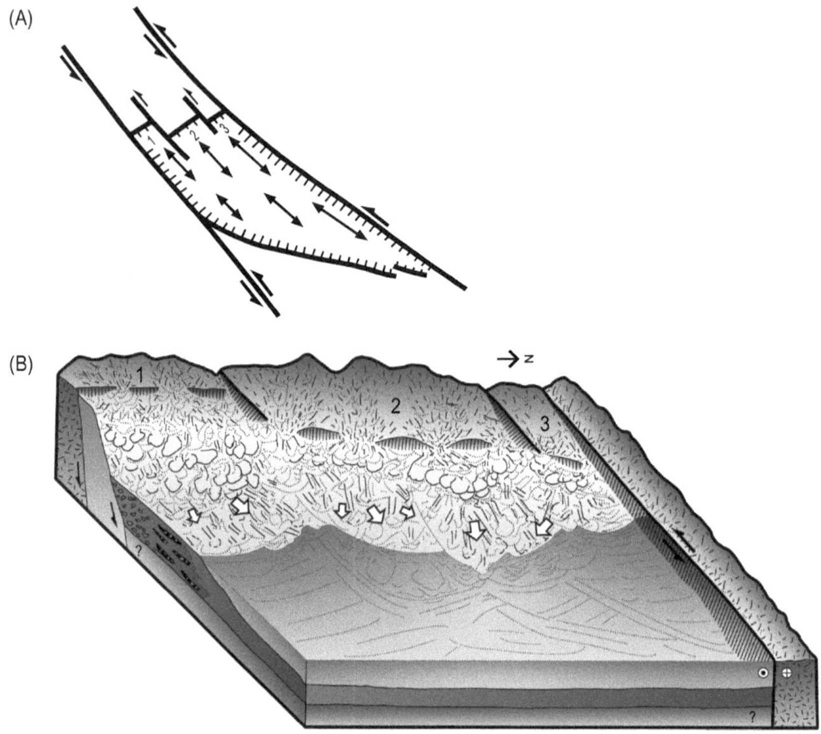

Figure 10.28 (A) Plan view of basin-margin development in the southwestern part of the Eumseong Basin. Note three basinal slivers developed in the pull-apart margin. (B) Basin-filling model for the development of sedimentary facies and stratigraphic architecture in the southwestern part. Note three alluvial-to-lacustrine systems (1, Ochang; 2, Baeti; 3, Kugokri systems) responding to each basinal sliver in the pull-apart margin. Open arrows show dominant directions of sediment transport. Solid arrows indicate transtensional faulting with left-slip motion.
Source: Ryang and Chough (1999) by permission of the Elsevier.

downstream change in clast size and sedimentary facies, characterized by relatively synchronous development of the alluvial–lacustrine system. It reflects the lithology of the adjacent source rocks and subsurface structure of normal-slip-dominant faults.

10.6 Haenam Basin

10.6.1 Introduction

The Haenam Basin was filled with a succession of andesitic tuff, Uhangri Formation, acidic tuff, and rhyolite in ascending order (Figure 10.29). The Uhangri Formation consists of interlayered conglomerate, gravelly sandstone, sandstone, siltstone, black shale, and chert (Table 10.4) (Figure 10.30) (Chun and Chough, 1995).

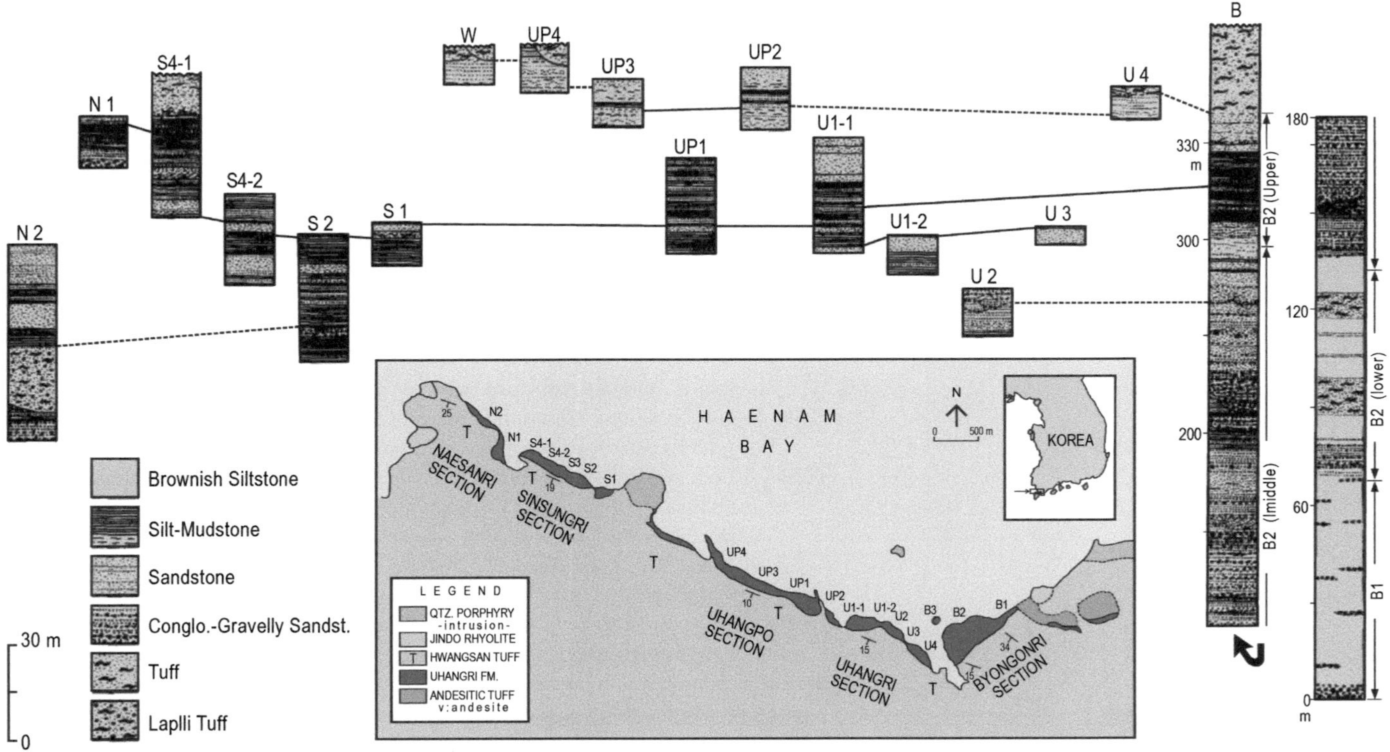

Figure 10.29 Simplified geologic map of the study area and locations and relative stratigraphic positions of measured sections of the Uhangri Formation. Site of the Haenam Basin is marked by an arrow in the inset. For facies codes, see Table 10.4.
Source: Chough et al. (1996) by permission of the Elsevier.

Table 10.4 Description, Dominant Transition, and Interpretation of Individual Sedimentary Facies in the Uhangri Formation

Facies (Code)	Description	Dominant Transition (%)	Interpretation
Disorganized conglomerate–gravelly sandstone (G1)	Very thin to thick (2–75 cm); matrix-supported (partly clast-supported); subangular to subrounded granule- to pebble-size clasts commonly including outsized gravels; poorly sorted medium to coarse sand matrix; erosive lower boundary. Thick to very thick (up to 4 m); mud-matrix rich; clast-supported (partly matrix supported); wedged and wavy or irregular geometry; brownish mud or volcanic ash matrix; nonerosive base; common in lowermost part of formation	G1→S1 (56) G1→4 (12)	Cohesionless debris flow or high-concentration turbidity current (Hampton, 1970, 1972, 1975; Hein, 1982; Lowe, 1976, 1982; Pickering et al., 1986); subaerial cohesive debris flow (Pierson, 1981; Shultz, 1984)
Normally graded conglomerated –gravelly sandstone (G2)	Variable in thickness (3–140 cm); either matrix- or clast-supported; angular to subangular; poorly sorted sand matrix, but well sorted in lower part (facies association I); conglomerate generally wedged and scoured; gravelly sandstone laterally continuous	G2→S4 (1 7) G2→S1 (17) G2→M1 (13)	Rapid settling in high-concentration turbidity current (Gloppen and Steel, 1981; Hein, 1982; Kleinspehn et al., 1984; Marzo and Anadon, 1988; Pickering et al., 1986)
Inversely (to normally) graded conglomerate–gravelly sandstone (G3)	Very thin to thick (3–80 cm); poorly to moderately sorted; either clast- or matrix-supported (partly clast-supported); subangular pebbles rarely including cobble to boulder clasts; inversely graded unit laterally continuous (without scoured base), but inverse-to-normally graded unit generally wedged; some thick units deeply scour underlying finer-grained beds	G3→S1 (42) G3→S4 (21)	Frictional freezing of traction carpets in high-concentration turbidity current (Lowe, 1982; Massari, 1984; Postma et al., 1988)

(*Continued*)

Table 10.4 (Continued)

Facies (Code)	Description	Dominant Transition (%)	Interpretation
Stratified conglomerate–gravelly sandstone (G4)	Thin to very thick (4–180 cm); poorly to moderately sorted; matrix-supported (partly clast-supported); subangular to subrounded pebble clasts (rare outsized clasts); sometimes normally or inverse-to-normally graded; rip-up clasts (parallel to bedding plane) in lowermost part of each unit; sharp lower boundary without scour	G4→S4 (50) G4→M1 (12)	Progressive freezing of traction carpet in high-concentration turbidity current (Pickering et al., 1986; Chun and Chough, 1992)
Cross-stratified conglomerate–gravelly sandstone (G5)	Planar cross-stratified unit (5–205 cm thick); clast-supported; subangular to subrounded clasts; large clasts concentrated in lower part (fining-upward); wedge shape and erosive lower boundary. Trough cross-stratified unit (27–347 cm thick); clast-supported; moderate to well sorted; angular to subangular clasts; scoured base; concave-up lower and flat upper boundaries	Not evident	Migration of gravel bars or channel fills (Miall, 1977, 1978; Middleton and Trujillo, 1984; Rust, 1978, 1979)
Massive sandstone (Sl)	Variable in thickness (< 0.5 cm, >l0 m), but mostly very thin; dominantly very fine to medium sands in grain size; poorly to moderately sorted; abundant feldspar-rich volcanic ash particles; very thin, discontinuous siltstone layers commonly interlayered (amalgamated); common shale chips in thick units; ubiquitous in all parts of Uhangri Formation (most abundant facies: c. 20%)	Not evident	Rapid suspension sedimentation of density and turbidity currents; liquefied, grain or cohesionless debris flows (Middleton and Hampton, 1976)
Normally graded sandstone (S2)	Variable in thickness (1–210 cm), but generally thin to medium; generally medium to coarse sands in grain	S2→61 (18) S Z→P S(∼15)	High- or low-concentration turbidity current (Hein, 1982;

(Continued)

Table 10.4 (Continued)

Facies (Code)	Description	Dominant Transition (%)	Interpretation
	size; distribution normal grading; moderately to well sorted; commonly alternating with massive sandstone (S2/S1)	S2→S/MI (18)	Lowe, 1982; Middleton and Hampton, 1973, 1976)
Inversely graded sandstone (53)	Thin to medium thick (5–18 cm); generally medium-to-coarse sands; poorly to moderately sorted; slightly fining in topmost part of each unit (inverse-to-normally graded); common repetitive occurrence; rip-up clasts; generally common in upper part of formation	Not evident	Traction carpet of high-concentration turbidity current (Lowe, 1982)
Stratified sandstone (S4)	Very thin to thick (1–83 cm), but mostly thin (2–11 cm); medium to coarse sands; moderately to well sorted; some discontinuous mud layers interlayered; some associated and interlayered with ripple cross-laminated sandstone units; generally continuous; common scour surfaces and rip-up clasts; each laminae laterally continuous up to a few decameters	S4→S1 (13) S4→S/M1 (18) S4→M2 (11)	Migration of low-amplitude bed forms or by low-concentration turbidity current (upperflow-regime condition) (Allen, 1964; Bridge, 1981; Middleton, 1966, 1967)
Cross-stratified (laminated) sandstone (S5)	Planar cross-stratified unit (5–96 cm thick): fine to coarse sands; well-sorted; angular or tangential basal contact and generally fining-upward; including granules and some pebbles; discontinuous and rapidly thinned out; flat lower and upper boundaries. Ripple cross-laminated unit (1–22 cm); fine to medium sands; well sorted; two styles of ripples	Not evident	Avalanching by traction-dominant flows and/or migration of small bedforms (Allen, 1968; Cant and Walker, 1976; Costello and Southard, 1981; Feary, 1984; Miall, 1977; Reineck and

(Continued)

Table 10.4 (Continued)

Facies (Code)	Description	Dominant Transition (%)	Interpretation
	(symmetrical and asymmetrical forms); well associated with stratified sandstone (54)		Singh, 1980; Rust, 1978; Southard, 1971)
Sandstone/mudstone couplets (S/M1)	Alternation of thin sandstone and mudstone; variable in thickness (0.5–485 cm), but each couplet generally less than 1 cm; sandstone laminae laterally continuous, but mudstone laminae either lenticular, scoured or slightly deformed; fining- and thinning-upward	S4→S/M1 (18) G4→SIM1 (12)	Deposition by fluctuations of shear stress and average velocity in boundary layers of low-concentration turbidity current (Kuenen and Menard, 1952; Lombard, 1963; Pickering et al., 1986; Sanders, 1965; Walker, 1965)
Sandstone/cherty mudstone couplets (S/M2)	Alternation of sandstone and cherty mudstone; medium to thick-bedded (10–55 cm); commonly convoluted; sandstone layers massive or crudely laminated; cherty mudstone layers homogeneous or slightly normally graded with transitional lower and sharp upper boundaries	Not evident	Chemical fluctuation (pH and silica concentration) of lacustrine water caused by episodic density under- and interflows and by volcanic activity
Homogeneous mudstone (M1)	Three types: (i) brownish mudstone (9–474 cm) commonly including caliche nodules, dominant in lower part of formation and associated with massive sandstone (S1); (ii) greenish mudstone (2–30 cm thick) mainly composed of volcanic ash, common in middle and upper parts of formation and associated with gravelly sandstone; and (iii) dark gray mudstone (1–18 cm thick), dominant in upper part and associated with black shale (M2)	Not evident	Suspension sedimentation; rapid mass deposition from high-concentration mud-dominated turbidity current (Pickering et al., 1986)

(Continued)

Table 10.4 (Continued)

Facies (Code)	Description	Dominant Transition (%)	Interpretation
Laminated mudstone (M2)	Alternation of silt-rich laminae and mud-rich dark gray laminae (2 cm→10 m thick); upwardly fines and thins within one facies unit; some lenticular (streaky) siltstone and fine sandstone layers intercalated; calareous nodules, syneresis cracks, pisolitic caliche nodules and pyrite framboids included; ostracods of brackish or freshwater origin concentrated in some layers; prevail in upper part of formation	S4→M2 (11)	Suspension sedimentation alternately affected by pulses of riverine inflows; some by velocity fluctuations in turbidity current (Hesse and Chough, 1980; Lambert et al., 1976; Normark and Dickson, 1976; Sturm and Matter, 1978)
Cherty mudstone/dark grey mudstone couplets (M3)	Alternation of dark grey mudstone and cherty mudstone laminae; thin-bedded (4–9 cm); individual laminae 0.1 mm to 1 cm in thickness; cherty mudstone laminae progressively thinning upward and commonly wavy and deformed	M3→M2 (100)	Chemical fluctuation of lake water by episodic riverine inputs, resulting in alternating precipitation under high and low pH conditions
Gravelly mudstone (GM1)	Medium to very thick-bedded (10–130 cm); matrix supported; wedged and/or irregular geometries; slightly erosive lower boundary, but some deeply scoured base; either massive, normally graded or inverse-to-normally graded	Not evident	Cohesive debris flow deposits (Hampton, 1973; Johnson, 1970; Middleton and Hampton, 1979; Naylor, 1980)

Source: Chun and Chough (1995) by permission of the Wiley-Blackwell.

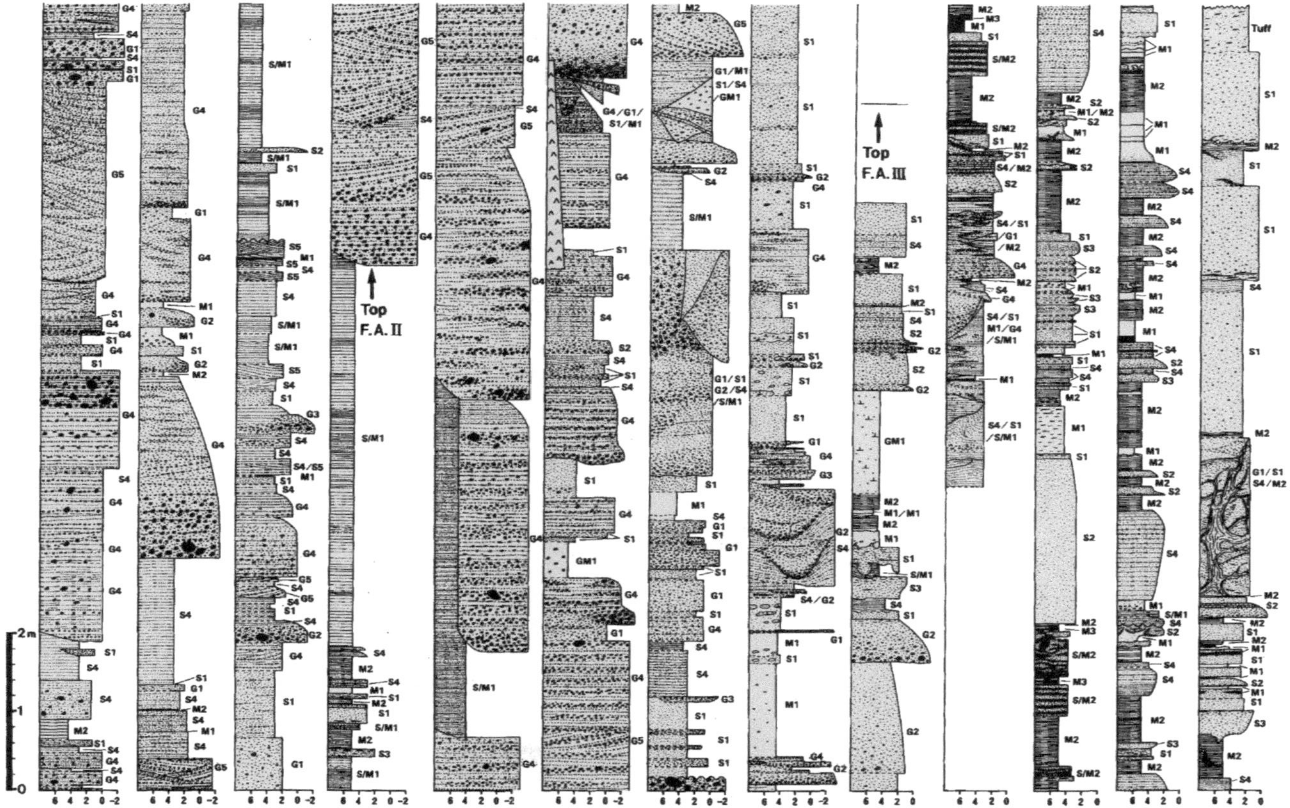

Figure 10.30 Columnar sections of the Byongonri section, representative of the Uhangri Formation. The columns are arranged in ascending order from left to right and are labeled with facies symbols classified as 16 distinct sedmentary facies (Table 10.4) based on lithology and primary sedimentary structures. The lowermost part (ca. 68 m in thickness) of this section, which has a similar pattern to the overlying succession, is omitted.
Source: Chun (1990).

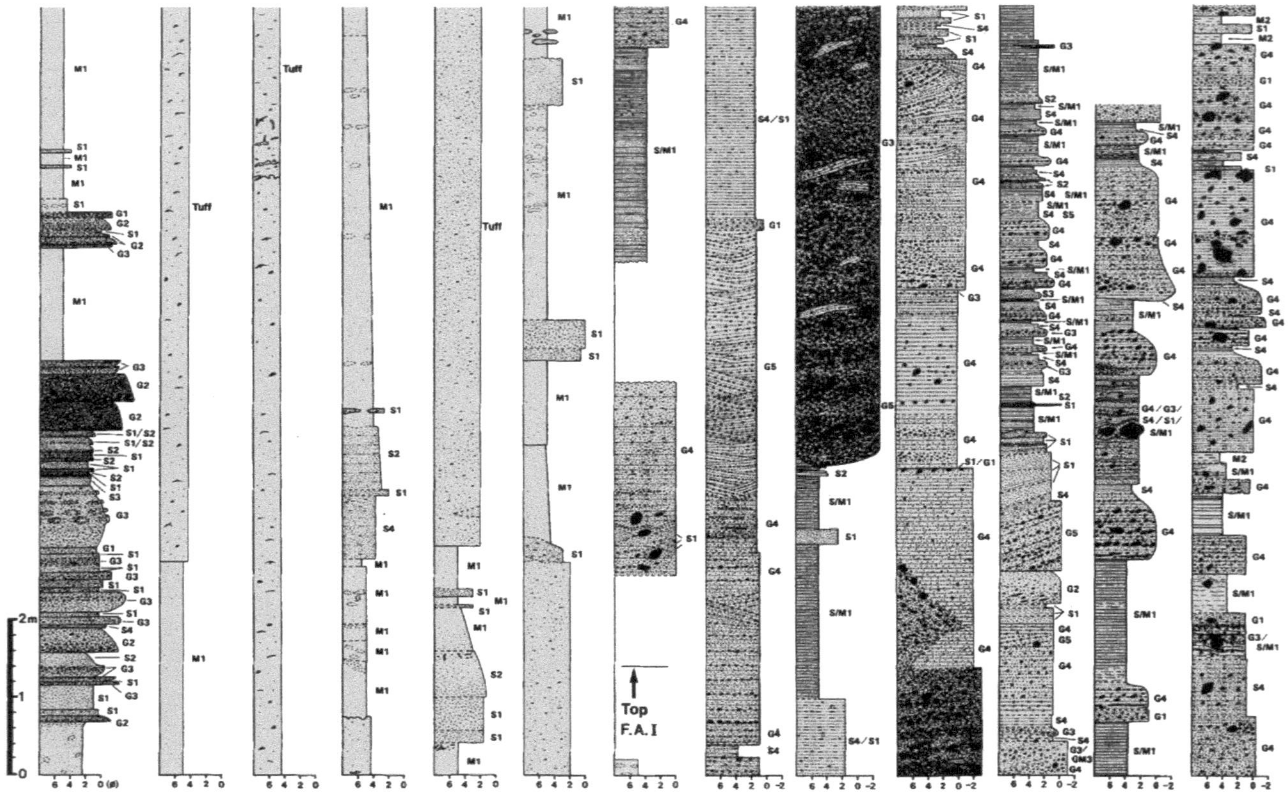

Figure 10.30 (Continued)

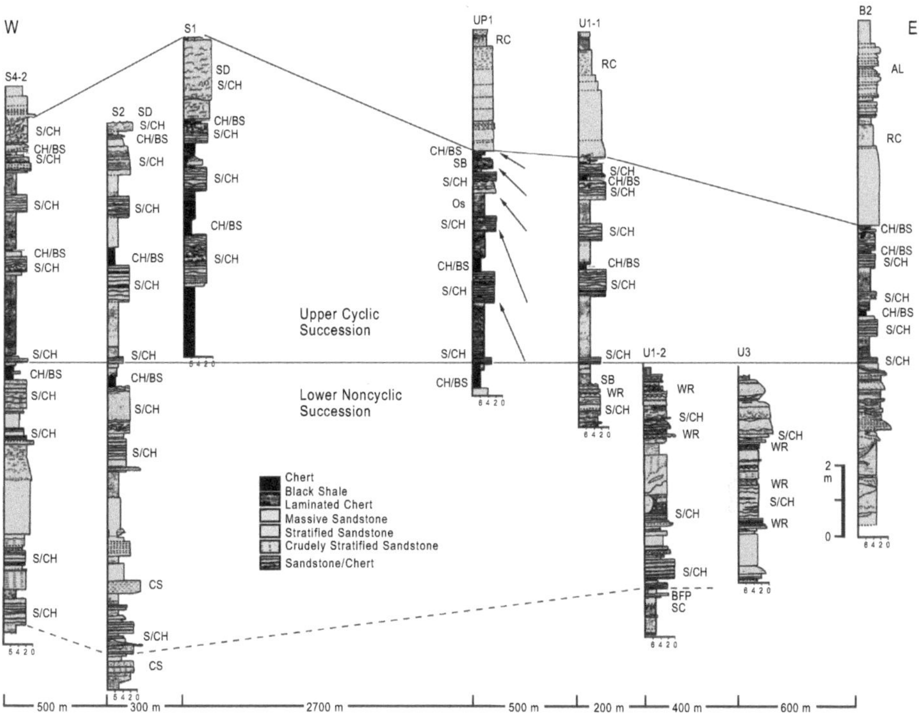

Figure 10.31 Correlation of individual layers of sandstone/chert, chert/black shale, and others, Uhangri Formation.
Source: Chough et al. (1996) by permission of the Elsevier.

In five outcrop sections, bed-by-bed correlation was made by using sequential orders and key beds, such as siltstone layers with accretionary lapilli or rip-down clasts, and sandstone/chert and laminated chert/black shale couplets (Chun, 1990) (Figure 10.29).

10.6.2 Sedimentary Facies

The lower part (ca. 130 m thick) of the succession consists mainly of brown siltstone, wedge-shaped disorganized conglomerates, and thin gravelly sandstones, representing floodplain environments, influenced by debris flows and sheetfloods (Figure 10.30). The lower-middle part (ca. 110 m thick) comprises fining-upward cycles of stratified and graded conglomerate–gravelly sandstones, laminated sandstones, and sandstone/siltstone couplets, formed in subaqueous lobes. The upper-middle part (ca. 60 m thick) consists of wedge-shaped conglomerates and gravelly sandstones bounded by scoured bases, which formed in distributary channels and mouth bars of delta front. The upper part (ca. 35–75 m thick) consists of a laterally

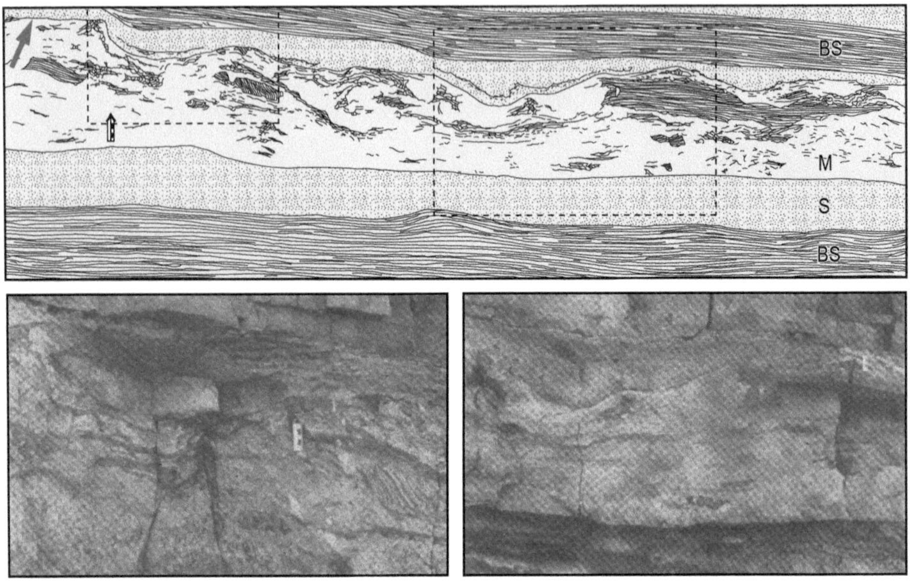

Figure 10.32 Line drawing of Uhangri section outcrop with photographs (insets) showing shale clasts (rip-down clasts) in mudstone. The shale layer from which the clasts originated is now sheared and completely consumed after the ripping process; it should have been in the position of the large arrow. Note extensive shearing of laminated shale clasts.
S, sandstone; M, mudstone; BS, black shale. Scale in centimeter.
Source: Chough and Chun (1988) by permission of the SEPM.

continuous succession of finely laminated mudstones, crudely stratified sandstones, and convoluted sandstone/chert and laminated chert/black shale couplets (Figure 10.30). It contains symmetrical ripple marks, tracks of shorebirds, syneresis cracks, plant fossils, and brackish or freshwater ostracodes, which collectively indicate deposition in shallow lacustrine environments.

The deposits formed in alkaline lake environments surrounded by volcanic terrain. The upper part of the formation comprises sandstone/chert and laminated chert/black shale couplets (Figure 10.31). The chert is inorganic in origin, which is composed of silica (74–81%), alumina (10–12%), and potash (6–7%). The impurity is due to the presence of feldspars and clays in microcrystalline quartz matrix. In sandstone/chert alternations, the sandstone layer is massive, graded, stratified, rippled, or ripple-cross-laminated, whereas the chert layer is homogeneous. However, the laminated chert/black shale couplets consist of alternating dark- and light-colored laminae of chert and the overlying black shale (Figure 10.31). These couplets formed by episodic riverine inflows into a silica-rich alkaline lake (Chough et al., 1996). The freshwater input triggered inorganic precipitation of silica through an abrupt decrease in lake-water pH. The sandstone/chert couplets formed by underflows, whereas the laminated chert/black shale couplets by

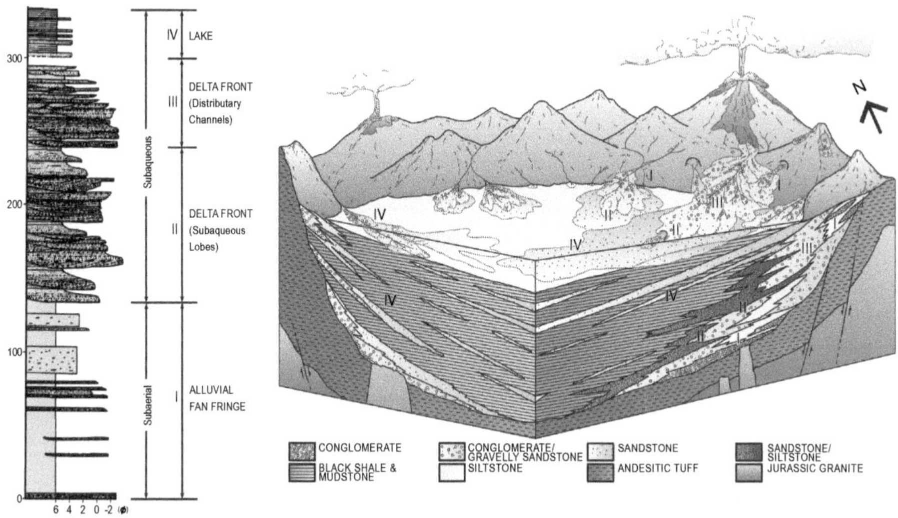

Figure 10.33 Summarized sequential pattern and depositional model of the Uhangri Formation. The broad channelized lobes probably formed on the northern margin, radiating generally southward. In the western part, the lacustrine margin was relatively deep and steep with sedimentation mainly controlled by vertical settling and episodic mass flows. *Source*: Chun and Chough (1995) by permission of the Wiley-Blackwell.

overflows. In some tuffaceous mudstone, cherty mudstone, fine sandstone beds, and black-shale clasts of various size and shape occur (Figure 10.32). The rip-down clasts originated from the overlying bed by liquefied intrastratal flow of tuffaceous and cherty muds and fine sand during penecontemporaneous deformation (Chough and Chun, 1988).

The Uhangri succession generally shows a fining-upward trend with a transition from an alluvial-fan fringe, coarse-grained subaqueous delta, to a shallow lake. The retrogradation was due to continuous subsidence, caused by movement of oblique-slip faults. It also shows an abrupt change from subaerial to subaqueous sedimentary facies, which is indicative of transgression of lake water. These features suggest that the basin formed in a waning stage of fluvial discharge during the period of volcanic interruption (Figure 10.33).

10.7 Jinan Basin

10.7.1 Geological Setting

The Jinan Basin contains a nonmarine succession of conglomerate, gravelly sandstone, sandstone, and dark gray mudstone, more than 10 km thick (Figure 10.34). Sixteen sedimentary facies have been identified in the northeastern part of the basin

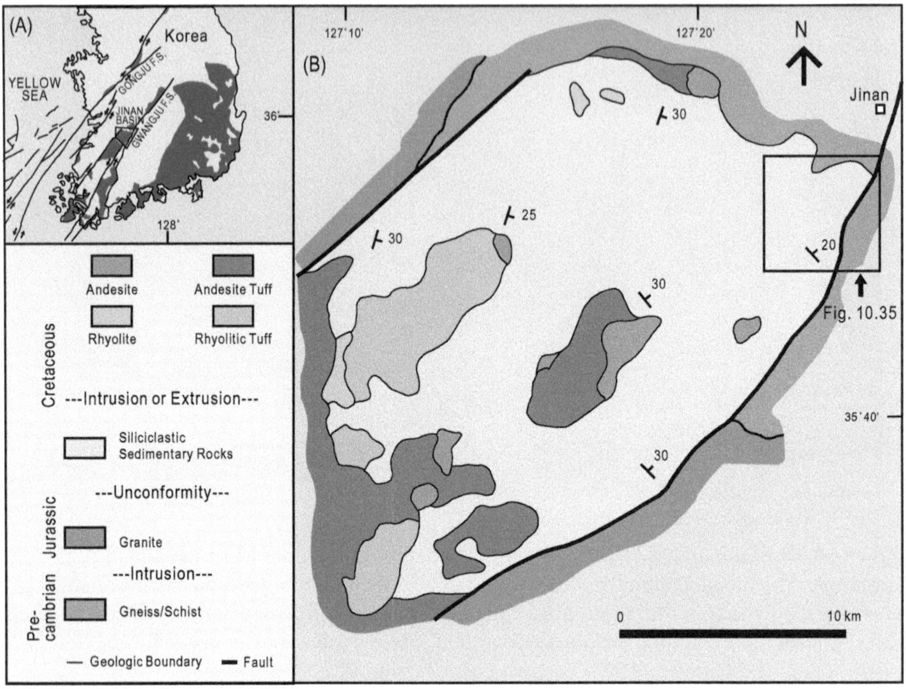

Figure 10.34 (A) Cretaceous nonmarine basins and major faults of the southern Korean Peninsula. Arrows show directions of fault movement. (B) Geologic map of the northern part of Jinan Basin.
Source: Lee and Chough (1999) by permission of the Elsevier.

(Table 10.5) (Figure 10.35). They can be grouped into five facies associations, which represent the depositional environments of alluvial fan, steep delta slope, base of delta slope and prodelta, and lacustrine plain (Table 10.4) (Lee and Chough, 1999). Based on the facies associations, depositional architecture, and paleocurrent directions, the sedimentary successions can be divided into a marginal fill (about 3.2 km thick) and a longitudinal fill (about 2 km thick).

10.7.2 Sedimentary Facies

The marginal fill developed northward from the active strike-slip margin in the southeastern part. The lower part consists of an alluvial fan and Gilbert-type delta, onlapped by lacustrine plain deposits, successively. The overlying upper part is dominated by large-scale, steep delta and prodelta successions. The successive alluvial fans, small-scale Gilbert-type deltas, and large-scale, steep delta and prodelta deposits are overlapped northeastward. The shingled geometry comprises sinistral strike-slip displacement along the basin-bounding fault. The progressive changes in

Table 10.5 Description and Interpretation of Individual Sedimentary Facies in the Jinan Basin

Facies	Description	Occurrence	Interpretation
Matrix-supported, disorganized conglomerate (Facies Gmd)	Matrix-supported, (very) poorly sorted, pebble- to boulder-grade clasts (up to 1.5 m in size); (very) poorly sorted, sand to small pebble matrix; randomly oriented clast fabric; some bouldery protruding clasts; some basal inverse grading; sharp, nonerosional base; tabular geometry; purple sand matrix in some units; greater than 70 cm in thickness	M-1, M-5, M-10, L-1, L-5	Debris flows (Costa, 1988; Nemec and Steel, 1984; Shultz, 1984)
Clast-supported, disorganized conglomerate (Facies Gcd)	Clast-supported, (very) poorly sorted, pebble- to boulder-grade clasts (up to 3 m in size); (very) poorly sorted, sand to small pebble matrix; either randomly oriented or imbricated clast fabric; common bouldery uptruding clasts; some basal inverse grading; either nonerosional or erosional base with a relief of a few decimeters; irregular, undulatory upper boundary in some units; either wedge-shaped with blunt margin or channelized; purple sand matrix in some units; variable in thickness ranging from 30 to 150 cm	M-1, M-2, M-4, M-5, M-6, M-8, M-9, M-10, L-2, L-3, L-4, L-5, L-6	Cohesionless debris flows (Nemec, 1990; Nemec and Steel, 1984; Shultz, 1984) or hyper-concentrated/ high-magnitude flood flows (Allen, 1981; Jo et al., 1997; Smith, 1986)
Clast-supported, clast parallel-oriented conglomerate (Facies Gcp-1)	Clast-supported, boulder- to cobble-grade clasts (up to 2–2.5 m in size) with rare matrix; parallel-oriented clast fabric to bedding plane; diffuse lower and upper boundaries; wedge-shaped with some blunt margins; 50–150 cm in thickness	M-8	Grain flows or debris falls (Hwang and Chough, 1990; Nemec, 1990; Sohn et al., 1997)
	Matrix-rich, (very) poorly sorted, pebble- to cobble-grade clasts (5–20 cm in size) with rare boulder-	M-10, L-4, L-5, L-6	Cohesionless debris flows or density-modified grain flows (Fisher, 1971;

(*Continued*)

Table 10.5 (Continued)

Facies	Description	Occurrence	Interpretation
Matrix-rich, clast parallel-oriented conglomerate (Facies Gcp-2)	grade clasts; coarse sand to granule matrix; parallel-oriented clast fabric; diffuse lower and upper boundaries; laterally amalgamated; mostly less than 40 to 50 cm thick		Hubert and Filipov, 1989; Lowe, 1982; Nemec and Steel, 1984)
Inversely graded conglomerate (Facies Gi)	Poorly to very poorly sorted, clast-supported, pebble- to boulder-grade clasts; poorly sorted, medium sand to small pebble matrix; common bouldery protruding clasts; some upward increases in clast size and matrix content; either randomly or parallel oriented clast fabric; irregular, undulatory upper boundary; planar-convex geometry; 20–70 cm in thickness	M-6, M-7, M-8, M-9, M-10, L-4, L-5	Cohesionless debris flows (Middleton and Hampton, 1976; Nemec and Steel, 1984; Shultz, 1984)
Normally graded conglomerate (Facies Gn)	Poorly sorted, pebble- to cobble-grade clasts; medium sand to granule matrix; normally graded; clasts either imbricated or parallel-oriented to bedding plane; sharp, erosional base ranging from 5 to 70 cm; purple sand matrix in some units; variable in thickness ranging from 15 to 100 cm	M-2, M-4, M-5, M-6, M-7, M-9, M-10, L-3, L-5	High-concentration turbidity currents (Chun and Chough, 1992b; Lowe, 1982; Postma and Roep, 1985;) or hyper-concentrated/high-magnitude flood flows (Allen, 1981; Wells and Harvey, 1987)
Crudely stratified conglomerate (Facies Gcs)	Alternation of diffuse, gravel- and sand-rich layers; gravel-rich layer: clast-supported, pebble- to cobble-grade clasts with medium sand to granule matrix, 5–20 cm in thickness, either randomly oriented or parallel-oriented clast fabric, some inversely graded; sand-rich layer: laterally discontinuous, less than 10 cm thick; variable in thickness ranging from 30 to 70 cm	M-2, M-7, M-9, M-10, L-3, L-6	Cohesionless debris flows, density-modified grain flows or high-concentration turbidity currents (Lowe, 1979; Middleton and Hampton, 1976; Nemec, 1990; Postma, 1986)

(Continued)

Table 10.5 (Continued)

Facies	Description	Occurrence	Interpretation
Openwork gravel lens (Facies Gol)	Moderately to well sorted, openwork, cobble to boulder-grade clasts; single-grain to a few clasts thick layer; (broad) lenticular geometry; clasts either randomly oriented or imbricated; terminated abruptly in flow-parallel and flow-transverse directions; several clasts thick layer laterally passes into single-grain thick layer	M-6, M-8, M-9, L-1, L-2, L-3, L-4, L-5, L-6	Debris falls; several clasts thick layer: grain-assemblage debris falls; single-grain thick layer: single-grain debris falls (Kim et al., 1995; Nemec, 1990; Sohn et al., 1997)
Stratified (gravelly) sandstone (Facies Ss)	Alternation of a few centimeters thick granular to pebbly sandstone and medium to coarse sandstone layers; individual layers: diffuse boundary; variable in thickness ranging from 15 to 50 cm; sometimes, undulatory drapes over uptruding clasts in the underlying conglomerate	M-2, M-7, M-9, L-2	High-concentration turbidity currents (Hiscott and Middleton, 1979; Lowe, 1982; Pickering et al., 1986)
Inversely graded (gravelly) sandstone (Facies Si)	Poorly to moderately sorted, medium to granular sandstone and pebbly sandstone; inversely graded; mostly less than 10 cm thick; single or repetitive units	M-10, M-7, L-1, L-2	High-concentration turbidity currents (Chun and Chough, 1995; Lowe, 1982)
Massive (gravelly) sandstone (Facies Sm)	Poorly to moderately sorted, massive medium to granular sandstone and pebbly sandstone; some diffuse, discontinuous pebble to granule trains; sharp lower and upper boundaries; purple color in some units; variable in thickness ranging from 5 to 50 cm	M-2, M-3, M-4, M-7, M-10, L-1, L-2, L-3, L-5	Rapid suspension settling from either high-/low-concentration turbidity currents (Lowe, 1982; Surlyk, 1984) or subaerial sand-laden flows with weak tractive transport (Arnott and Hand, 1989; Maizels, 1993)
Normally graded (gravelly) sandstone (Facies Sn)	Poorly to moderately sorted, medium to granular sandstone and pebbly sandstone; distribution normal grading; some diffuse, discontinuous	M-2, M-3, M-4, M-7, M-10, L-1, L-2, L-3, L-5	Turbulent suspension of either high-/low-concentration turbidity currents (Chun and Chough, 1995; Lowe,

(Continued)

Table 10.5 (Continued)

Facies	Description	Occurrence	Interpretation
	granule to pebble trains; sharp lower boundary; purple color in some units; variable in thickness ranging from 5 to 50 cm		1982) or subaerial sand-laden flows with weak tractive transport (Maizels, 1993; Smith and Lowe, 1991)
Laminated sandstone (Facies Sl)	Moderately or well sorted, horizontally laminated, fine to coarse sandstone; individual laminae: a few millimeters to centimeters thick; 5–30 cm in thickness	M-2, M-3, M-7, M-10, L-1, L-3, L-5	Low-concentration turbidity currents (Chun and Chough, 1995; Tb division; Lowe, 1988)
Laminated sandstone–mudstone couplets (Facies SMl)	Thin sandstone layers alternating with dark-gray to black mudstone layers; 5–10 cm in thickness; each laminae: a few millimeters thick; fining- and thinning-upward trend	M-3	Low-concentration turbidity currents (Tbd division; Bouma, 1962; Pickering et al., 1986)
Laminated mudstone (Facies Ml)	Dark-gray to black, laminated mudstone; mostly less than 10 cm thick; each lamina: a few millimeters thick; fining- and thinning-upward trend	M-3	Low-concentration turbidity currents (Td division; Chun and Chough, 1995; Hesse and Chough, 1980)
Homogeneous mudstone (Facies Mh)	Homogeneous texture; grayish olive green in color; less than 2 cm thick; drapes irregular surface of underlying conglomerates	M-6	Suspension sedimentation in subaqueous condition

Source: Lee and Chough (1999) by permission of the Springer-Verlag.

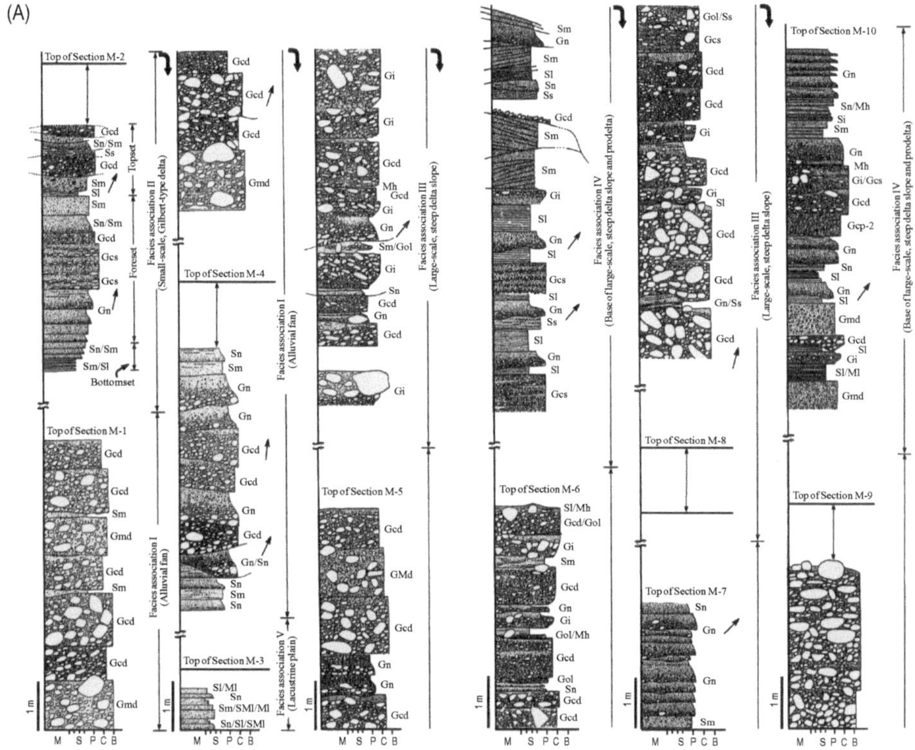

Figure 10.35 (A) Representative columnar sections M-1 to M-10 showing marginal fill deposits. (B) Representative columnar sections L-1 to L-6 showing longitudinal fill deposits. For location, see Figure 10.34. For facies codes, see Table 10.5. Solid straight arrows indicate paleoflow direction with north toward the top of the figure. M, mud; S, sand; P, pebble; C, cobble; B, boulder.
Source: Lee and Chough (1999) by permission of the Elsevier.

sedimentary facies and depositional architecture reflect an increase in basin subsidence, which was closely related to the changes in displacement on the basin-bounding fault.

The longitudinal fill represents an axial system, prograded from east to west. It is characterized by eastward overlapping of large-scale, steep delta and prodelta succession, which is opposite to the paleocurrent direction. It is indicative of eastward migration of the depocenter along the basin axis. However, the slightly oblique (i.e., northward) progradation and the northward syndepositional normal faults in the marginal fill imply a northward basin-floor tilting which led to the deposition of large scale, steep delta slope and prodelta in the basin axis.

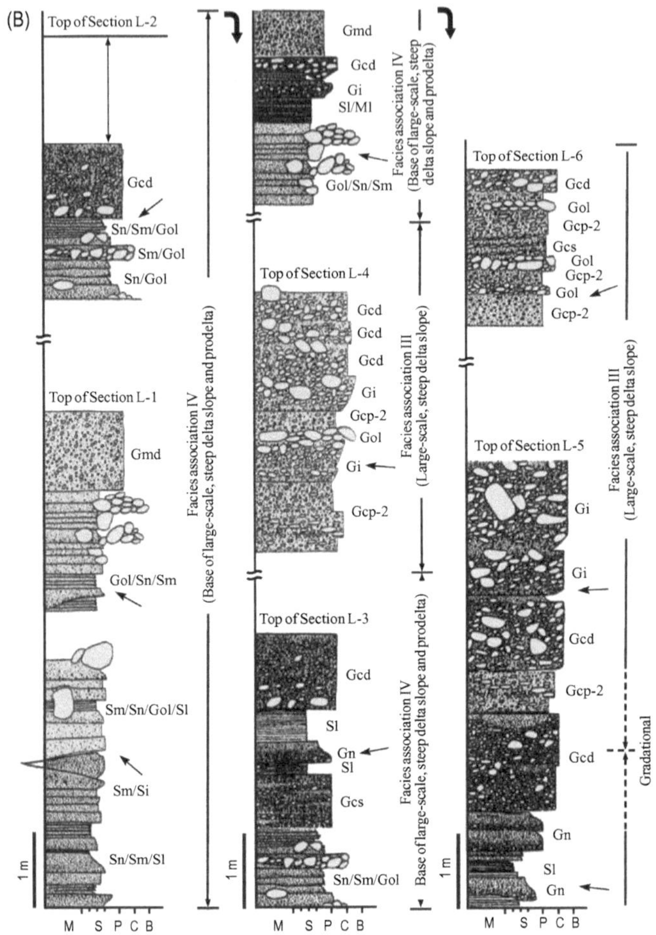

Figure 10.35 (Continued)

11 Tertiary Basins

11.1 Background

The southeastern margin of the Korean Peninsula was subject to dextral shear stress from the NNW–SSE opening of the East Sea during the Early to Middle Miocene (Yoon and Chough, 1995). A number of small-scale sedimentary basins formed in the area associated with NNW-trending dextral strike-slip faults and NNE- to NE-trending normal faults (Son, 1998). Most of these basins were rifted during the Early Miocene and filled with kilometer-thick successions of nonmarine to shallow marine sediments together with abundant dacitic and basaltic volcanic deposits (Bahk and Chough, 1996; Son et al., 2000, 2005). The normal faulting and strike-slip deformation continued until ca. 16–17 Ma. Afterward, major crustal extension occurred in the Pohang Basin, causing rapid subsidence of the basin from inner neritic (< 50 m deep) to middle-to-lower bathyal depths (Ingle, 1992).

11.2 Pohang Basin

The Pohang Basin occurs in the eastern part of the northeast-southwest-trending Yangsan Fault which has experienced right-lateral strike-slip deformation (Chough et al., 1990) (Figure 11.1). During the Miocene, the strike-slip movement was negligible and instead was under extensional regime (Hwang et al., 1995). Large-scale drainage systems developed at the junction between the strike-slip zone and the oblique transfer faults, acting as major sources for coarse-grained sediments, forming a number of fan deltas (Figure 11.1). High rate of sediment supply and rapid subsidence of the hanging wall resulted in the progradation of large-scale Gilbert-type fan deltas (Doumsan, Duksung, and Gohyun systems).

11.2.1 Fan Delta Systems

The fan deltas comprise a Gilbert-type topset, which is characterized by flat-bedded, cross-bedded, and wedged conglomerate formed in braided channels and bars (Figure 11.2). The foreset is dominated by stratified conglomerate formed by gravity sliding and avalanching. At the toeset, slope failure caused deposition of well-sorted and graded conglomerate, which were deformed during the deposition. Further downslope, at the bottomset, well-sorted graded conglomerate and

Geology and Sedimentology of the Korean Peninsula. DOI: http://dx.doi.org/10.1016/B978-0-12-405518-6.00011-5

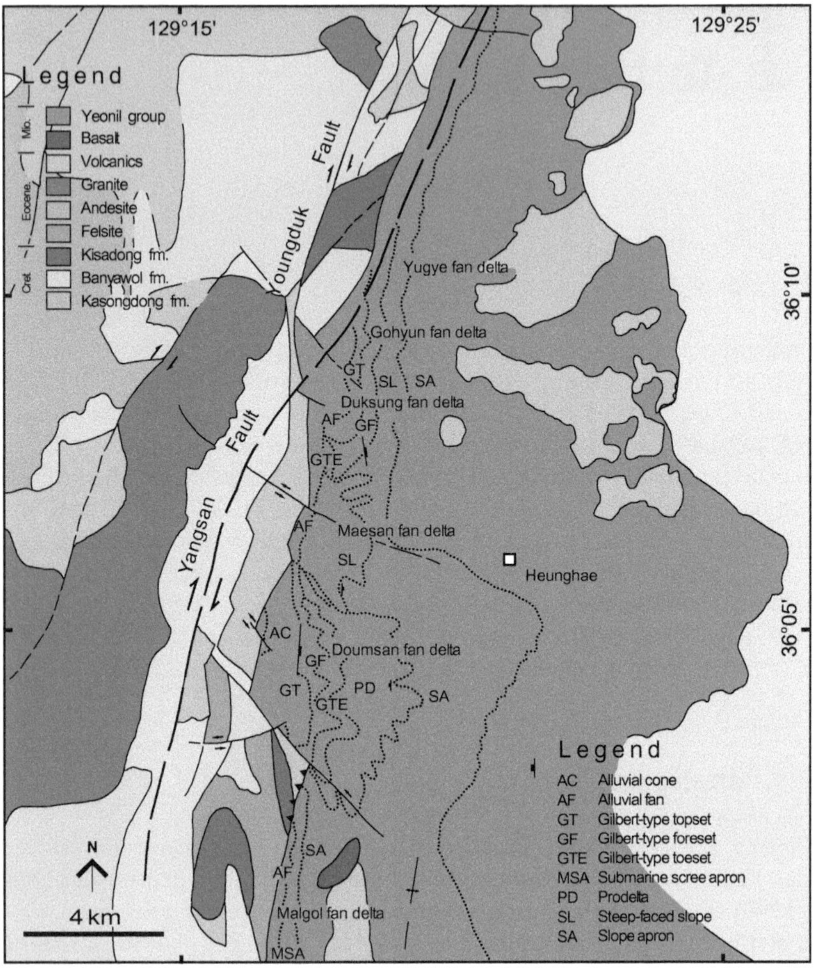

Figure 11.1 Simplified geologic map of the western part of Pohang Basin (Miocene) and outline of fan delta systems.
Source: Chough and Hwang (1997) by permission of the SEPM.

sandstone formed from turbidity currents in submarine channels and lobes, forming slope aprons (Choe and Chough, 1988). On the basin slope, high- and low-concentration turbidity currents and slide/slump as well as debris flows formed debris aprons. On the basin plain, hemipelagic sedimentation prevailed with scarce low-concentration turbidity currents.

Along the steeply inclined footwall scarps, however, the drainage systems were relatively small, forming scree-apron-type (Malgol and Yugye systems) and steep-slope-type (Maesan system) fan deltas (Figure 11.1). The Maesan fan delta comprises alluvial fan, steep-faced slope, and basin plain. Subaerial debris flows and

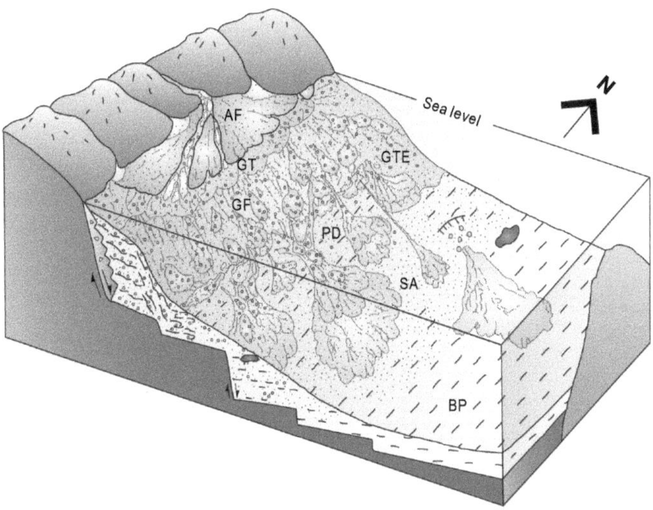

Figure 11.2 Simplified model of depositional environments in the Doumsan fan delta deep-water system. AF = alluvial fan, GT = Gilbert-type topset, GF = Gilbert-type foreset, GTE = Gilbert-type toeset, PD = prodelta, SA = slope apron, BP = basin plain. *Source:* Chough et al. (1990) by permission of the SEPM.

dense inertia-dominated currents were transformed into subaqueous sediment gravity flows in steep-faced environments (Hwang and Chough, 2000). Further downslope, these flows were channelized, forming lobate bodies of conglomerate and sandstone at the terminal edge of the channels, accompanied with large-scale slides and slumps. The channels, lobes, and splays are similar to those of steep slopes in modern fan delta systems.

11.2.2 Evolutionary Stages

Based on the variations in sedimentary facies and distribution, architecture, and overlapping patterns, the fan delta systems in the Pohang Basin can be divided into four stages of evolution (Figure 11.3). In the Doumsan and Duksung systems, the high rate of sediment supply and the rapid rise of the relative sea level resulted in the transformation of shoal-water-type to Gilbert-type foreset. In the Malgol and Maesan systems, rapid sea-level rise accompanied by a low rate of sediment supply caused the transformation of alluvial fans to scree-apron and steep-faced slope fan deltas, respectively.

In the second stage, subsidence of the hanging wall formed large-scale truncation surfaces in the Doumsan and Duksung fan deltas. Coarse-grained sediments were continuously supplied in the Doumsan system, whereas only fine-grained sediments were deposited in the Duksung system. In the Malgol system, sediment supply was almost terminated at this stage. In the third stage (Middle Miocene), the

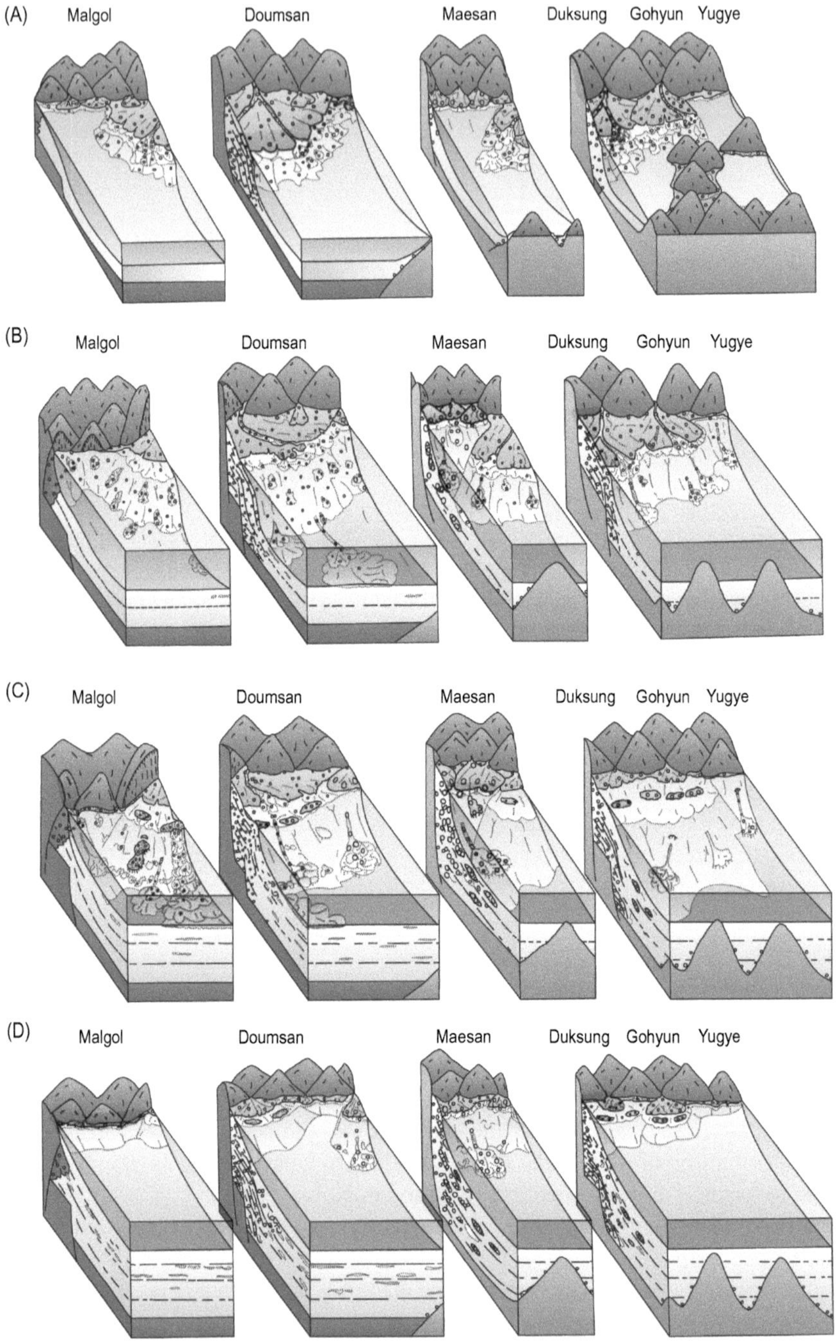

Figure 11.3 Depositional model for fan delta systems in the Miocene Pohang Basin: (A) stage 1 (Early Miocene); (B) stage 2 (Middle Miocene); (C) stage 3 (Middle-Late Miocene); and (D) stage 4 (Late Miocene).
Source: Hwang et al. (1995) by permission of the Elsevier.

sediment supply rate of the Doumsan fan delta decreased abruptly. The Duksung fan delta was overlain by the Maesan and Gohyun systems. During the last stage (Late Miocene), the sediment supply of the Doumsan and Maesan systems was also terminated.

11.3 Other Tertiary Basins

A number of small-scale early Miocene sedimentary basins occur along the southeastern margin of the Korean Peninsula (Figure 11.4). These basins formed by

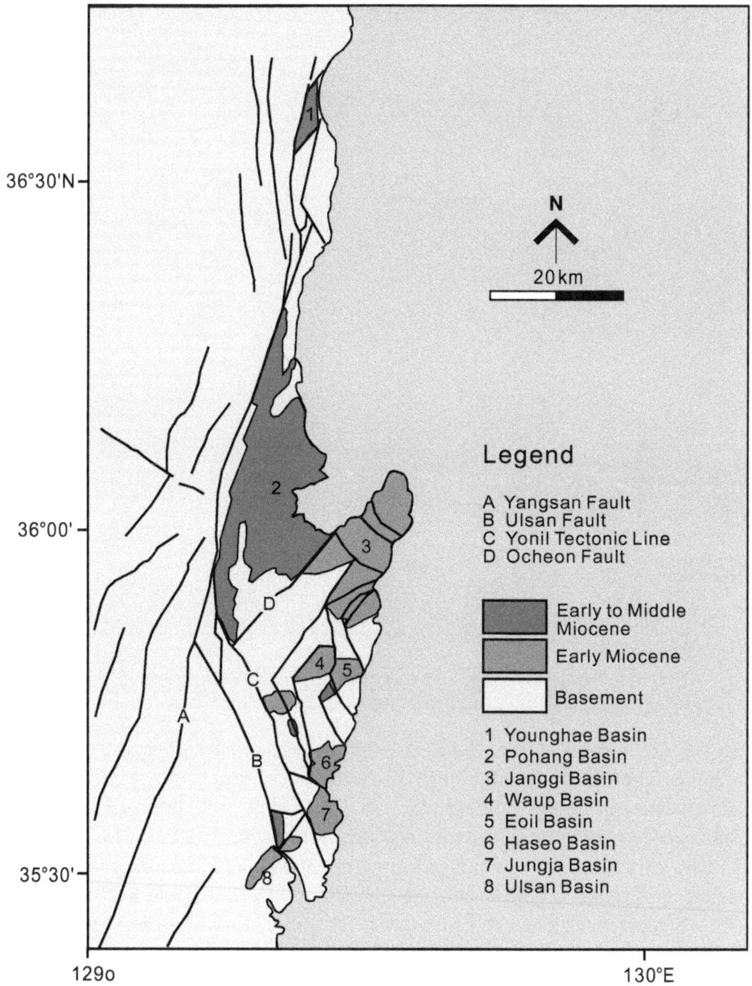

Figure 11.4 The distribution of Tertiary sedimentary basins and major faults in the southeastern part of the Korean Peninsula.
Source: Son (1998).

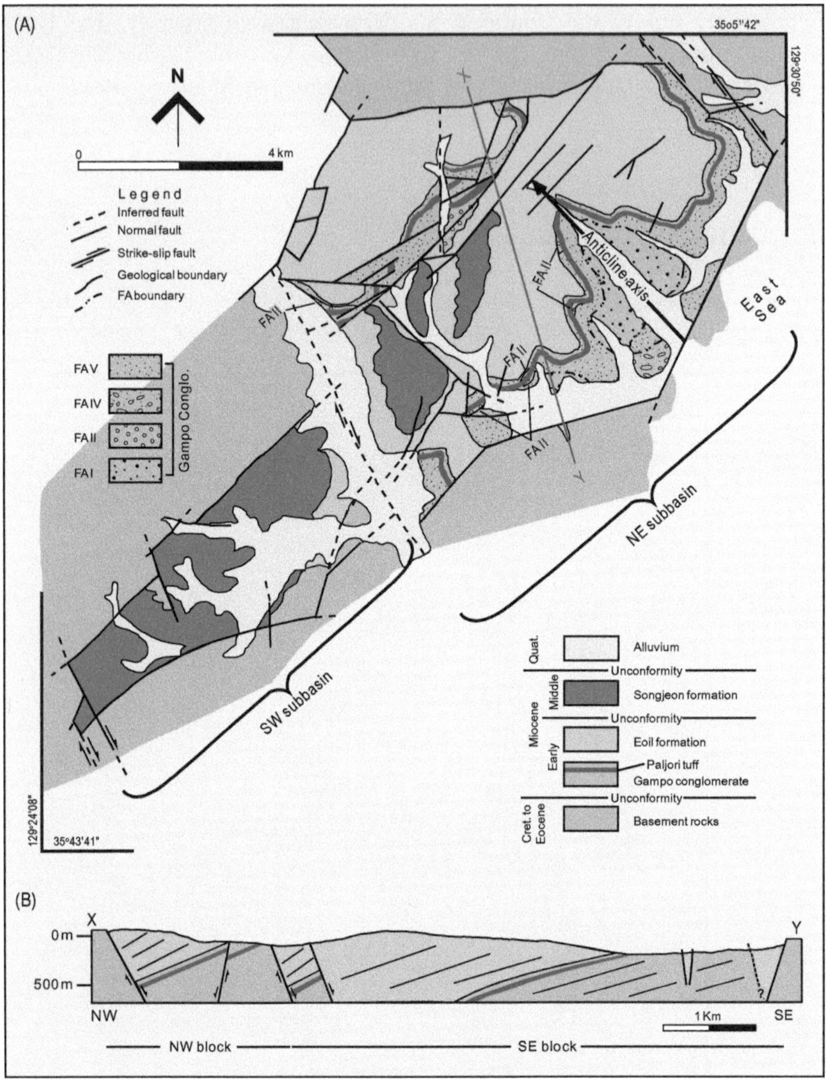

Figure 11.5 (A) Geological map of the Eoil Basin showing distribution of facies associations (Son et al., 2000). The basin is divided into the northeastern and southwestern subbasins by a NNW-trending strike-slip fault in the middle part. The Paljori tuff, intercalated in the topmost part of the Gampo Conglomerate, provides a good stratigraphic marker. The basement rocks of the Eoil Basin are composed of Cretaceous sedimentary rocks, Eocene granite (Kim et al., 1995a), Paleocene–Eocene acidic volcanic rocks (Jin et al., 1988), and Early Miocene andesites (the Hyodongri Volcanics; Tateiwa, 1924) and acidic tuffs (the Waupri Tuff; Tateiwa, 1924; Jin et al., 1989). FA I: Alluvial fan deposits, FA II = Gravelly braided-stream deposits, FA IV = Delta front deposits, FA V = Sandy mouth-bar deposits. (B) Geological cross section of the northeastern subbasin, showing the bed attitude of the basin fill strata dipping progressively steeper toward the northwest.
Source: Kwon et al. (2011) by permission of the Elsevier.

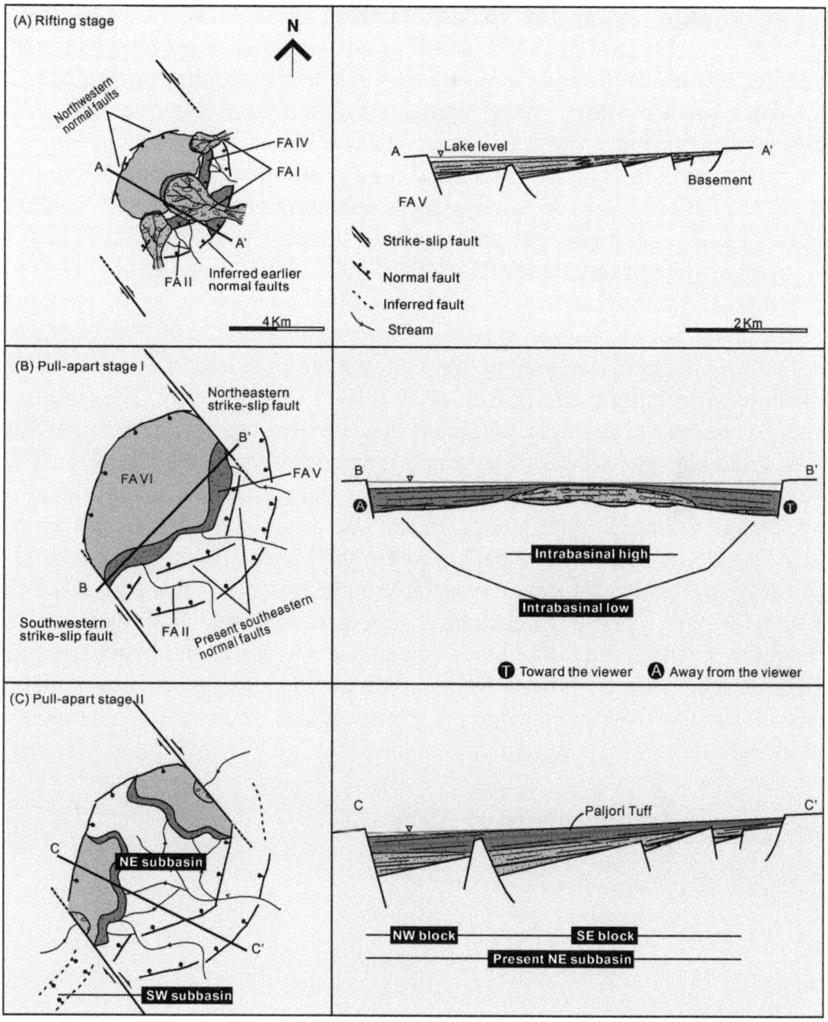

Figure 11.6 Cartoon illustrating the evolution of the Eoil Basin. (A) The early basin formation is predominantly controlled by the normal faults between the incipiently developing strike-slip faults. Major displacement along the northwestern normal faults caused deepening of the basin toward the northwest, resulting in half-graben geometry. (B) The role of strike-slip faults on the southwestern and northeastern margins began to be important after the deposition of FA VI. The intrabasinal lows started to form on the southwestern and northeastern basin margins. The intrabasinal high, which formed during the early rifting stage, developed in the central part of the basin and was accentuated by the later strike-slip fault activity. (C) The strike-slip faults continued to act as master faults of the basin together with the northwestern normal faults. The overall basin floor inclined toward the northwest. The depocenters developed near the basin corners, resulting in thick accumulation of deposits there.
Source: Kwon et al. (2011) by permission of the Elsevier.

dextral deformation during the backarc opening of the East Sea (Kwon et al., 2011). They are bounded by NNW-trending dextral strike-slip faults and NNE- to NE-trending normal faults and have generally northwest-dipping half-graben geometry of the basin fills (Son, 1998). Paleomagnetic studies indicate that the fault blocks and the overlying basin fills were rotated about 30° clockwise due to the dextral deformation in contrast to those of the Pohang Basin (Kim and Kang, 1989, 1996). Most of these basins were filled with nonmarine conglomerates, sandstones, mudstones, and a variety of volcaniclastic deposits in alluvial fans, delta slopes, braided streams, and lakes (Son et al., 2000, 2005; Jeong et al., 2008; Kwon et al., 2011). Basaltic diatremes, i.e., tephra-filled volcanic necks of phreatomagmatic volcanoes, are also found in places (Kwon and Sohn, 2008; Son et al., 2012). A detailed architectural analysis of the conglomerates (Gampo Formation) of the Eoil Basin was recently made (Kwon et al., 2011) (Figure 11.5). The results demonstrate that normal faults were dominant in the early stage of the basin evolution, whereas strike-slip faults became more pronounced in the late stage (Figure 11.6). It is suggestive of dynamic changes of the basin margin faults through time.

The Janggi Group in the Janggi Basin comprises pyroclastic and epiclastic deposits (conglomerate, trachyte tuff, basaltic tuff, andesitic tuff, and coal seams) of the Early to Middle Miocene, associated with basaltic flows and dikes (Bahk and Chough, 1996). These volcaniclastic deposits represent both syn- and inter-eruption events. The former comprises volcanic debris flows and hyperconcentrated flood flows, which were initiated from turbulent flood surges. However, the latter consists of channel and interchannel deposits, which show abrupt compositional variation from those of the syn-eruption deposits.

12 East Sea

12.1 Introduction

The East Sea is a semienclosed backarc basin surrounded by the Asian continent and the Japanese islands. It consists of three deep basins separated by submarine topographic highs such as the Korea Plateau, Oki Bank, and the Yamato Ridge (Figure 12.1). The topographic highs are made of relatively shallow-seated acoustic basements of rifted continental crust and the overlying thin sedimentary succession, in contrast to the adjacent deep basins of extended continental crust (Tamaki, 1988; Chough et al., 2000) (Figure 12.1). The continental backarc basins formed by multiaxial extensional opening since the Oligocene (Chough and Barg, 1987; Tamaki, 1988).

The southwestern part of the sea has recently been under particular attention, from which an array of both multichannel deep-seismic reflection profiles and high-resolution subbottom profiles (Chirp) was obtained (Figures 12.2 and 12.3) (Lee et al., 2002, 2004; Bahk et al., 2005; Kwon et al., 2009; Kim et al., 2011). The geology of the eastern continental margin and the Ulleung Basin was summarized at the turn of the century in *Marine Geology of Korean Seas* (Chough et al., 2000). In this volume, I focus on the new findings on the South Korea Plateau and the Ulleung Interplain Gap.

12.2 South Korea Plateau

12.2.1 Geological Setting

The South Korea Plateau is a submerged topographic high, separating the Ulleung and Japan basins (Figure 12.1). The basement of the plateau is interpreted as rifted continental crust formed in the early stage of backarc opening in the Late Oligocene to Early Miocene (Kwon et al., 2009). The plateau area is divided into the western and eastern parts by the north Ulleung escarpment (Figure 12.2). In the western South Korea Plateau, seismic reflection profiles commonly delineate shallow-seated (< 3.5 s in two-way travel time below sea level, twtt) acoustic basements with very rugged topographic features such as volcanic ridges and mounds and intervening troughs. However, the eastern South Korea Plateau is characterized by relatively monotonous topography except for a deep fault-controlled valley in

Geology and Sedimentology of the Korean Peninsula. DOI: http://dx.doi.org/10.1016/B978-0-12-405518-6.00012-7

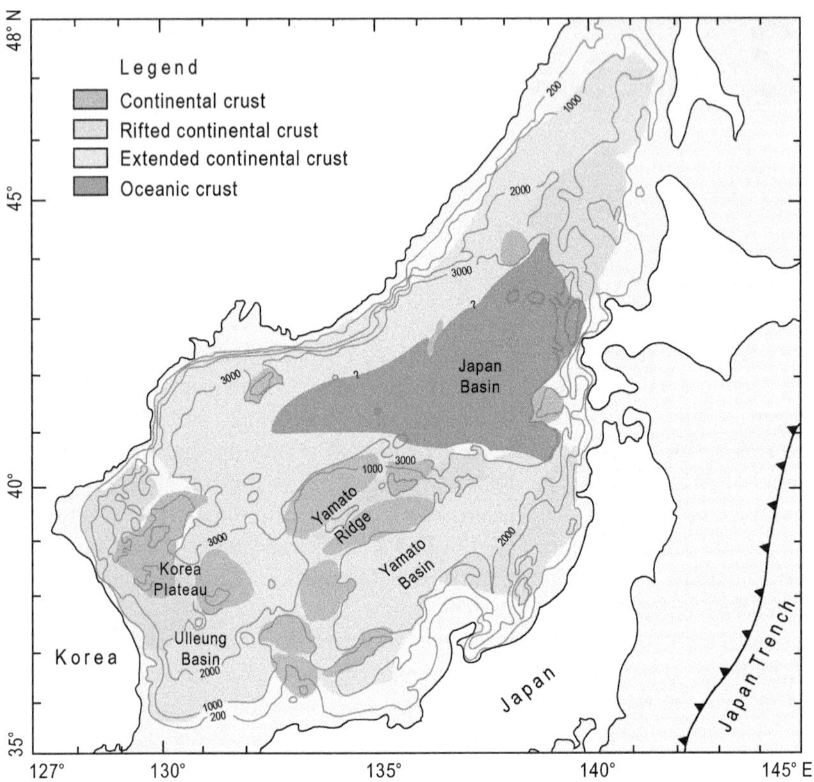

Figure 12.1 Crust types of the East Sea. Contours of water depth in meters.
Source: Tamaki (1988).

the southern margin and a seamount chain in the northern margin (Figure 12.2). The overlying Tertiary–Quaternary sedimentary succession shows considerable variation in thickness from very thin veneers on the ridges and seamounts to relatively thick (maximum of ca. 2 km) trough fills (Kwon et al., 2009). The ridges and seamount chains in the plateau occur at a water depth of about 600–1500 m (Figure 12.2). The ridges are relatively broad (> 30 km wide) and rarely exceed 1° in slope gradient, mostly ranging from 0.1° to 0.5° except near the ridge summits and small-scale topographic highs which rise about 200–300 m above the surrounding seafloor (Figure 12.2). In contrast, the seamount chains are mostly steep and narrow (< 10 km wide). The ridges and seamount chains commonly form troughs and basins (Figure 12.2).

12.2.2 Geologic Structures

In the western South Korea Plateau, the acoustic basement commonly appears as fault-bounded horsts (Figure 12.4). Some parts of the continental-margin terraces

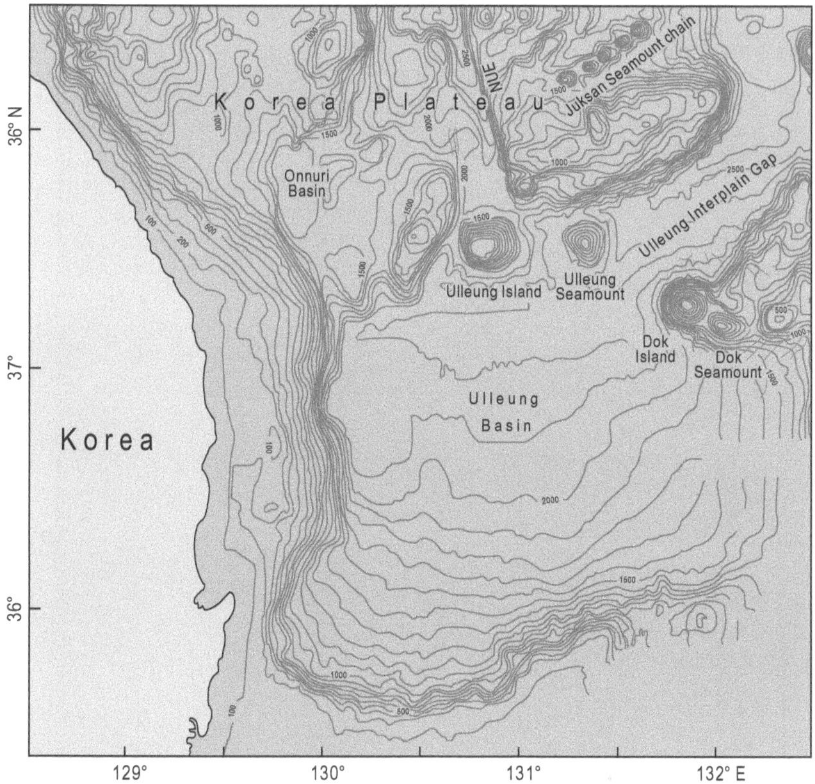

Figure 12.2 Bathymetric map of the Ulleung Basin and the southern part of the Korea Plateau. Contour interval is 100 m. NUE = North Ulleung Escarpment.
Source: Courtesy of the National Oceanographic Research Institute.

are also bounded by faults with a normal separation (Kwon et al., 2009). These features are associated with continental rifting during the early stage of backarc opening in the Late Oligocene to Early Miocene (Tamaki, 1988; Jolivet and Tamaki, 1992). The bounding faults show slightly curving fault planes, and the fault offset apparently reaches up to a few hundred meters. The strike of the fault planes is north to south on the plateau, and northeast to southwest along the eastern Korean continental margin (Figure 12.4). Along the western margin of the plateau, sedimentary successions appear to have been significantly compressed and locally uplifted, giving rise to two thrust belts (Figure 12.4). The reverse faults usually accompany anticlinal folds or fault drags. These contractile structures strike dominantly northwest to southeast and dip westward. Along the faults, thick successions are apparently uplifted and laterally juxtaposed against a thinner, structurally lower (down-dropped) succession. Some contractile structures seem to have developed in response to the reactivation of preexisting extensional basement faults. Yoon and

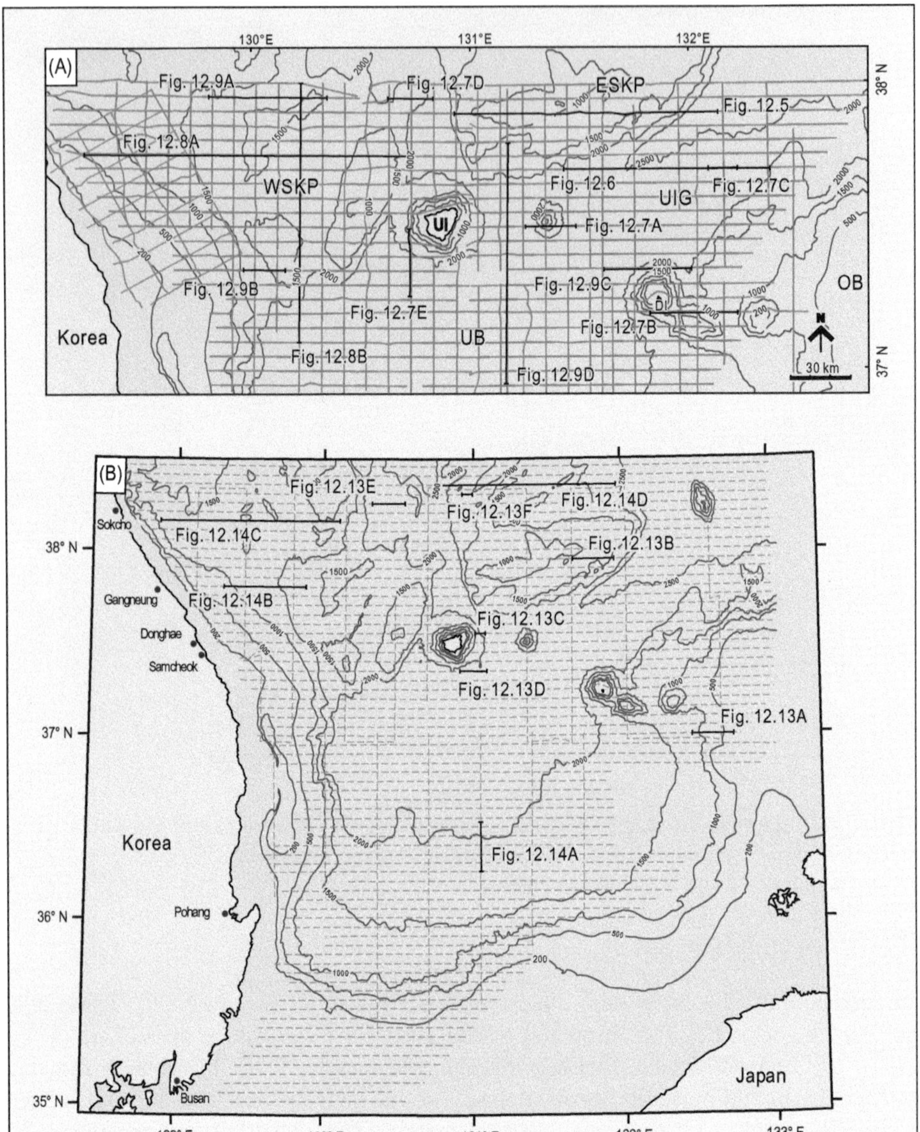

Figure 12.3 Tracklines of multichannel seismic profiles (A) by the KIGAM, and Chirp sonar survey (B) by the National Oceanographic Research Institute. WSKP = Western South Korea Plateau, ESKP = Eastern South Korea Plateau, UB = Ulleung Basin, UI = Ulleung Island, US = Ulleung Seamout, DI = Dok Island, OB = Oki Bank. Contours in meters.

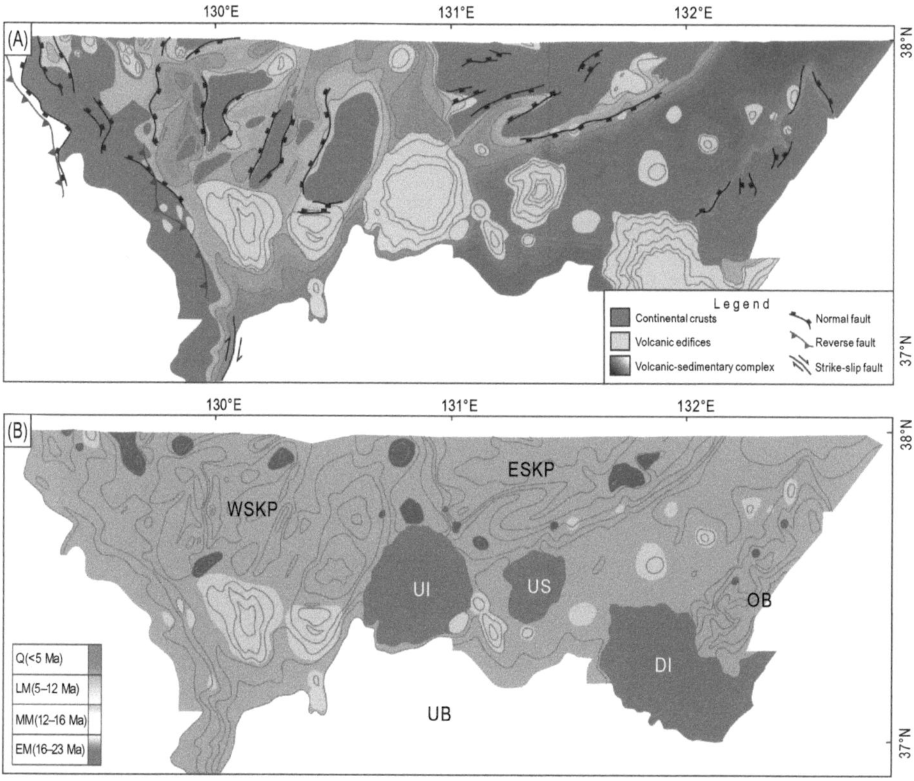

Figure 12.4 (A) Map showing the types of acoustic basement and its spatial distribution. Fault-lines are modified after Kim et al. (2007) and Kwon et al. (2009). (B) Map showing the spatio-temporal variation of volcanic edifices. Volcanic edifices colored in red, yellow, and blue conform to the stage 1, 2, and 3 volcanic events. WSKP = Western South Korea Plateau, ESKP = Eastern South Korea Plateau, UB = Ulleung Basin, UI = Ulleung Island, US = Ulleung Seamout, DI = Dok Island, OB = Oki Bank, Q = Quaternary, LM = Late Miocene, MM = Middle Miocene. (For interpretation of the references to colour in this figure legend, the reader is referred to the online version of this book.)
Source: Kim et al. (2011) by permission of the Elsevier.

Chough (1995) suggested that this contractile deformation has been associated with the backarc closing phase since the end of the Middle Miocene.

12.2.3 Seismic Stratigraphy

12.2.3.1 Acoustic Basement

According to the seismic reflection characteristics, the acoustic basement in the South Korea Plateau can be classified into three types (Kim et al., 2011)

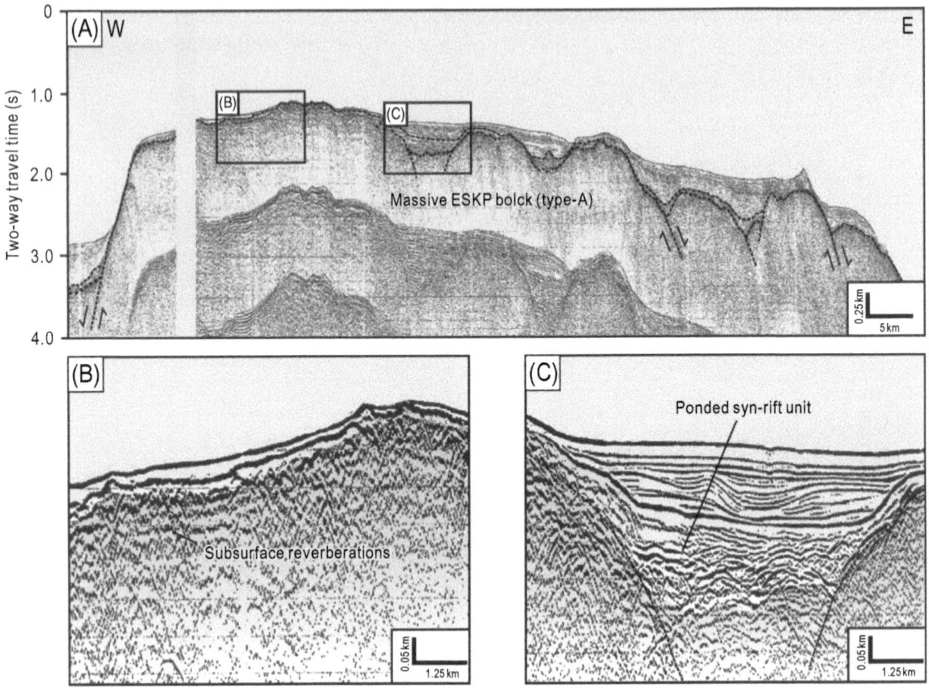

Figure 12.5 (A) Seismic reflection profile showing overall topography of the Eastern South Korea Plateau (for location, see Figure 12.3A). (B) Details of basement high in the massive ESKP block. Below high-amplitude and continuous surface reflector, subsurface reverberations are identified. (C) Details of small-offset fault-bounded trough in the ESKP block with pond-like syn-rift unit.
Source: Kim et al. (2011) by permission of the Elsevier.

(Figure 12.4). Type-A acoustic basement is a remnant of rifted continental crust comprising relatively shallow-seated continental-margin platforms, elongated ridges, and equidimensional blocks in the South Korea Plateau and its vicinities (Figures 12.4 and 12.5). Type-B acoustic basement consists of syn-rift volcanic sills/flows and sediment complexes, which infill the rift-controlled basement lows in the South Korea Plateau, the Ulleung Interplain Gap, and the northern Ulleung Basin (Figure 12.6). Type-C acoustic basement is a cluster or chain of volcanic edifices, which shows internal facies variation from the main body to apron slope facies (Figure 12.7).

12.2.3.2 Syn-Rift Sedimentary Unit

This unit is the lowermost succession that directly overlies the acoustic basement. It forms a wedge in some basement troughs, and is generally absent or very thin above the basement highs (Figure 12.8). The maximum thickness varies from about

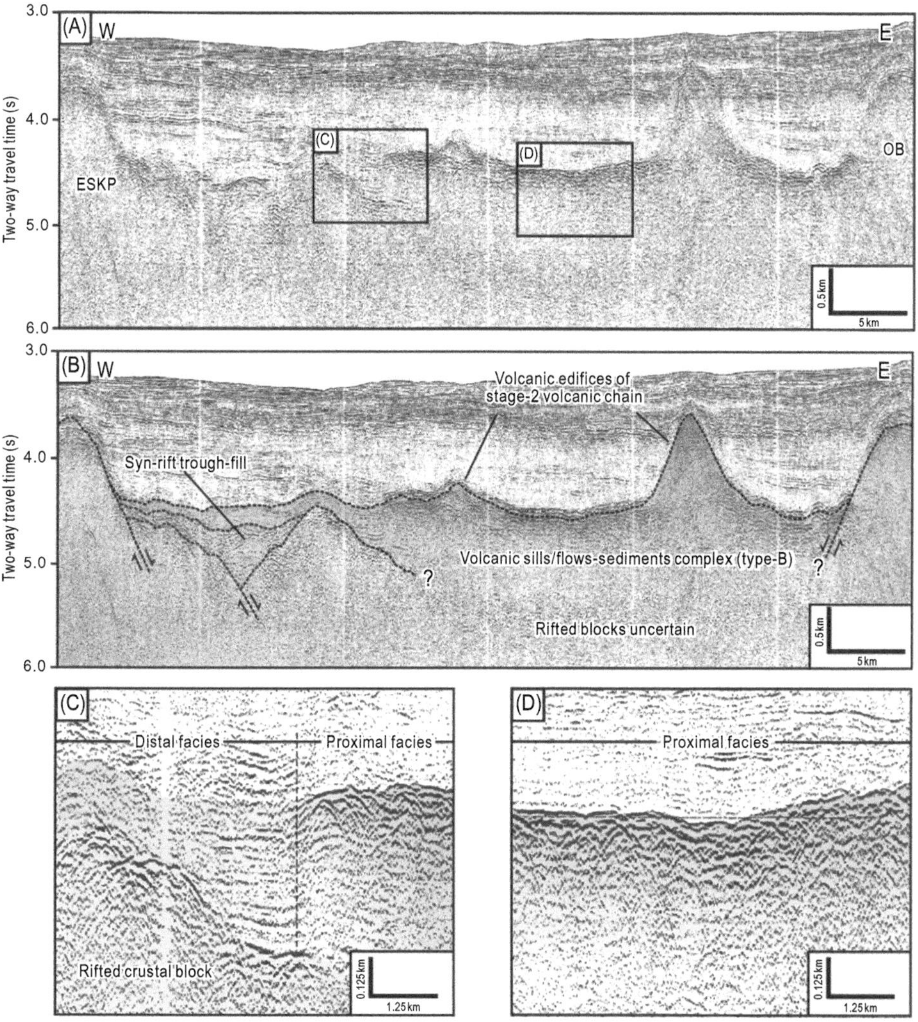

Figure 12.6 (A) Seismic reflection profile showing overall topography of the Ulleung Interplain Gap (UIG) (for location, see Figure 12.3A). (B) Details of highly rugged acoustic basement of the UIG. Irregularly overlapping high-amplitude hyperbolae with almost transparent internal reflection configurations occur near the volcanic eruption centers. (C) Details of crudely stratified seismic reflection facies of acoustic basement identified relatively far from the eruption centers. (D) Details of acoustic basement types in the UIG. Rifted crustal blocks of the Eastern South Korea Plateau (ESKP) and Oki Bank (OB) are bordering on the NW and SE of the Ulleung Interplain Gap (UIG). Volcanic edifices are scattered along the eruption center. Underlying rifted blocks are identified in the western boundary of the passage.
Source: Kim et al. (2011) by permission of the Elsevier.

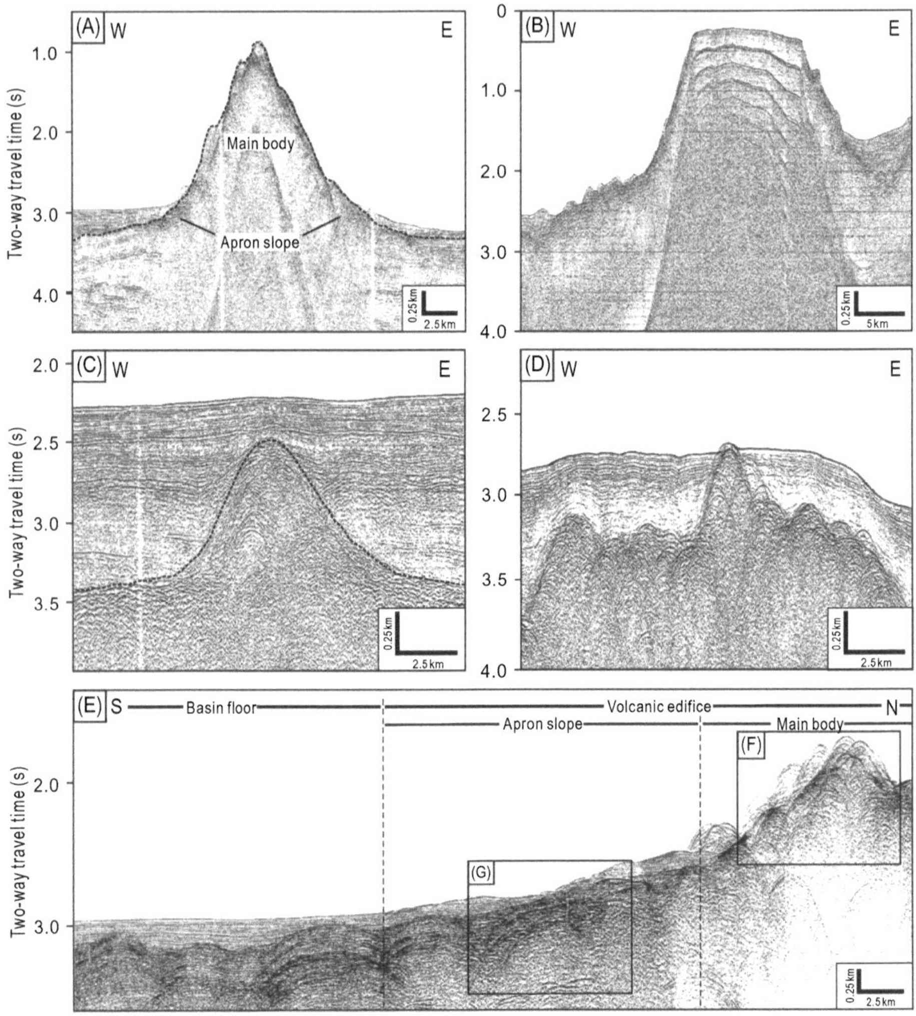

Figure 12.7 (A)–(D) Seismic reflection profiles showing details of various types of volcanic edifices. (A) and (B) A seamount. (C) One of the small volcanic cones scattered around the UIG. (D) Volcanic mound which supports small pinnacles with various vertex elevations. (E) Selected seismic reflection profile showing the variation of seismic reflection facies in a volcanic edifice. Profile is from southern slope side of the UI. (F) Detailed view of the main body facies. (G) Detailed view of apron slope facies. For location, see Figure 12.3A.
Source: Kim et al. (2011) by permission of the Elsevier.

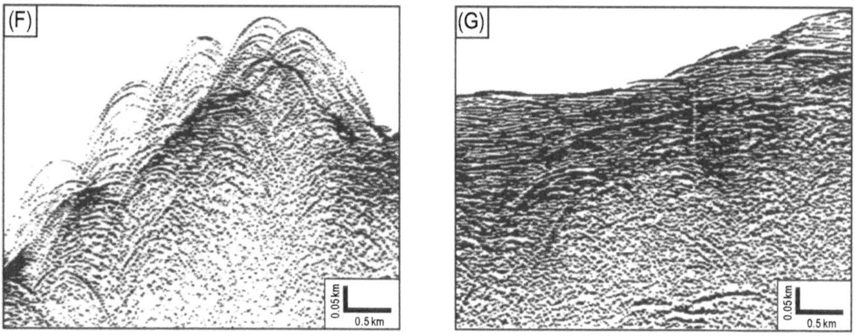

Figure 12.7 (Continued)

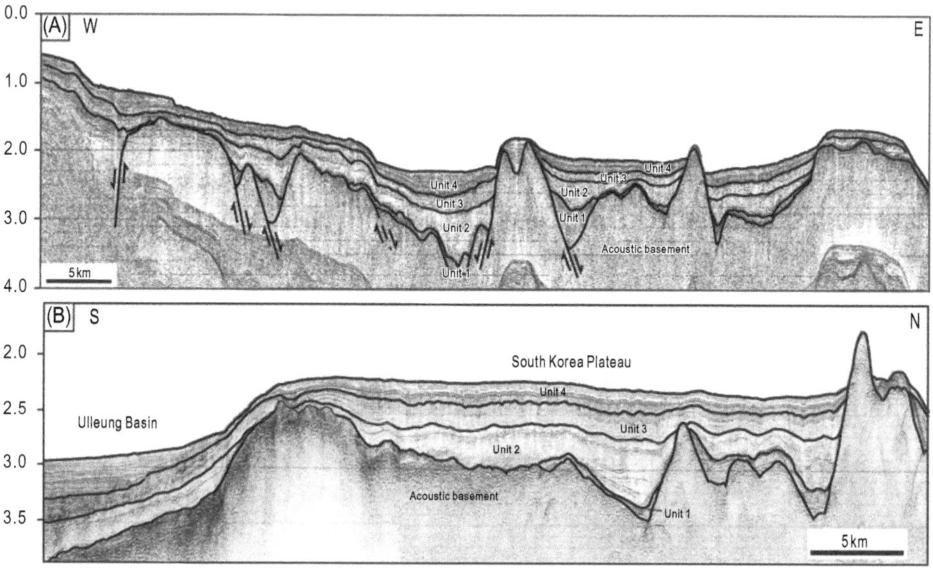

Figure 12.8 Seismic reflection profiles showing the overall distribution of stratigraphic units (for location, see Figure 12.3A). Unit 1 = syn-rift volcanic sills/flows-sediments complex, Unit 2–4 = post-rift turbidites and hemipelagic sediments.
Source: Kwon et al. (2009).

0.05 to more than 0.2 s twtt, depending on the basement topography. This unit is characterized by a wide range of reflection configurations, including short and tangle high-amplitude reflectors, contoured subparallel reflectors of low to moderate amplitude, and chaotic to transparent reflections. The trough-fill strata on the slope of the continental-margin terraces and the top of fault-bounded ridges often show onlap, divergent, and prograded fills. The upper boundary of the unit is rugged and invariably identified by a relatively continuous high-amplitude reflection showing erosional truncation. The syn-rift unit is interpreted as a volcanic sill/flow-sediment complex intercalated with fluvial to lacustrine deposits (Kwon et al., 2009).

12.2.3.3 Post-Rift Sedimentary Units

According to the distinctive features of seismic reflection configuration (Kwon et al., 2009), there are three sedimentary units in the post-rift phase. The lower unit (Late Miocene) consists of onlap-fill strata that occupy the basement troughs, which are traceable into the neighboring Ulleung Basin (Figure 12.8), suggesting that the plateau was progressively flooded during the deposition of this unit. Acoustically, the lower unit comprises stratified reflectors with low to intermediate amplitude. It locally shows chaotic to transparent configurations. These features are suggestive of turbiditic and hemipelagic sedimentation with episodic sediment input by slide/slump and debris flows.

The middle unit (Late Miocene to Early Pliocene) is characterized by overall sheet-drape geometry with a relatively uniform thickness, ranging from 0.1 to 0.3 s twtt (Figure 12.8). Internally, the unit shows a facies transition from the lower transparent or chaotic to the upper well-stratified reflections, suggesting a change from mass flows to turbidity current/hemipelagic processes.

The upper unit also shows a sheet-draping reflector of relatively uniform thickness, from 0.1 to 0.2 s twtt (Figure 12.8). It comprises well-stratified continuous reflectors with high to moderate amplitude. This reflection configuration is suggestive of turbidites and hemipelagic sediments formed by relatively uniform sedimentation (Lee and Suk, 1998). Sediments of two drill cores (25 and 28 m long), recovered from the upper part of the upper unit on the gently sloping Korean continental margin, comprise thin (<1 m) layers of bioturbated or laminated muds and massive sands (Kwon, 2005). Along the Korean continental margin, the upper unit includes sharp erosional scars, as well as blocky masses of hyperbolic or chaotic facies downslope, representing slope failures accompanied by slide/slump and debris flows (Kwon et al., 2009).

12.2.4 Volcanic Activities

Volcanic eruptions played an important role in the shaping of the plateau. Volcanic activities identified in the South Korea Plateau and its vicinity can be classified into three stages based on volcano-stratigraphic correlation with the existing stratigraphic framework (Kim et al., 2011) (Figures 12.4B and 12.9). Stage 1 volcanism is characterized by a fissure-type eruption represented by scattered volcanic cones and mounds, and volcanic sills and lava flows interlayered with syn-rift sedimentary units (Figure 12.10). Early-stage continental rifting prior to the Middle Miocene most likely controlled this volcanism (Figure 12.11A). Stage 2 volcanism is characterized by a northeast- to southwest-trending chain of volcanic edifices along the northern margin of the Ulleung Basin and the Ulleung Interplain Gap. During the Middle Miocene, weakening of the backarc extension might have resulted in limited volcanic eruptions, which formed the elongated volcanic fields across the Ulleung and Yamato basins (Figure 12.11B). Stage 3 volcanism represents highly peaked volcanic islands and seamounts, showing vertically stacked apron slopes in seismic profiles. This vertical stacking pattern is indicative of

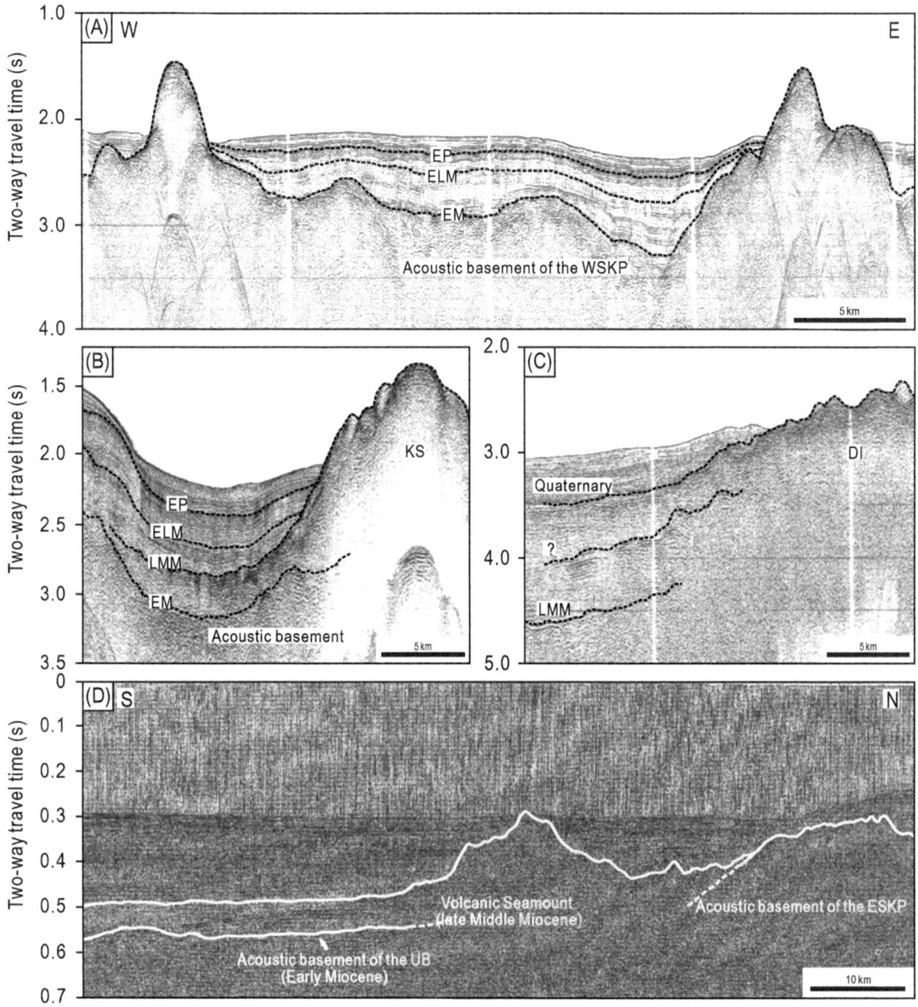

Figure 12.9 Seismic reflection profiles showing details of the acoustic basement and volcano-stratigraphic units (for locations, see Figure 12.3A). (A) Volcano-stratigraphic units in the northern WSKP. (B) Volcano-stratigraphic units near the Kiminu Seamount in the southern WSKP. (C) Volcano-stratigraphic units near the Dok Island. (D) Volcano-stratigraphic correlation between the southern ESKP and northern UB. EP: Early Pliocene, ELM: early Late Miocene, LMM: late Middle Miocene, EM: Early Miocene, KS: Kiminu Seamount, DI: Dok Island, UB: Ulleung Basin, WSKP: Western South Korea Plateau, ESKP: Eastern South Korea Plateau.

Source: Age data from Lee (1992b), Park (1998), and Kwon et al. (2009). Kim et al. (2011) by permission of the Elsevier.

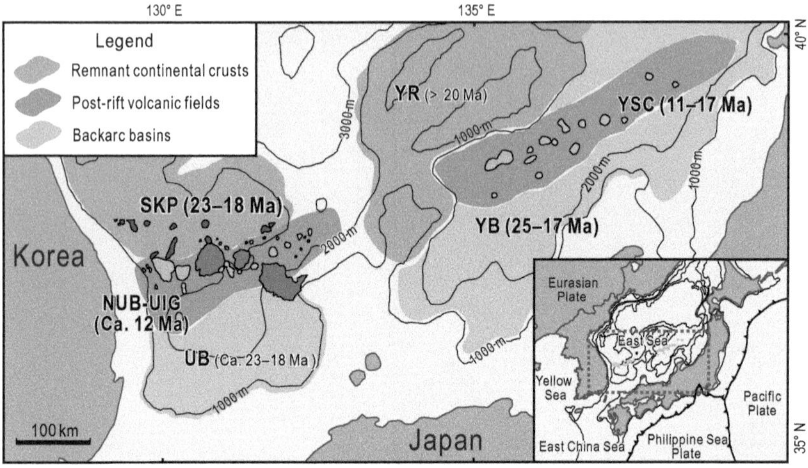

Figure 12.10 Map showing the types of acoustic basement and spatio-temporal variation of volcanic activities in the southern East Sea. For the age of volcanic edifices (indicated by red, yellow, and blue colors), see Figure 12.4B. LO = Late Oligocene, LMM = Late Middle Miocene, MM = Middle Miocene, NUB = northern Ulleung Basin, SKP = South Korea Plateau, UB = Ulleung Basin, UIG = Ulleung Interplain Gap, YB = Yamato Basin, YR = Yamato Ridge, YSC = Yamato Seamount Chain. (For interpretation of the references to colour in this figure legend, the reader is referred to the online version of this book.) *Source:* Age data from Jolivet and Tamaki (1992), Kaneoka (1990), Kaneoka et al. (1990), Kwon et al. (2009), Lee (1992b), Park (1998), and Tamaki et al. (1992). Kim et al. (2011) by permission of the Elsevier.

multiple eruption events from a single eruption center from at least the Late Miocene to the Quaternary (Figure 12.11C).

12.2.5 Echo Types

The shallow sedimentary successions can be subdivided into a few seismic units above the acoustic basement (Lee et al., 2004) (Figure 12.12). The deposits occur mainly in a series of basement troughs (grabens) bounded by faulted blocks and volcanic ridges, and are commonly floored by extrusive volcanic rocks. High-resolution subbottom (Chirp) profiles reveal that large masses of wavy stratified sediments (type III-4) cover gently sloping ridges (Figure 12.12) (Lee and Chough, 2001). They form wavy, continuous reflective layers with a basal deformed zone. On steep slopes, these layers are bounded by slide and slump deposits, formed by submarine creep (Lee and Chough, 2001). On the basis of clarity, continuity, and the geometry of subbottom echoes as well as the

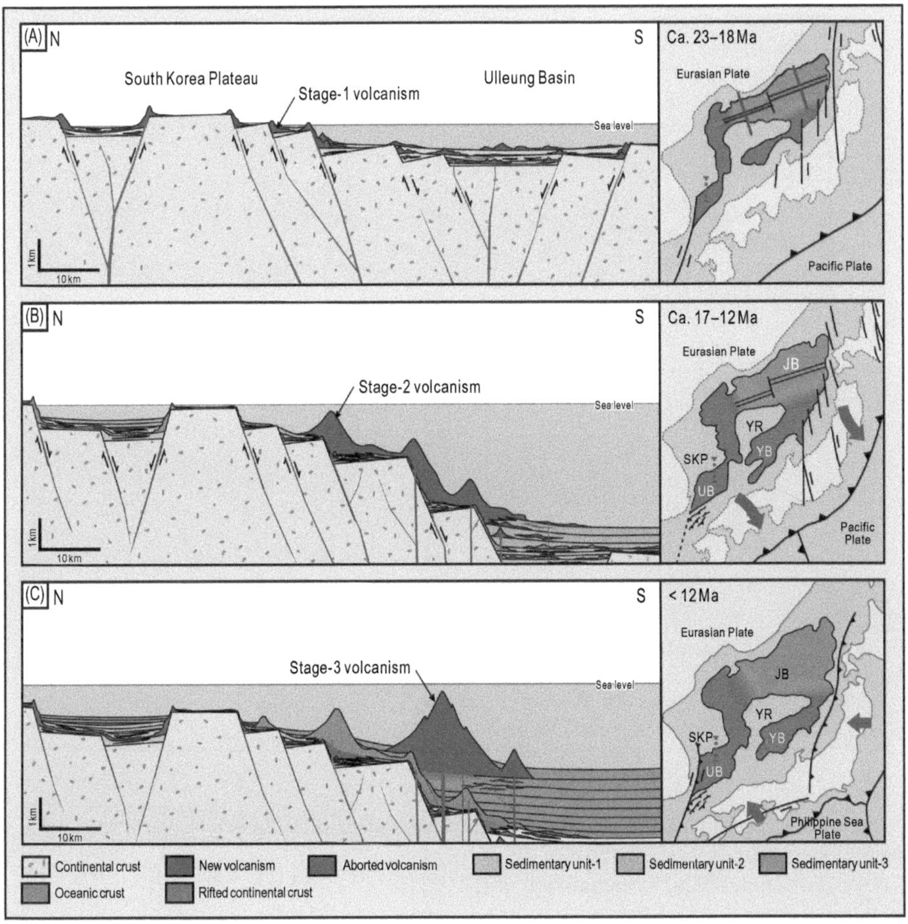

Figure 12.11 Three-stage, volcano-stratigraphic model of the South Korea Plateau and the Ulleung Basin in corporation with tectonic evolution of the East Sea.
Source: Kim et al. (2011) by permission of the Elsevier.

seafloor morphology, 16 echo types have been identified in the area (Table 12.1) (Lee et al., 2004).

Echo types in the South Korea Plateau exhibit a distinctive zonal distribution depending systematically on seafloor morphology. Pelagic and hemipelagic sediments (type I-2) and rock basement highs (type III-1) prevail on the ridge summits and seamount chains, and are commonly bounded downslope by submarine creeps (type III-4) and slide/slump (type IV-1) over the entire slope areas of the ridges and seamount chains (Figure 12.13). The mass-movement deposits

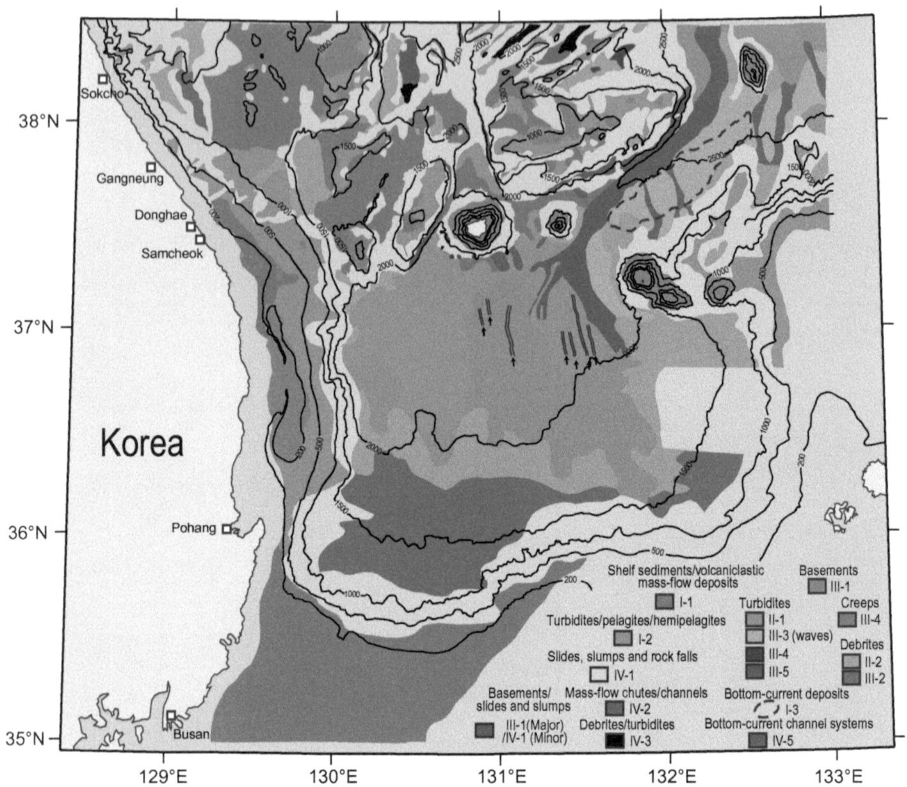

Figure 12.12 Distribution of echo types and interpretation of sedimentary processes in the southwestern part of the East Sea.
Source: Lee et al. (2004) by permission of the AGU.

change downslope to debris-flow and turbidite deposits (types II, III-2, IV-2, and IV-3) in the troughs and Onnuri Basin (Figure 12.14). The voluminous creeps, slides, and slumps over the entire slopes of the plateau, deeper than 300 m in water depth, were most likely generated by frequent seismic shaking (Lee et al., 2004).

12.3 Ulleung and Dok Islands

The Ulleung and Dok islands are separated by the Ulleung Interplain Gap, which connects the Ulleung Basin to the Japan Basin (Figure 12.2). These islands were formed by a series of volcanic eruptions during the late Tertiary and Quaternary. Ulleung Island consists largely of basalt and tuff as well as trachyte, andesitic

Table 12.1 Description and Interpretation of Echo Types in the East Sea.

Class	Type	Line Drawing	Description	Occurrence	Interpretation
I	I-1 I-2		Very sharp surface echoes with either no or rare diffuse internal reflectors; sometimes, strongly reflective, prolonged subbottom echoes below the surface echoes; nearly flat or slightly irregular topography; shallow (< 15 m) penetration depth	Eastern continental shelf of Korea; southern shelf of the Ulleung Basin (UB); lower slope of the Ulleung Seamount and Island and the escarpment of the South Korea Plateau (SKP)	Coarse-grained sediments on the shelf surface (Pratson and Laine, 1989); coarse-grained volcaniclastic mass-flow deposits (Chough et al., 1985)
	I-2	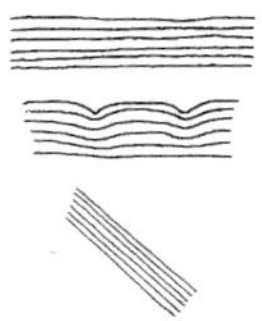	Distinct bottom echoes and several discrete, continuous internal reflectors; 15–80 m in penetration depth; nearly flat surface topography; a filling geometry in the topographic depressions; a few small-scale (0.7–1 km wide, <10 m deep), intermittent channel-levee systems showing a N–S or NNW–SSE orientation; northward decrease in channel relief; continuous aggrading reflectors in the levees; slightly to steeply inclined surface topography in the eastern continental slope of Korea and the ridge summits of the South Korea Plateau (SKP)	Basin plain of the UB; summits and broad, gentle sloping areas of the ridges in the SKP; upper to middle parts of the eastern continental slope of Korea; distal lower slopes of the northern Oki Bank (OB) and the volcanic islands and seamounts in the Ulleung Interplain Gap (UIG); upper slope and ridge summits in the OB	Turbidites with a few small-scale depositional channel-levee systems (Damuth, 1980; Flood and Damuth, 1987); pelagites or hemipelagites (Yoon et al., 1996; Lee et al., 2002).
	I-3		An uppermost transparent layer which is discordant to internal reflectors of the underlying sedimentary sequences; locally, vague or indistinct,	Southeastern flank of the Ulleung Interplain Channel	Bottom-current deposits (Howe, 1996; Faugères et al., 1999)

(*Continued*)

Table 12.1 (Continued)

Class	Type	Line Drawing	Description	Occurrence	Interpretation
			discontinuous internal reflectors in the transparent layer; 3–6 m in thickness; lateral decrease in thickness, forming a mound shape in cross section; a NE–SW-trending elongate distribution along the Ulleung Interplain Gap		
	II-1	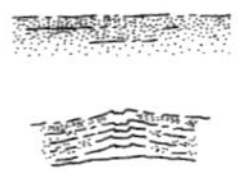	Indistinct to very prolonged bottom echoes with either no or a few sub-parallel, intermittent internal reflectors; 20–40 m in penetration depth; nearly flat surface topography; a filling geometry in the topographic depressions; a few small-scale (0.3–1 km wide, less than 5 m deep), intermittent channel-levee systems showing lobate geometry in cross section; a N–S or NNW–SSW orientation of channel-levee systems; northward reduction in channel relief; continuous aggrading reflectors in the levees	Basin plain of the UB; distal lower slopes of the northern SKP and the northern OB	Turbidites with a few small-scale depositional channel-levee systems (Damuth and Hayes, 1977; Clark and Pickering, 1996; Chough et al., 1997)
II	II-2		Laterally wedged, acoustically transparent masses; variable surface echoes ranging from seafloor-tangent hyperbolae to weak or very prolonged echoes; seafloor-tangent surface echoes changing downslope to weak prolonged bottom echoes; convex up or nearly flat	Lower slope of the UB; lower continental slope of Korean peninsula; lower slope of the ridges/seamount chains of the SKP: intervening troughs of the SKP;	Debrites (Embley, 1976; Damuth, 1980; Lee et al., 1999b)

			upper surface with concave-upward base; downslope decrease in thickness	Onnuri Basin; lower slopes of the northern OB and the volcanic seamounts and islands in the UIG	
III	III-1		Large single or irregular overlapping hyperbolae with widely varying vertex elevations (up to 200–300 m) above the seafloor; highly variable dimensions of the hyperbolae; strongly reflective surface echo and prolonged subbottom echoes of the hyperbolae; locally covered by thin blocky or hyperbolic masses (type IV-1)	Ridge summits, seamount chains and escarpments of the SKP; volcanic seamounts and islands in the UIG	Basement highs or outcrops (Damuth, 1980; Laine et al., 1986).
	III-2		Regular, intense overlapping hyperbolae with little or slightly varying vertex elevations; 50–250 m in wavelength and less than 10–15 m in hyperbola relief; laterally wedged geometry; convex-up upper surface and concave-upward base; a progressive downslope decrease in spacing, size, and vertex elevation of the hyperbolae	Southern and eastern lower slopes of the UB; middle to lower part of the eastern continental slope of Korea; lower slope of the ridges and seamount chains of the SKP; intervening troughs of the SKP	Debrites (Nardin et al., 1979; Damuth and Embley, 1981; Lee et al., 1999b)
	III-3		Regular to slightly irregular wavy bottom echoes with upslope migrating subbottom reflectors in downslope-parallel section; in contour-parallel section, irregular and less wavy forms; parallel orientation of the wave crests to the regional slope; asymmetric wave	Lower slopes of the OB and Dok Island	Turbidity-current sediment waves (Normark et al., 1980; Wynn et al., 2000; Wynn and Stow, 2002)

(Continued)

Table 12.1 (Continued)

Class	Type	Line Drawing	Description	Occurrence	Interpretation
			geometry with steeper, thicker upslope and gentler, thinner downslope flanks; waves: 0.3–2 km in wavelength, less than 15 m in wave height; gradual downslope reduction in wave height, changing to nearly flat surface topography; progressive downslope decrease in thickness and wave asymmetry; mostly overlain by type I-3		
	III-4	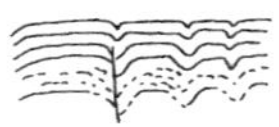	Irregular, wavy surface echo and several continuous internal reflectors; basal mingled reflections overlying the undeformed layers; wavelength of 0.2–5 km and wave height of 2–15 m; systematic variations in both wave dimensions and thickness of internal reflectors with change in slope gradient; common internal faults in the wave troughs	Broad, gentle-sloping ($<0.5°$) areas of the ridges of the SKP	Creep deposits (Syvitski et al., 1987; Lee and Chough, 2001)
	III-5		Wavy, prolonged bottom echo with a few indistinct, parallel subbottom echoes; 0.3–2 km in wavelength and <5 m in wave height; internal faults in wave troughs; slightly higher (5–10 m) topographic position than the adjacent echo types II-2 and IV-1	Intervening troughs of the SKP	Deformed turbidites by shearing of debris flows and slides/slumps (Damuth and Hayes, 1977; Gardner et al., 1999)
IV	IV-1	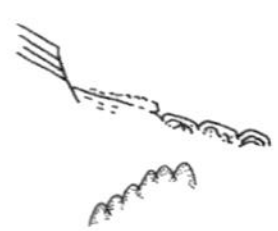	Irregular blocky, lumpy or hyperbolic masses bounded upslope by scarps or scars; a few meters to 90 m in scarp	Upper to middle slopes of the UB; upper to lower slope of the ridges and	Slide/slump deposits and mass-failure scars (Embley and Jacobi,

		relief; highly variable degrees of internal deformation; scars marked by sharp glide plane or irregular drape of thin acoustically transparent or hyperbolic masses; several scars below the head scarp; step-like geometry of failed masses; locally, strongly reflective, slightly irregular to regular overlapping hyperbolae (0.3–1.2 km wide) with slightly varying vertex elevation (< 10 m) above the seafloor	seamount chains and escarpments of the SKP; upper to lower part of the eastern continental slope of Korea; upper to lower slopes of the northern OB and volcanic seamounts and islands in the UIG	1977; Chough et al., 1985)
IV-2		Various echo characters on the channelized geometry; channel floors: nearly flat to convex-up, sharp, prolonged or small-scale hummocky bottom echoes with either transparent or rare intermittent, discontinuous subbottom echoes; channel walls or levees: either eroded or aggrading internal reflectors; channels: 2–8 km wide, 7–15 m deep, downslope decrease in relief	Lower slopes of the northern Oki Bank and volcanic islands and seamounts, and the escarpment of the South Korea Plateau	Mass-flow chutes and channels (Flood and Damuth, 1987; Clark and Pickering, 1996)
IV-3	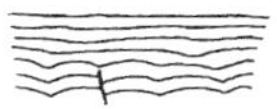	Flat, distinct to indistinct surface echoes with several continuous subbottom echoes; filling geometry on the depressions; nearly flat internal reflectors in the upper part and wavy internal reflectors with a subbottomward increase of wave amplitude in the middle to lower part; small-scale internal faults in the wavy part	Intervening troughs of the SKP	Turbidites (Pratson and Laine, 1989; Chough et al., 1997) wavy internal reflectors in the middle to lower part: deformation (O'Leary and Laine, 1996)

(*Continued*)

Table 12.1 (Continued)

Class	Type	Line Drawing	Description	Occurrence	Interpretation
	IV-4		Laterally wedged transparent masses and indistinct stratified masses; filling geometry in the depressions; transparent masses: convex-up to nearly flat upper surface and concave-up base, often bounded below by the eroded indistinct stratified masses; indistinct stratified masses: partly disrupted reflections	Intervening troughs of the SKP	Interlayered debrites (transparent masses) and turbidites (indistinct stratified masses) (Embley, 1976; Chough et al., 1997).
	IV-5		Large-scale, slightly to strongly channelized surface topography with various bottom and subbottom echoes; 1.5–13.6 km wide, 15–85 m deep; asymmetric flank geometry in cross section; split into several branches in the southern part; channel floor: sharp to very prolonged bottom echoes with either no or a few intermittent, prolonged subbottom reflectors that are often truncated by the present channel-floor surface; channel margins or walls: slightly irregular topographic, indistinct bottom echoes which truncate distinct to indistinct internal reflectors; in the southern branches, local draping of small-scale, thin transparent layers on the channel floor	Axis of the Ulleung Interplain Gap	Bottom-current channel system (Johnson and Damuth, 1979; Johnson, 1984)

Source: Lee et al. (2004) by permission of the AGU.

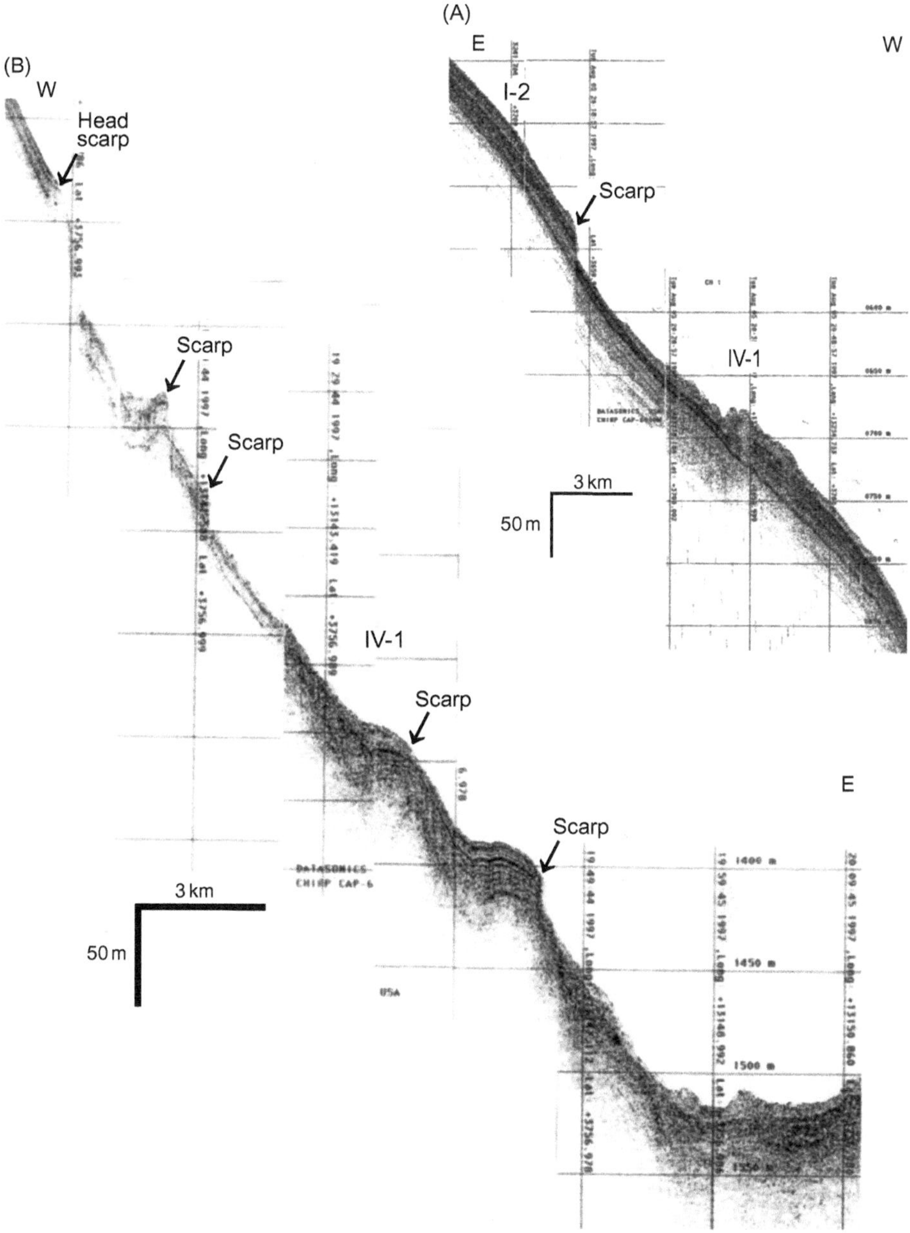

Figure 12.13 High-resolution (2–7 kHz) subbottom profiles of combined echoes (IV). For location of each profile, see Figure 12.3B. (A) and (B) Irregular blocky, lumpy, or hyperbolic masses bounded upslope by scars (type IV-1). Note several subsidiary scarps below the head scarps. (C) Local occurrence of strongly reflective, regular overlapping hyperbolae with slightly varying vertex elevation in type IV-1. (D) Channelized topography with various bottom and subbottom echoes (type IV-2). (E) Flat, distinct to indistinct surface echoes with several continuous internal reflectors (type IV-3). Wavy internal relectors in the middle to lower part with a subbottomward increase in wave amplitude. Note internal faults (open arrows). (F) Combined laterally wedged, transparent masses and indistinct stratified masses (type IV-4). *Source:* Lee et al. (2004) by permission of the AGU.

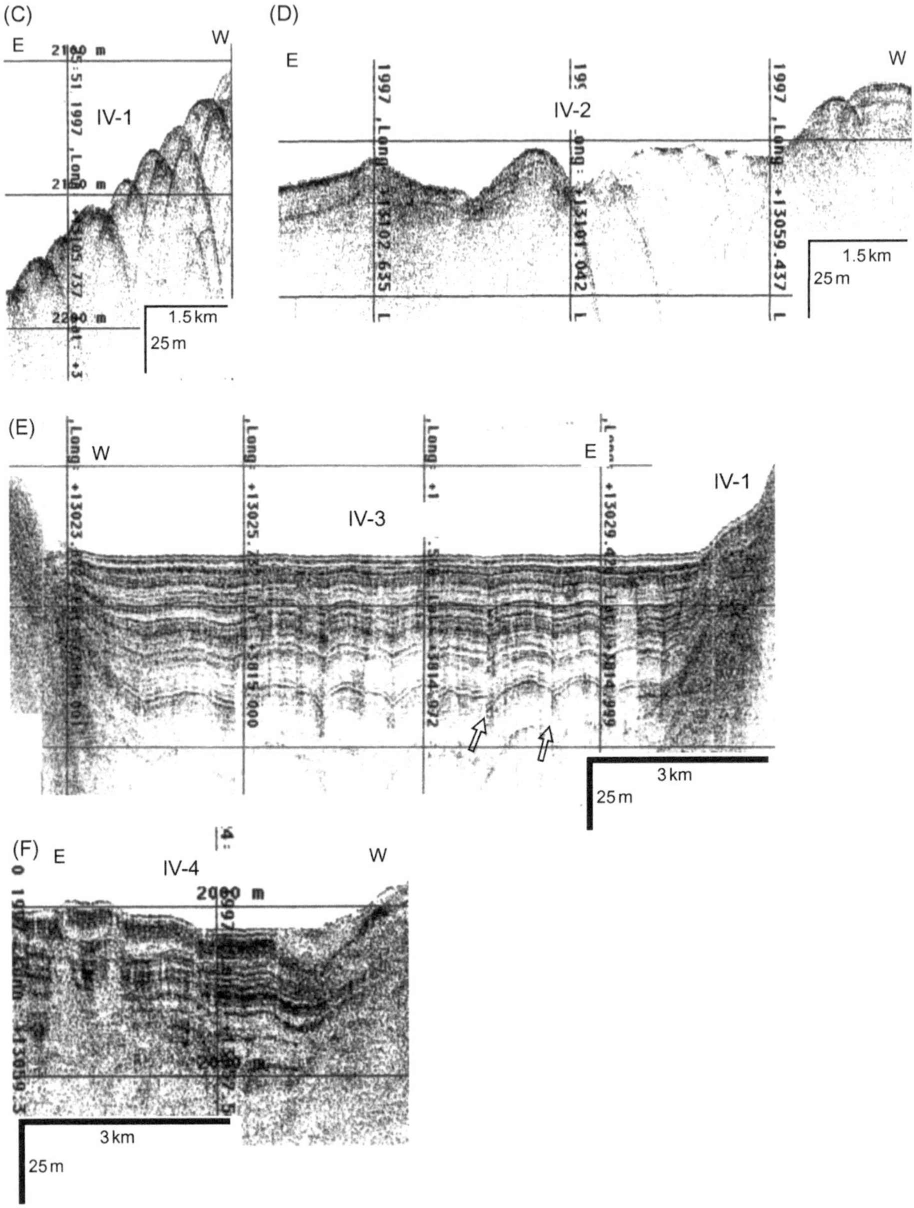

Figure 12.13 (Continued).

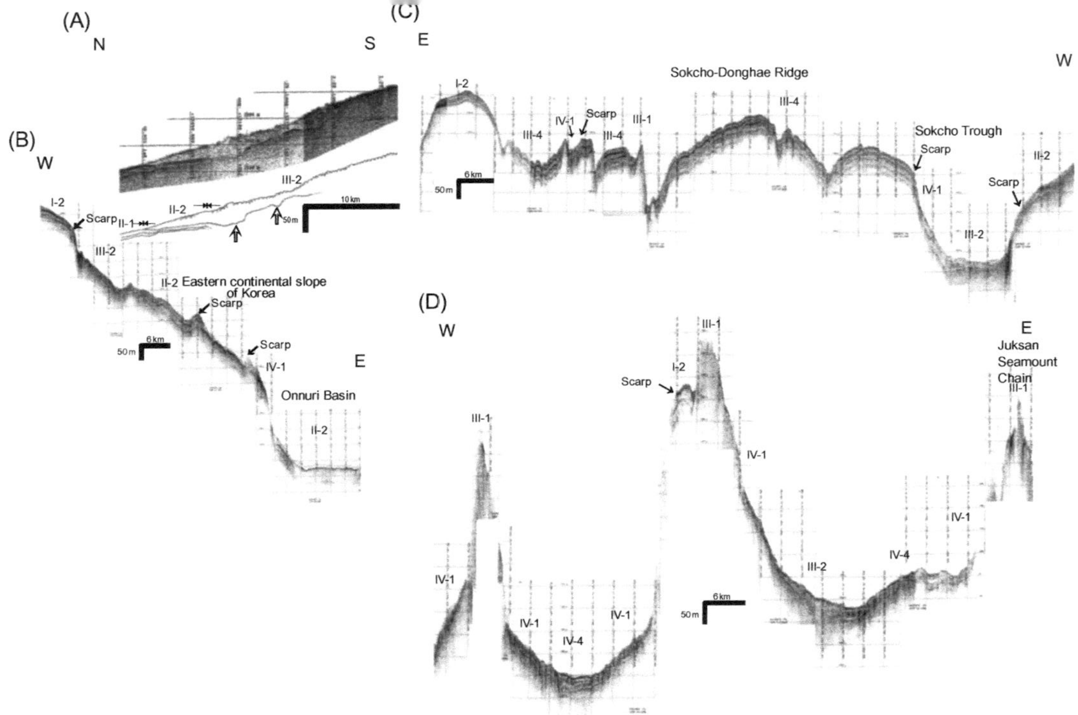

Figure 12.14 Typical Chirp subbottom profiles showing occurrence and association of echo types according to the topographic provinces. For location of each profile, see Figure 12.3B. (A) Longitudinal seismic profile of a debris lobe consisting of types II-2 and III-2 in the southern lower slope of the Ulleung Basin. Note (1) gradual downslope change in echo characters from hummocky bottom echo to seafloor-tangent hyperbolic and smooth prolonged surface echo and (2) gradual downslope decrease in size, spacing, and relief of the hummocks. (B) Eastern continental slope of Korea and Onnuri Basin in the South Korea Plateau. Type I-2 on the upper to middle continental slope changing downslope to types III-2 and II-2 with the head scarp. At the lower continental slope, type II-2 is dissected by scarps and remolded into echo type IV-1. Note type II-2 (debrites) draping over the entire floor of the Onnuri Basin. (C) Ridges and troughs in the South Korea Plateau. Extensive occurrence of type III-4 on broad, gentle sloping areas of the ridges. Type III-4 is dissected by scarps at the upper slope of the ridges and transitional downslope to echo types IV-1 and III-2. (D) Seamount chains and troughs in the South Korea Plateau. Dominant occurrence of type IV-1 on the middle to lower slope of the seamount chains and types III-2 and IV-4 on the troughs.
Source: Lee et al. (2004) by permission of the AGU.

basalt, and phonolite (Won and Lee, 1984). Volcanic rocks of Ulleung Island are dated from 8.07 ± 0.39 to 0.51 ± 0.07 Ma. Dok Island comprises basalt, trachyte, andesite, and tuff (Sohn and Park, 1994). The volcanic rocks consist of a number of units, depending on the eruptive and depositional styles and on whether subaqueous or subaerial effusive conditions prevailed (Sohn and Park, 1994). These rocks range in age from 4.6 to 1 Ma (Sohn and Park, 1994; Song et al., 2006).

13 Jeju Island

13.1 Geological Setting

Jeju Island is an intraplate volcanic island situated on the continental crust off the southern coast of the Korean Peninsula, about 1830 km^2 in area (74 km east to west and 32 km north to south). The mountain peak (Halla Mountain) is 1950 m high with a central crater (Baekrokdam crater, 575 m in diameter and 100 m in depth). The entire island is a shield volcano of basaltic lava flows and pyroclastic rocks (Figure 13.1). It contains approximately 360 scoria cones in inland areas and a dozen tuff cones and rings along the coastal regions. The volcanism occurred throughout the Quaternary and continued into historic times. The volcanism was formerly interpreted to have originated from plume-related hotspot magmatism (Lee, 1982b; Tatsumi et al., 2005). Recent studies suggest, however, that the volcanism was caused by decompression melting of the shallow asthenosphere (Choi et al., 2006), possibly in association with rotation in a subduction direction of the Philippine Sea plate at around 2 Ma (Brenna et al., 2012). A geological cross-section shows that the island was built on the granite basement and the overlying continental shelf sediments of the Pliocene age, named the U Formation (Figure 13.1). The early-stage volcanism of Jeju Island was characterized by widespread and prolonged hydrovolcanic activity and attendant volcaniclastic sedimentation on the continental shelf, which resulted in the Seoguipo Formation. Lava effusion became dominant in the middle to late Pleistocene after the island emerged above fluctuating Quaternary sea levels (Sohn et al., 2008).

The late-stage lava effusion produced a broad and gently sloping lava shield and over 300 volcanic cones, rings, domes, and craters (Figure 13.2A). Molten magma surfaces in various ways, depending on the chemical composition, temperature, and volatile component. A shield volcano, such as the Jeju Island, originates from high-temperature basaltic magma that flows downslope for tens of kilometers. Lava is either relatively high in temperature (low in viscosity) (pahoehoe lava) or low in temperature (high in viscosity) (aa lava) (Figure 13.2). The pahoehoe lava forms a smooth surface topography with a gentle slope (about 5° in slope gradient), whereas the aa lava is characterized by a rough surface with klinkers (broken pieces of hardened lava). The pahoehoe lava dominated the low-gradient east and west slopes of Jeju Island, whereas the aa lava formed the steep north and south slopes (Park et al., 2006a) (Figure 13.1).

Geology and Sedimentology of the Korean Peninsula. DOI: http://dx.doi.org/10.1016/B978-0-12-405518-6.00013-9

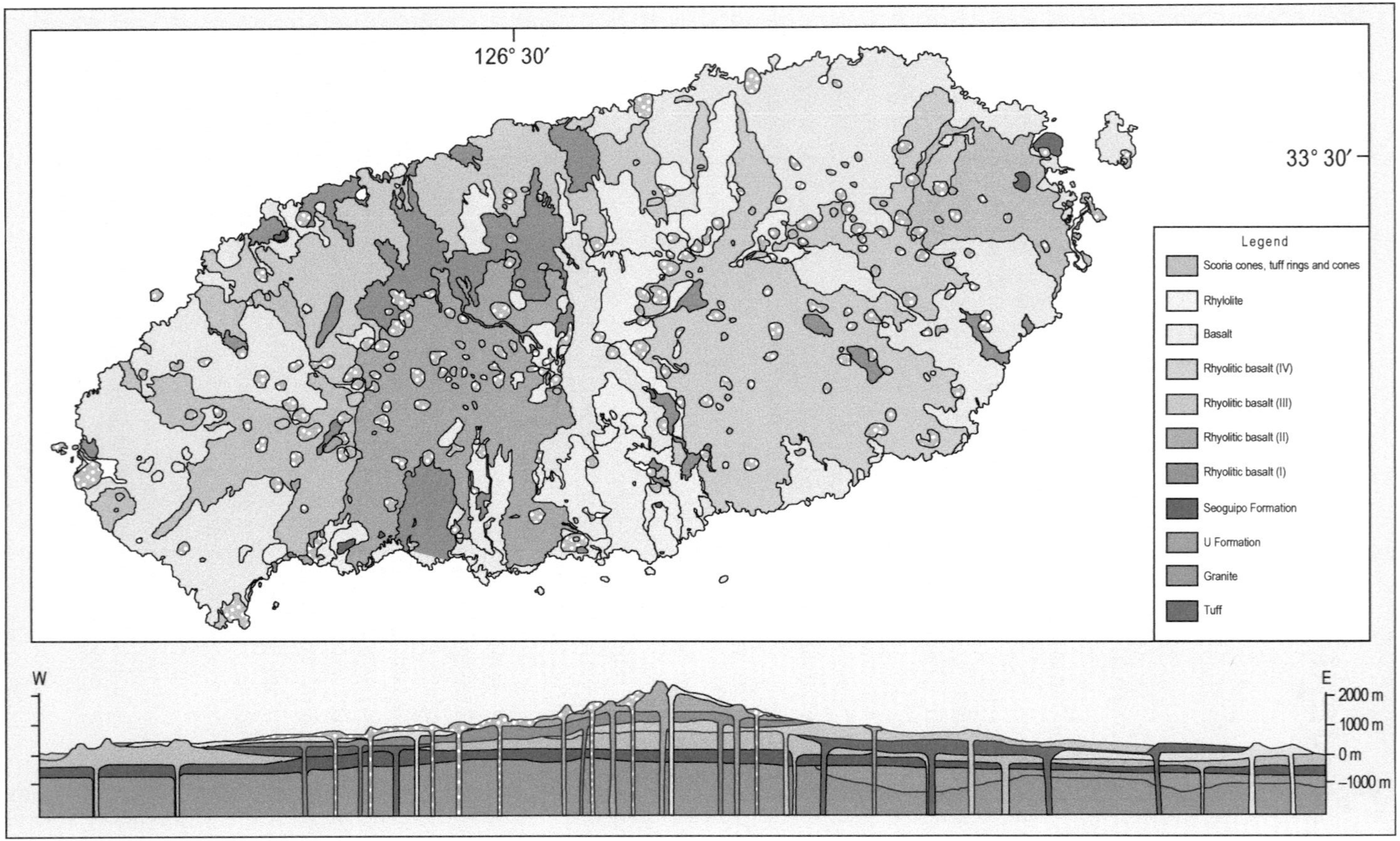

Figure 13.1 Simplified geologic map of Jeju Island.
Source: Park et al. (2006a).

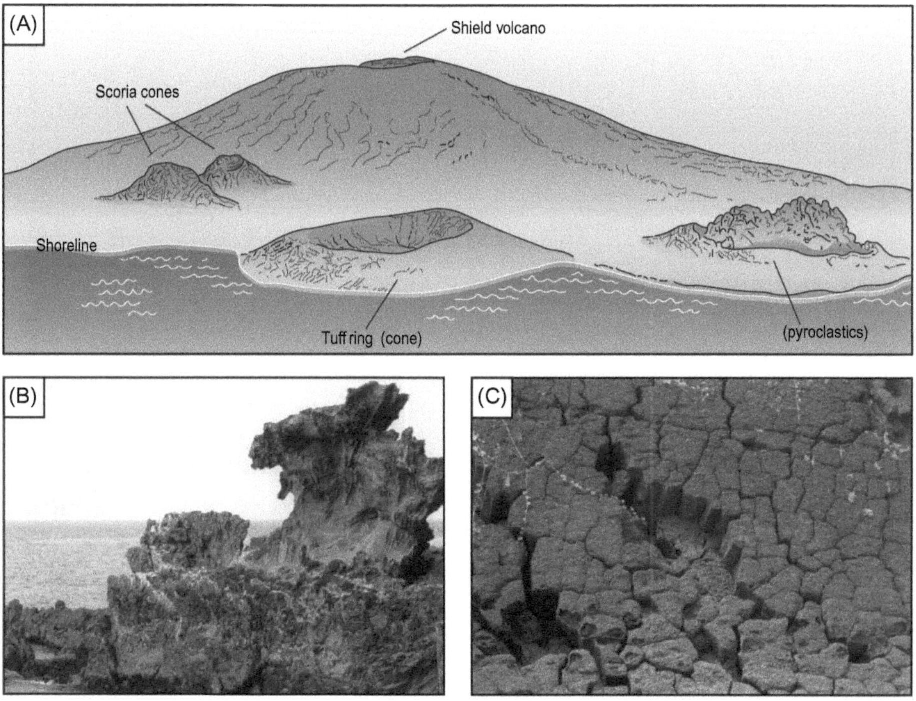

Figure 13.2 (A) Simplified morphology of volcanic features in Jeju Island. (B) Aa lava. (C) Pahoehoe lava.
Source: (A) Park et al. (2006a).

13.2 Scoria Cones

Jeju Island was dominated by monogenetic basaltic volcanism, which produced numerous small volcanic vents. The most abundant type of volcanoes produced by this volcanism is the scoria cone, which results from an isolated explosion of basaltic magma with a volatile component. Because of the sudden release of pressure near the surface, the lava fragments (scoria) are tossed into the air, forming a few hundred-meters-high chimney (column), and deposited within a radius tens of meters wide. The scoria contains abundant gas holes (vesicles). The deposit forms a cone-shaped mound (cinder cone). This type of explosion, first identified in Italy and named "Strombolian" volcanic eruption, prevailed in all of Jeju Island (Figure 13.3).

13.3 Tuff Rings and Tuff Cones

Tuff ring and tuff cone are small monogenetic volcanoes produced by explosive magma–water interactions. Tuff ring is a broad, low-relief volcano of gentle outer

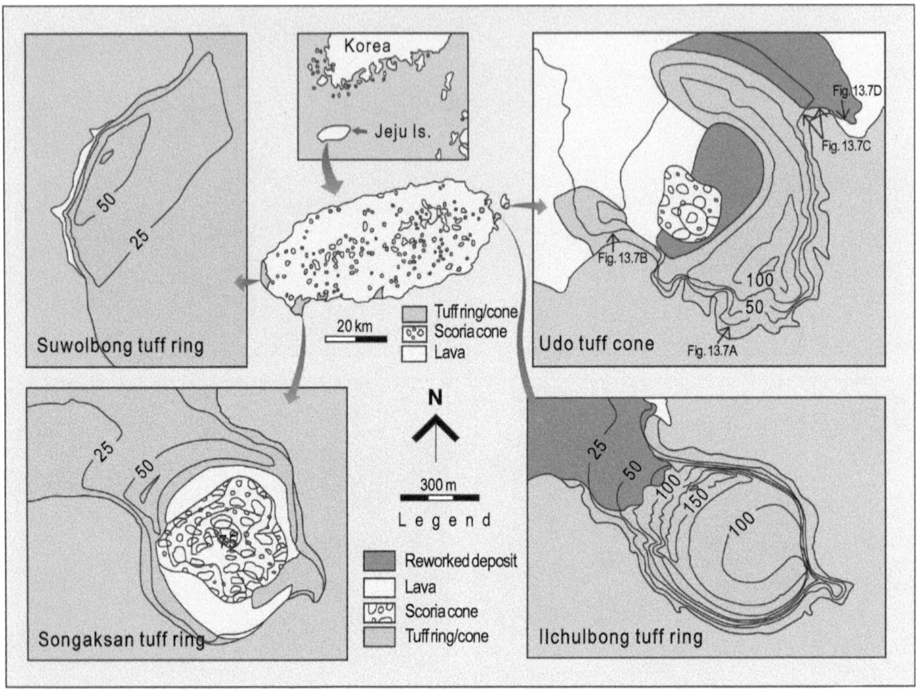

Figure 13.3 Geology of Jeju Island and location of tuff rings and cones as well as scoria cones. Contours in meters.
Source: Sohn (1996) by permission of the Geological Society of America.

slopes and a wide crater, whereas tuff cone is a narrow, steep volcano with high-angle outer slopes. In Jeju Island, the former includes the Suwolbong and Songaksan tuff rings in the western part, whereas the latter includes the Ilchulbong and Udo tuff cones in the eastern part (Figure 13.3). The tuff rings were largely formed by pyroclastic surges, whereas the tuff cones were dominated by fallout of pyroclastic ashes. The morphological variations of the rings and the cones are caused by the difference in depositional processes, which are further controlled by depositional setting, type, level, and lithology of aquifers, strength of country rocks, ground water behavior, and property and behavior of magma (Sohn, 1996).

The Suwolbong and Songaksan tuff rings were produced by relatively powerful explosions at deeper levels and were mainly emplaced by dry or wet pyroclastic surges (Table 13.1) (Figure 13.4). A detailed analysis of sedimentary facies suggests that the Suwolbong tuff comprises eighteen sedimentary facies, which are organized into six lateral and seven vertical facies successions (Figure 13.4). Depositional processes were dominated by a relatively dry base surge (Sohn and Chough, 1989). The base surge comprises turbulent and high concentration suspension near the vent whose deposits are generally unstratified due to the lack of

Table 13.1 Sedimentary Facies in Tuff Rings and Cones of Jeju Island, Korea

Facies code	Brief Description	Occurrence	Interpretation
Bm	Thin disorganized breccia layers and isolated clasts with common bedding sags	Ubiquitous in tuff rings and cones	Ballistic fall
Lm	Massive or crudely stratified lapilli deposits; well-sorted and openwork; occasionally scoriaceous	Several beds intercalated in tuff rings	Fallout from dry tephra jets or Strombolian eruptions
LTm and Tm	Massive, thick-bedded lapilli tuff and tuff; LTm_1 and Tm have coarse ash matrix, outsized clasts, amalgamated boundaries, and scoured bases;	LTm_1 and Tm: common in proximal regions of tuff rings	LTm_1 and Tm: highly concentrated near-vent surge
	LTm_2 has muddy matrix, intraformational clasts, lenticular geometry, and positive relief.	LTm_2: common in lower flanks of tuff cones	LTm_2: debris flow generated by slope failure
LTg	Normally graded lapilli tuff; thick-bedded, coarse-tail graded, scoured bases; beds with inverse-graded base denoted by LTig	Proximal regions of tuff rings	Rapid deposition from highly concentrated near-vent surge
Li and LTi	Inversely graded lapilli deposits and lapilli tuff; less than a few decimeters thick, clast supported; planar (Li_1 and LTi_1) or lenticular geometry (Li_2 and LTi_2)	Steep rim regions of tuff rings	Grain flow transfomed from dry tephra jets
Ti	Inversely graded tuff, thin-bedded	Proximal regions of tuff rings	Traction carpet formed under near-vent
LTs	Stratified lapilli tuff; LTs_1 is composed of alternating coarse- and fine-grained layers; LTs_2 is crudely bedded, poorly sorted	LTm_1: medial regions of tuff rings LTm_2: steep rim beds of tuff cones	LTs_1: relatively dry surge LTs_2: fallout from wet tephra jets
Ts	Thinly stratified tuff; Ts_1 shows plane-parallel bedding; Ts_2 shows mantle bedding	Medial to distal regions of tuff rings and cones	Ts_1: dilute surge Ts_2: air fall
LTx and Tx	Cross-stratified lapilli tuff and tuff; low-angle cross-stratification or backset stratification, scour-fill bedding	Proximal regions of tuff rings	Traction deposition from erosive near-vent surge

(Continued)

Table 13.1 (Continued)

Facies code	Brief Description	Occurrence	Interpretation
LTx and Tw	Wavy- or megaripple-bedded lapilli tuff and tuff; down current crest migration, continuous bedding a cross crest and trough	Most common in medial to distal regions of tuff rings	Traction and suspension deposition from dilute surge
LTc	Chaotic lapilli tuff; backset-stacked and brecciated strata	Ubiquitous in tuff cones	Slide and/or slump due to slope failure
La	Accretionary lapilli (La_1) and armored lapilli (La_2)	Common or absent in either tuff rings or tuff cones	Adhesion of sticky wet ash in eruption columns or surge clouds
Rm	Massive lenticular-bedded basaltic gravelly deposits bounded by irregular scour surfaces	Volcaniclastic apron around tuff cones	Debris flow generated by reworking of steep cone deposits
Rm	Trough or ripple cross-bedded basaltic sand in small channels or pockets	Channels or pockets	Reworking by rill flows and marine processes

Note: Summary of descriptive codes for facies classification: B = block deposit or breccia, L = lapilli deposit or lapillistone, LT = lapilli ash or lapilli tuff, T = ash or tuff, R = reworked deposits, m = massive or disorganized, g = normally graded, i = inversely graded, s = stratified, x = cross-bedded or scour-fill-bedded, w = wavy- or megaripple-bedded, c = chaotic, a = accretionary or armored.
Source: Sohn (1996) by permission of the Geological Society of America.

traction. As the base surge becomes diluted downcurrent through the fallout of clasts and mixing of ambient air, it develops large-scale turbulent eddies, segregated into coarse-grained bedload and overlying suspension load (Figure 13.5). The explosion depths were much deeper, since there was virtually no surface water or shallow-level ground water (Sohn, 1996). Magma interacted with the interstitial water in wet sediments and permeable rocks and was fragmented by steam explosion. The Songaksan tuff ring comprises mainly thin-bedded and gently dipping tuff with abundant megaripples (Figure 13.6) (Chough and Sohn, 1990). The deposits consist mainly of sideromelane ash and poorly vesicular lapilli as well as accidental components, which are indicative of deep excavation and incorporation of abundant country rocks by hydroexplosion.

The Ilchulbong and Udo tuff cones were formed by relatively weak explosions at shallow depth and emplaced by the fallout of dry to wet tephra (mostly lapilli) (Figures 13.6) (Sohn, 1996). The presence of surface water acted as an inexhaustible fuel. A sedimentological study of the Ilchulbong tuff cone reveals that the cone grew by an alternation of vertical and lateral accumulation (Sohn and Chough, 1992). The vertical buildup formed by plastering of wet tephra, resulting in oversteepening and the subsequent slope failure. This process of resedimentation contributed to the lateral growth. After the eruption ceased, the cone underwent

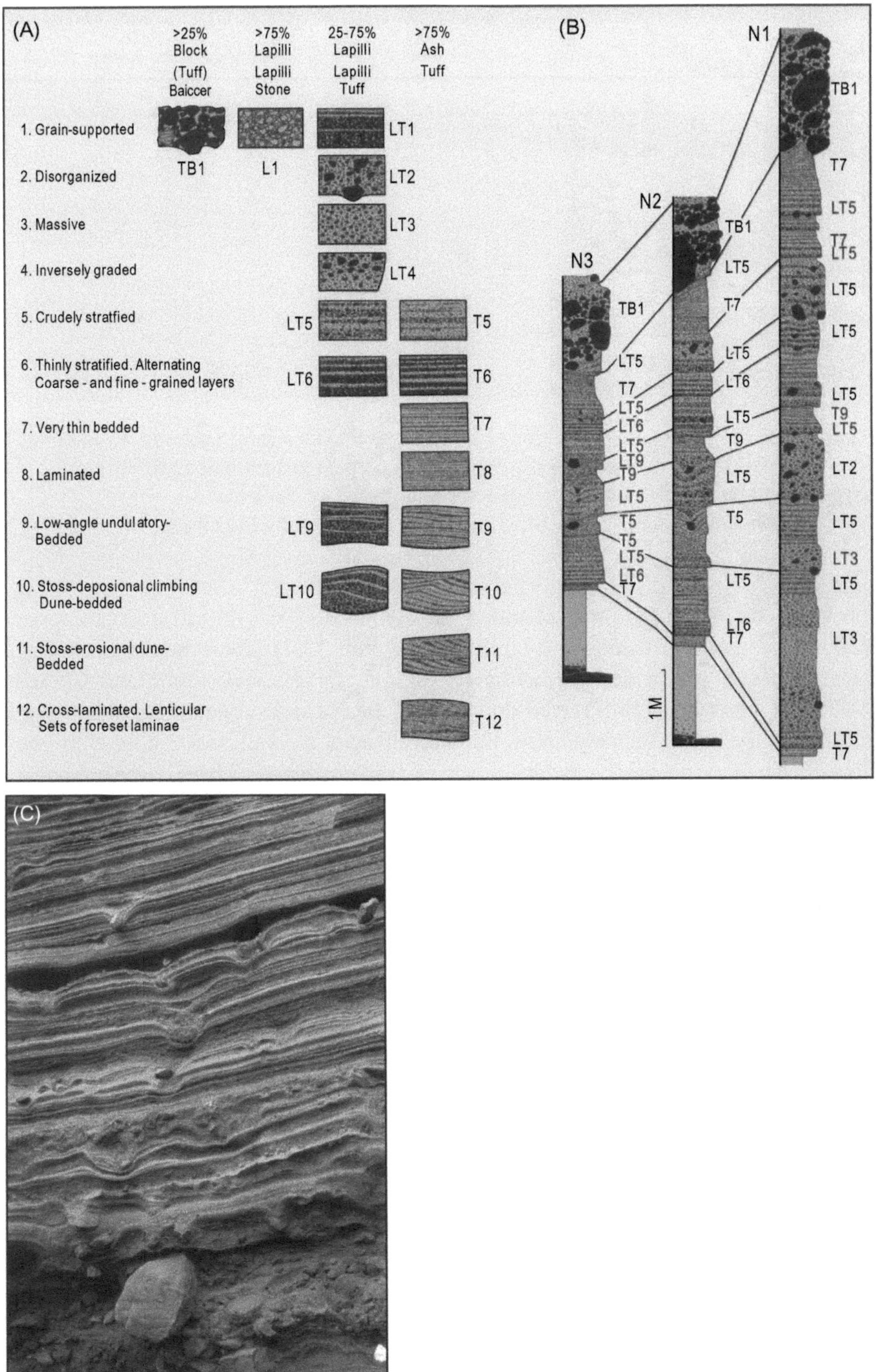

Figure 13.4 (A) Facies classification scheme based on grain size and sedimentary structures. (B) Bed-by-bed correlation of measured sections in Suwolbong tuff ring, comprising mostly crudely stratified units (LT5, T5). (C) Photograph of Suwolbong succession, showing laminated and graded layers with prominent sags.
Source: (A) and (B) Sohn and Chough (1989) by permission of the Wiley-Blackwell.

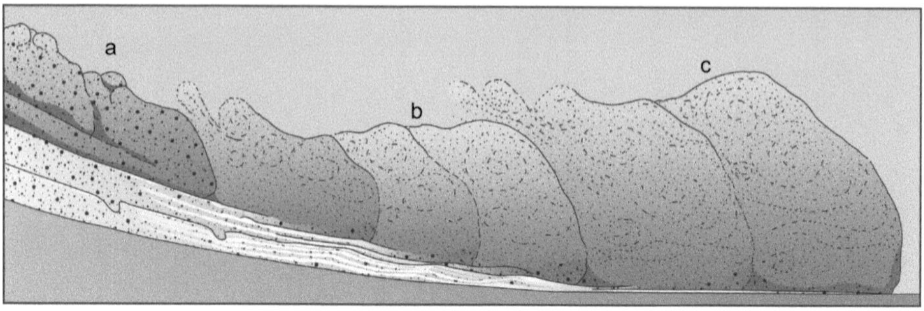

Figure 13.5 Diagrammatic illustration of a base surge cloud at three stages (a, b, and c) of development. The base surge expands, decreases in particle concentration and develops large-scale turbulent structures. For detailed explanation, see the text.
Source: Sohn and Chough (1989) by permission of the Wiley-Blackwell.

subaerial erosion and faulting of intracrater deposits. The Udo tuff cone comprises inclined beds of lapillistone and lapilli tuff (Figure 13.7). These beds were formed by grain flows which were transformed from pyroclastic fall (Sohn and Chough, 1993). As the debris fall traveled down the slope, particle concentration increased, and hence the mode of momentum transfer changed to collisional, similar to grain flow.

13.4 Seoguipo Formation

The Seoguipo Formation is exposed along the coastal cliff (about 50 m high) of Seoguipo city, overlain by hawaiite lava (0.4 Ma) (Figure 13.8). Examination of numerous borehole data suggests that the formation is part of volcaniclastic and fossiliferous deposits (approximately 100 m thick), which occur mostly below the present sea level beneath the shield-forming lavas (Koh, 1997). The formation consists of numerous superposed phreatomagmatic volcanoes intercalated with marine (or nonmarine) and volcaniclastic deposits, indicating that there was widespread hydrovolcanic activity and associated volcaniclastic sedimentation prior to the effusion of the shield-forming lavas (Sohn and Park, 2004; Sohn et al., 2008). A number of paleontological, paleomagnetic, stable isotopic, and geochronological studies (Tamanyu, 1990; Yi et al., 1998; Li et al., 1999; Khim et al., 2001; Kang et al., 2002; Kang, 2003) reveal that deposition began at about 1.8 Ma and continued until about 0.8–0.4 Ma. The formation deposited over a million years throughout the Early Pleistocene.

The formation consists of mixed basaltic volcaniclastic and siliciclastic gravelly sandstone, sandstone, and sandy mudstone with abundant fossils of mollusks and brachiopods (Kang et al., 1999; Li et al., 1999; Khim et al., 2001) (Figure 13.8). According to a detailed description of the sedimentary facies, it consists of 14

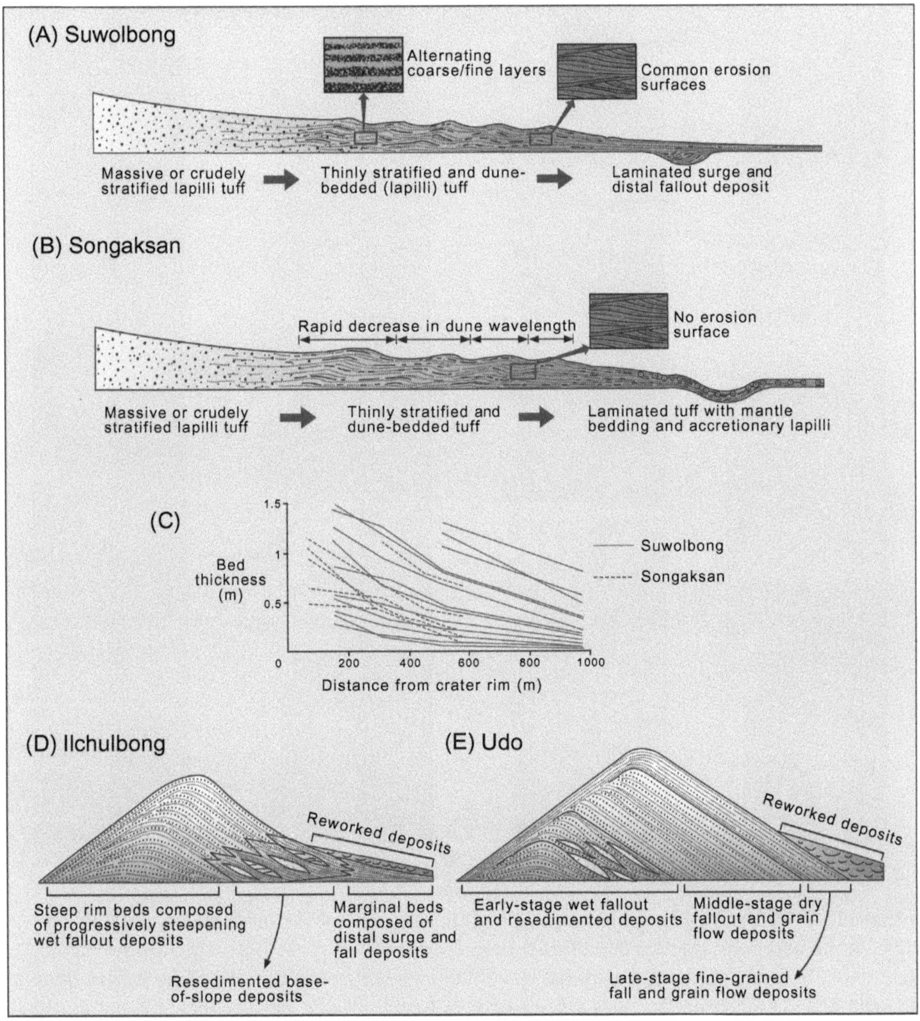

Figure 13.6 (A) and (B) Lateral (downflow) facies variations in pyroclastic-surge deposits of the Suwolbong and Songaksan tuff rings. (C) Plot of pyroclastic-surge bed thickness vs. distance from the crater rim in the tuff rings. (D) and (E) Spatial or temporal facies variations in the Ilchulbong and Udo tuff cones.
Source: Sohn (1996) by permission of the Geological Society of America.

sedimentary facies association units distinguished by sharp erosional surfaces, distinct changes in lithology, and fossil contents as well as the degree of bioturbation, representing distinct changes in depositional environments (Figure 13.9). The succession largely formed in nearshore and inner shelf areas with skeletal banks (channels, lobes, and ridges) shaped by tidal currents as well as storms in the outer shelf.

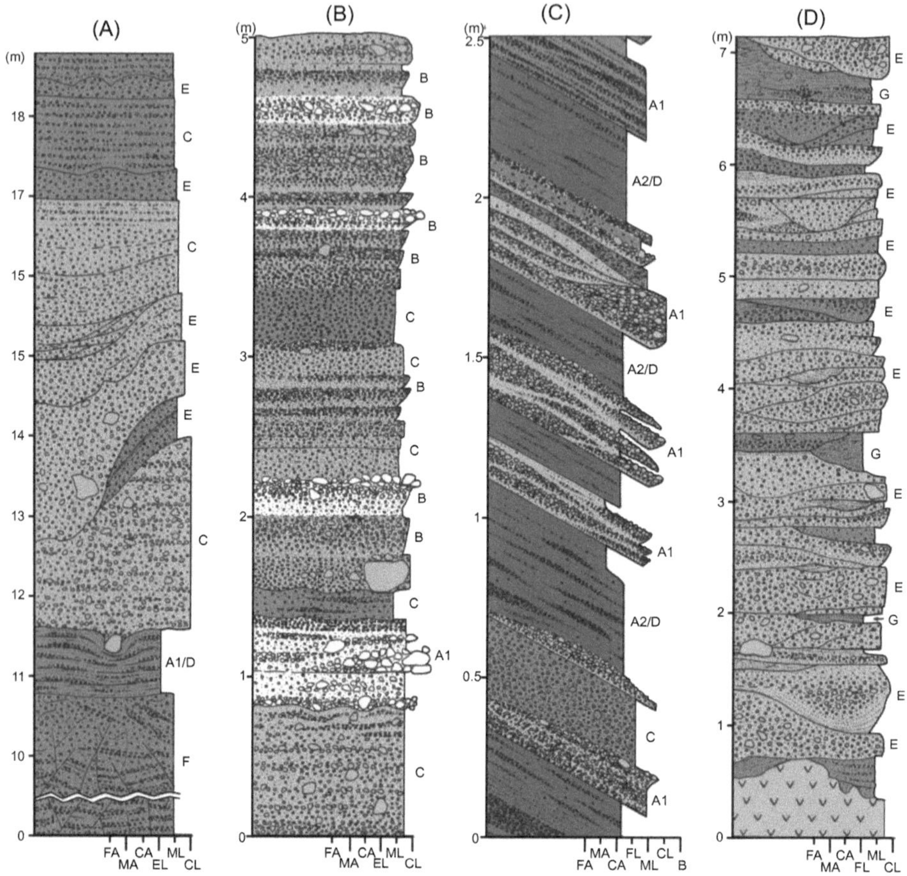

Figure 13.7 Part of stratigraphic sections in Udo tuff cone. (A) Lower part dominated by slide/slump and debris flows. (B) Middle part dominated by grain flows. (C) Upper part dominated by grain flows and ashfalls. (D) Volcaniclastic apron dominated by debris flows and rill flows. Letters on the right side of each column indicate facies codes. FA = fine ash, MA = medium ash, CA = coarse ash, FL = fine lapilli, ML = medium lapilli, CL = coarse lapilli, B = block. For locations, see Figure 13.3.
Source: Sohn and Chough (1993) by permission of the Wiley-Blackwell.

However, there are several volcaniclastics-dominant units in which no fossils or bioturbations occur. Sohn and Yoon (2010) made new observations of these deposits, interpreting them as having resulted from syneruption reworking of volcanic debris derived from pyroclastic clouds passing over water in a shallow sea. They mainly consist of normally graded, parallel-stratified, or low- to high-angle cross-stratified units with common intercalations of mm- to cm-thick mud flasers, formed by combined suspension and traction sedimentation under alternating currents and still waters of a tidal regime (Figure 13.10). Mud flasers in the deposits, which

Figure 13.8 (A) Location and geological maps of the Seoguipo city. CTA(s) = Cheongiyeon trachyandesite or scoria, JTB = Jeongbangdon trachyandesite, SF = Seoguipo Formation. (B) Sea-cliff exposure of Seoguipo Formation. (C) Part of the Seoguipo Formation, showing abundant fossils (mollusks and brachiopods).

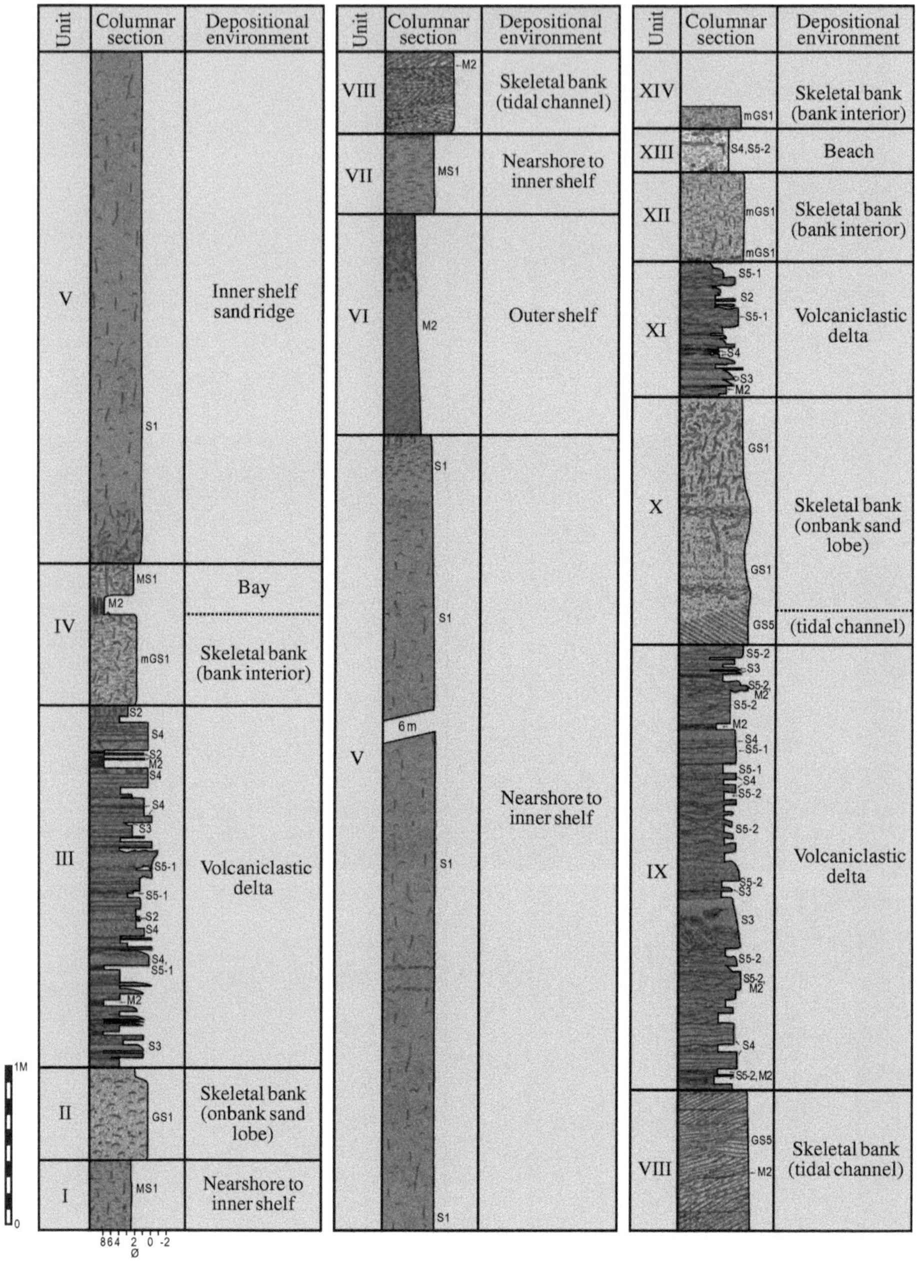

Figure 13.9 Sedimentary log and interpreted depositional environments of the Seoguipo Formation. GS1 = bioturbated gravelly sandstone, mGS1 = bioturbated muddy gravelly sandstone, GS4 = parallel-stratified gravelly sandstone, GS5 = cross-stratified gravelly sandstone, S1 = bioturbated sandstone, S2 = massive sandstone, S3 = graded sandstone, S4 = parallel-laminated sandstone, S5-1 = ripple cross-laminated sandstone, S5-2 = hummocky cross-laminated sandstone, Ms1 = bioturbated muddy sandstone, M2 = homogeneous mudstone.

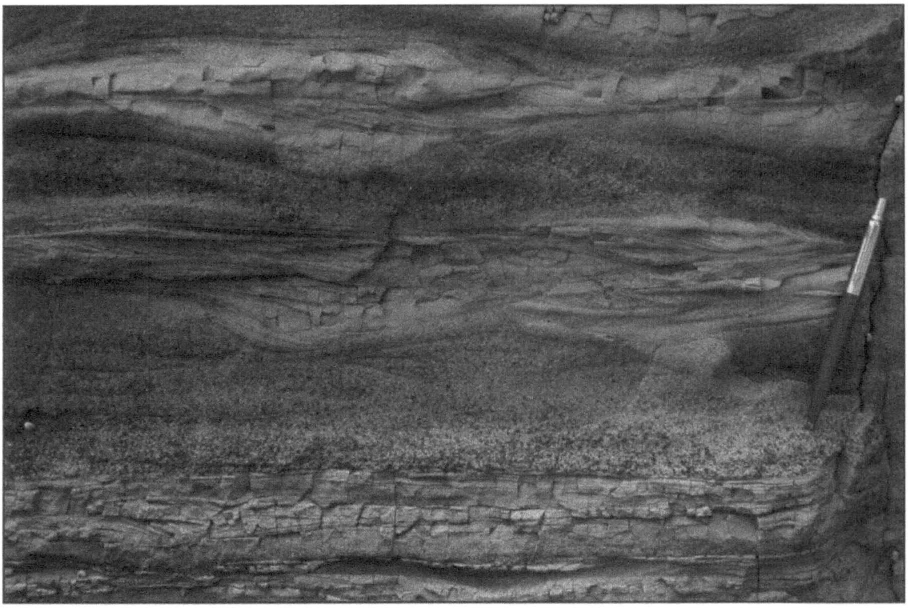

Figure 13.10 Volcaniclastics-dominant unit (Unit XI) of the Seoguipo Formation.

indicate periods of slack water during tidal cycles, suggest that the deposition occurred in a very short period of a fortnight or a month, about a million times faster than the adjacent siliciclastic sedimentary strata. Because of the unusually high sedimentation rate, the volcaniclastic deposit could record the fair-weather processes in the depositional site, i.e., the tidal processes, at a resolution that is almost never provided by ordinary sedimentary deposits.

14 Yellow Sea

14.1 Introduction

The Yellow Sea is a distinct epicontinental sea, characterized by a flat and broad seafloor (Figure 14.1). The western part of the sea includes deltas of the Huanghe River; off the Jiangsu coast with the shallow seafloor (about 50 m) gently sloping eastward. The eastern part forms numerous islands, estuaries, and tidal flats. Tidal sand ridges are ubiquitous in isobaths of about 70 m. Figure 14.2 shows surface sediment distribution. Depositional processes and environmental changes in the sea were summarized in the past (Chough et al., 2000). In this book, I will focus on new findings since then, based on data of high-resolution subbottom profiles, drill cores, and age dating.

14.2 High-Resolution Profiles

Tremendous amounts of closely spaced (2.2–2.4 km interval) high-resolution subbottom profiles (Chirp, 2–7 kHz) were obtained in the eastern part of the Yellow Sea by the National Oceanographic Research Institute (Figure 14.3). Analysis of these profiles, based on seafloor morphology, surface bedforms, and subsurface acoustic characteristics, reveals that sedimentary deposits comprise 11 echo types (Table 14.1; Figure 14.4). Figure 14.5 shows a synthesis of the distribution of echo types in the eastern part of the Yellow Sea (Chough et al., 2002).

The central part of the sea is occupied by acoustically transparent, sheetlike mud (echo type 1-3a), which was deposited during transgression. It is partly overlain by wedge-shaped mud (echo type 1-3b) in the southeastern part (Figure 14.4M). The eastern part (east of 125°E) comprises elongated mounds (echo types 2-1, 2-2, and 2-3) of tidal ridges. The inter-mound areas comprise echo types 1-1 and 1-2. Here, the tidal ridges are up to 20 m high, a few km wide, and 10–50 km long, trending northeast to southwest (Figure 14.5). Further south (35°–45°N), the seafloor forms subaqueous dunes (echo type 1-4) and tidal ridges with superposed dunes (echo type 2-2). The ridges west of 125°E are 10–30 km wide and 50–150 km long. The southeastern nearshore area around the islands consists of eroded seafloor (echo type 3-2) and large-scale banks (echo type 2-4). Echo type 3-2 is dominant in the area between islands and submerged basement highs.

Geology and Sedimentology of the Korean Peninsula. DOI: http://dx.doi.org/10.1016/B978-0-12-405518-6.00014-0

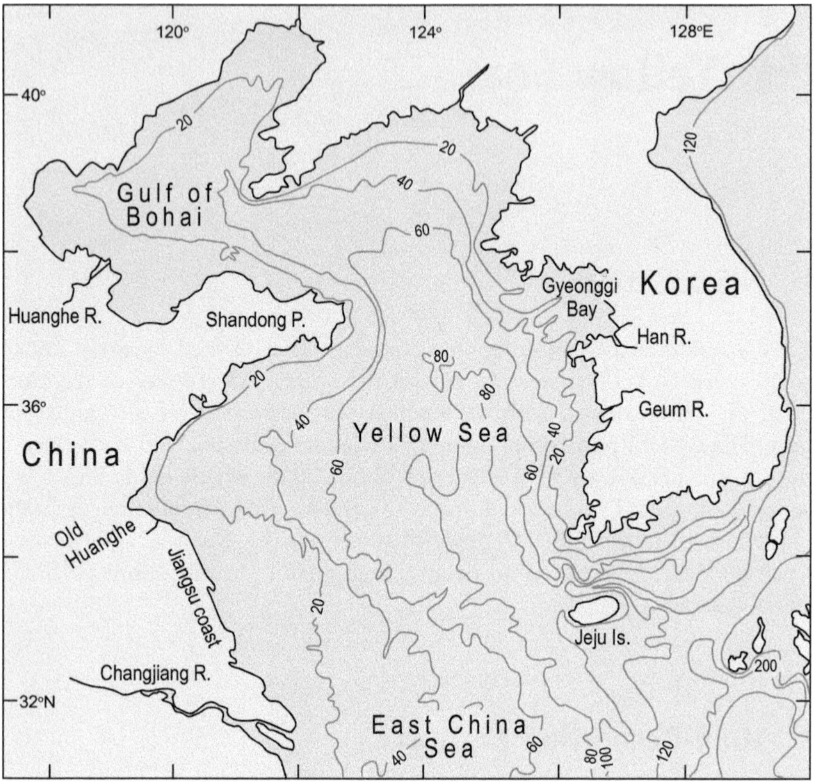

Figure 14.1 Bathymetry of the Yellow Sea. Contours in meters.
Source: Chough et al. (2004) by permission of *Geosciences Journal*.

In the north of 37°N, the seafloor is scoured (echo types 3-1 and 3-2) with minor occurrence of echo types 1-1, 1-2, and 1-4 (Figure 14.5). Small-scale channels are ubiquitous as well as large-scale valleys with a high topographic relief. On the other hand, flat seafloor and large-scale dune fields with peaked crests occur in intervening areas in the northernmost part of the area. These features most likely developed during the later phase of sea level rise and highstand. The narrow seaway between the Shandong Peninsula and the Jangsan promontory caused intensified strong tidal and nearshore currents to incise the seafloor on which very thin sediments accumulated.

The echo characteristics reflect dynamic processes of transgressive and highstand deposits during sea-level rise (Chough et al., 2002). Transgression commenced along the topographic lows of the eastern part of the sea, subjected to strong tidal currents. Tidal sand ridges and relict sands formed extensively in the inner shelf/nearshore area by way of landward retreats of the shoreface. The rounded profiles, subdued relief, and lower slope gradient of the ridges suggest that they were temporarily dormant.

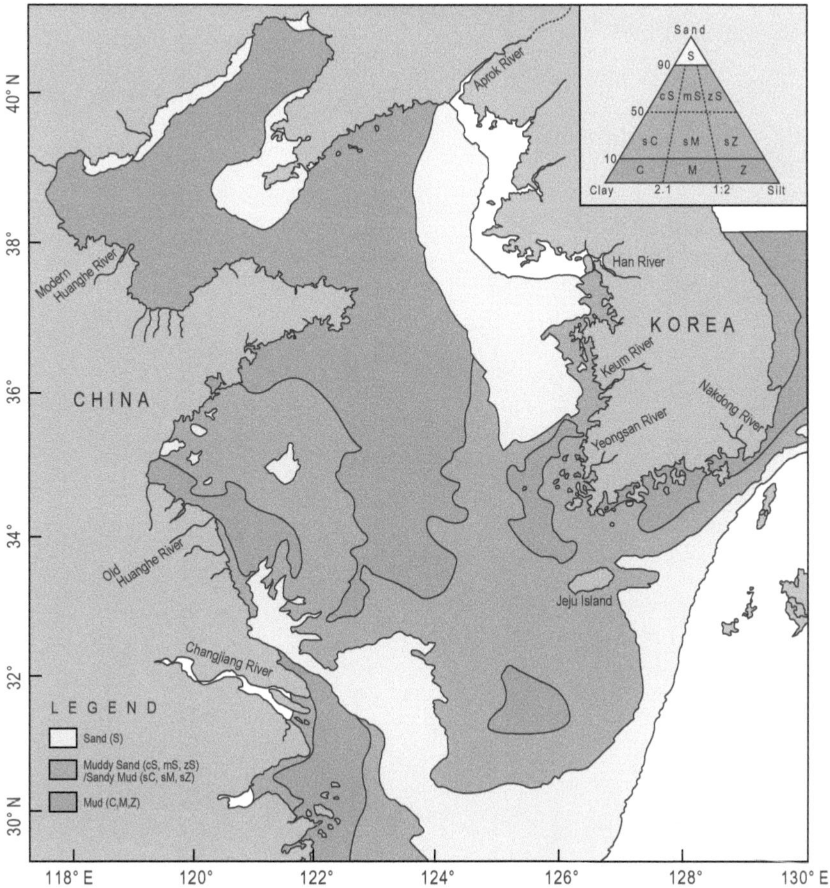

Figure 14.2 Map showing surface sediment distribution in the Yellow Sea and adjacent area. Sediment classification according to grain size using Folk's (1954) scheme. *Source*: Lee and Chough (1989) by permission of the Elsevier.

Some nearshore ridges undergo active growth, whereas others degrade by strong currents (tides and storms). In the offshore area, the transgressive sediment sheet of mud (echo type 1-3a) was deposited on the erosional surface. With continued sea-level rise, the transgressive ravinement and remoulding processes propagated to the northern part of the sea. The sheetlike mud wedge (echo type 1-3b), which occupies a large part of the central Yellow Sea, represents deposition of suspended mud derived from the Huanghe River (Lee and Chough, 1989). On the other hand, a mud bank (Heuksan mud belt) originated from the Geum River, was controlled by tidal and thermohaline fronts (Figure 14.6) (Lee and Chu, 2001). The mud bank locally overlies large-scale dunes. It is suggestive of partitioning of sedimentary facies controlled by sea-level rise and dispersal systems during the Holocene (Jin and Chough, 1998).

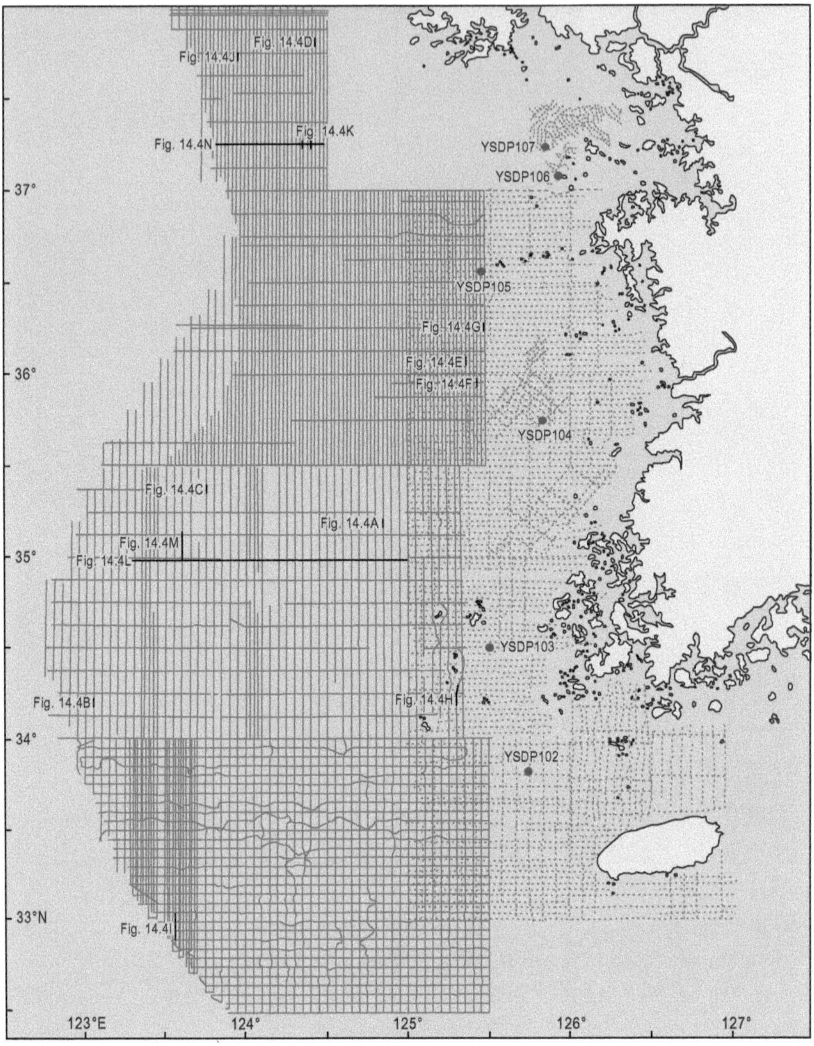

Figure 14.3 Tracklines of Chirp subbottom profiles (solid lines) acquired by National Oceanographic Research Institute (NORI), and air-gun, sparker, and 3.5 kHz seismic profiles (dotted lines) acquired by Korea Institute of Geoscience and Mineral Resources (KIGAM). Numbered thick lines are tracks of the cited figures. Note locations of drill cores.

14.3 Sediment Cores

Six drill cores were retrieved from the eastern part of the Yellow Sea by the Korea Institute of Geoscience and Mineral Resources, including the Heuksan mud belt (YSDP 102 and 103), the crest and trough of the tidal ridges (YSDP 104 and 105),

Table 14.1 Summary of Echo Types

Class	Type	Line Drawing	Description	
1	1-1		Relatively flat seafloor with either no or few subbottom reflectors	Seatfloor covered by coarse-grained sediments; relict sands
	1-2		Flat seafloor with moderately to well-developed subbottom reflectors	Seafloor covered by relatively fine-grained surface sediments
	1-3		Laterally extensive acoustically transparent unit of either sheet (1-3a) or wedge (1-3b) form	Transgressive sediment sheet (1-3a) and Holocene Huanghe-derived muds (1-3b)
	1-4		Flat seafloor covered by regularly spaced, wavy bedforms	Large-scale dunes formed by tidal currents
2	2-1		Mounds with no bedforms; either absent or well-developed internal reflectors	Tidal ridges; inactive
	2-2		Mounds covered by wavy bedforms	Tidal ridges with large-scale dunes; actively maintained
	2-3		Mounds accompanying acoustically transparent wedges on the flanks and smaller bedforms on the crest	Tidal ridges, degraded by tidal currents and storms
	2-4	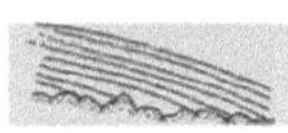	Large-scale mounds with distinct, continuous internal reflectors downlapping onto the substrate	Holocene mud belt
	2-5		Mounds with well-developed clinoforms	Relict subaqueous mud wedge
3	3-1		Regionally flat seafloor incised by shallow troughs	Shallow channel incision by strong tidal currents
	3-2		Seafloors of great topographic relief and deeply incised valleys	Extensive channel erosion or acoustic basement

Source: Chough et al. (2004) by permission of *Geosciences Journal*.

and the sediment lobes of Gyeonggi Bay (YSDP 106 and 107) (Jin, 2001) (Figure 14.3). The data have provided unprecedented information on the sedimentary facies and environmental changes in stratigraphic framework at the time scales of millennia and beyond.

YSDP 102 and 103 penetrated three discontinuity-bounded units, including the Holocene sequence in the Heuksan mud belt (Figure 14.7). The lowermost unit consists of gravelly sands formed in topographic lows and incised valleys during

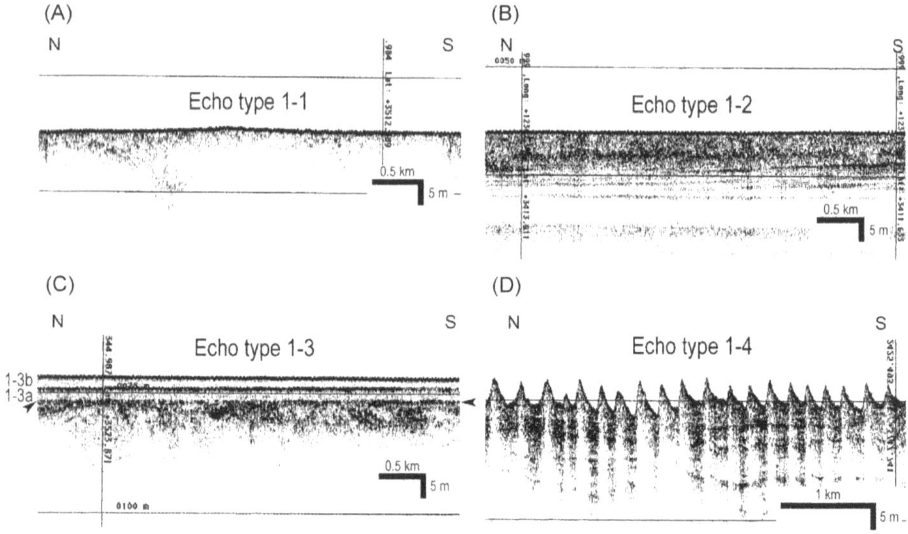

Figure 14.4 Chirp (2–7 kHz) seismic profiles. For location of each profile, see Figure 14.3. (A) Flat seafloor with no internal reflectors (echo type 1-1). (B) Flat seafloor with moderately to well-developed subbottom reflectors (echo type 1-2). (C) Laterally extensive acoustically transparent unit of either sheet form (echo type 1-3a) or wedge form (echo type 1-3b). (D) Regionally flat seafloor with regularly spaced, similar-sized, wavy bedforms (echo type 1-4). (E) Mounds with no bedforms (echo type 2-1). (F) Mounds with wavy bedforms (echo type 2-2). (G) Mounds with acoustically transparent wedge on the flanks and smaller-scale bedforms on the crest (echo type 2-3). (H) Large-scale mounds with parallel-stratified internal reflectors downlapping onto the substrate (echo type 2-4). (I) Mound with clinoforms prograding onto the erosional surface (echo type 2-5). (J) Regionally flat seafloor with small-scale troughs (echo type 3-1). (K) Erosional seafloor of greater topographic relief (echo type 3-2). (L) Lateral association of echo types across the central-eastern Yellow Sea. Note that the deepened eastern part of the sea is occupied by a series of tidal ridges. (M) Transgressive sediment sheet (echo type 1-3a) overlies extensively flat erosional surface. Note lateral wedging out of superficial prodeltaic mud unit (echo type 1-3b). (N) Eroded seafloor of northern Yellow Sea. Note the areal separation of the western shallow-incised seafloor (echo type 3-1) and the eastern high-relief erosional seafloor (echo type 3-2).
Source: (D) Chough et al. (2004) by permission of *Geosciences Journal.*

the last regression and early phase of the subsequent transgression (Jin and Chough, 1998). The middle unit exhibits ridge-and-swale topography and comprises gravelly sand and well-sorted massive sand. The overlying topmost unit consists of laminated and bioturbated mud, bounded by a sharp erosional surface, i.e., transgressive surface of erosion. The base of the mud unit was dated at 6000 or 14,000 years (Figure 14.8).

YSDP 104 was drilled on the crest of a sediment ridge (Figure 14.7). The lowermost unit (substratum older than 40,000 years BP) consists of disorganized sandy

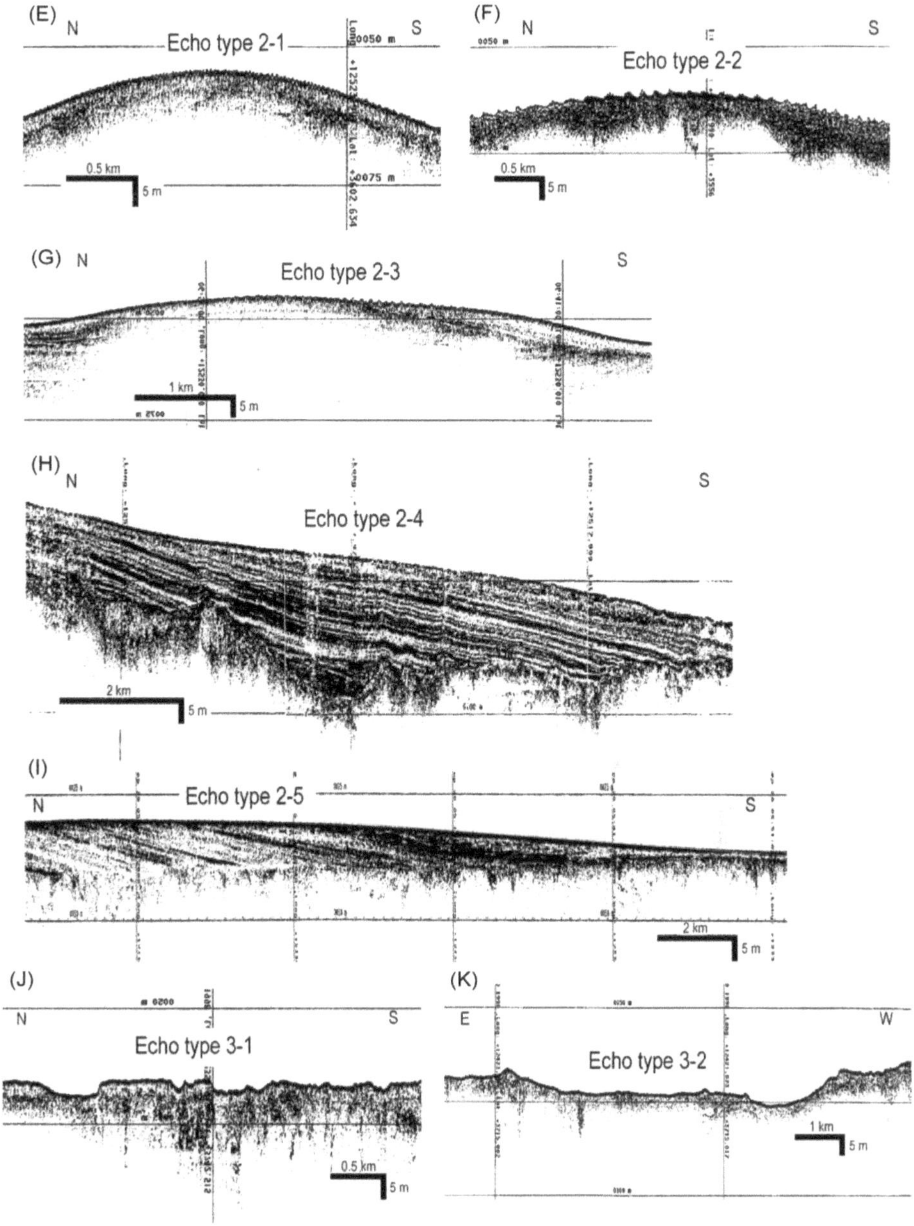

Figure 14.4 (Continued)

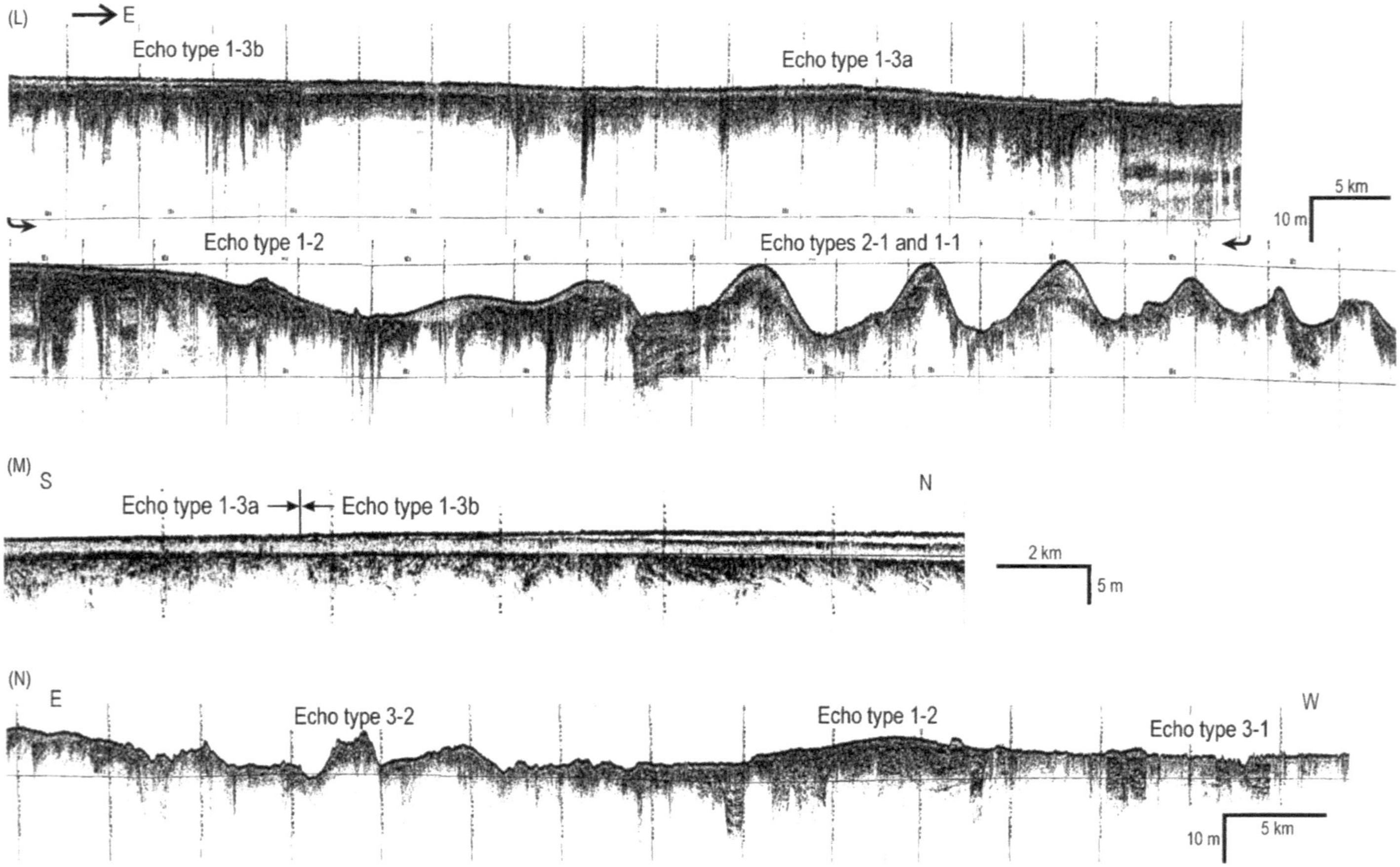

Figure 14.4 (Continued)

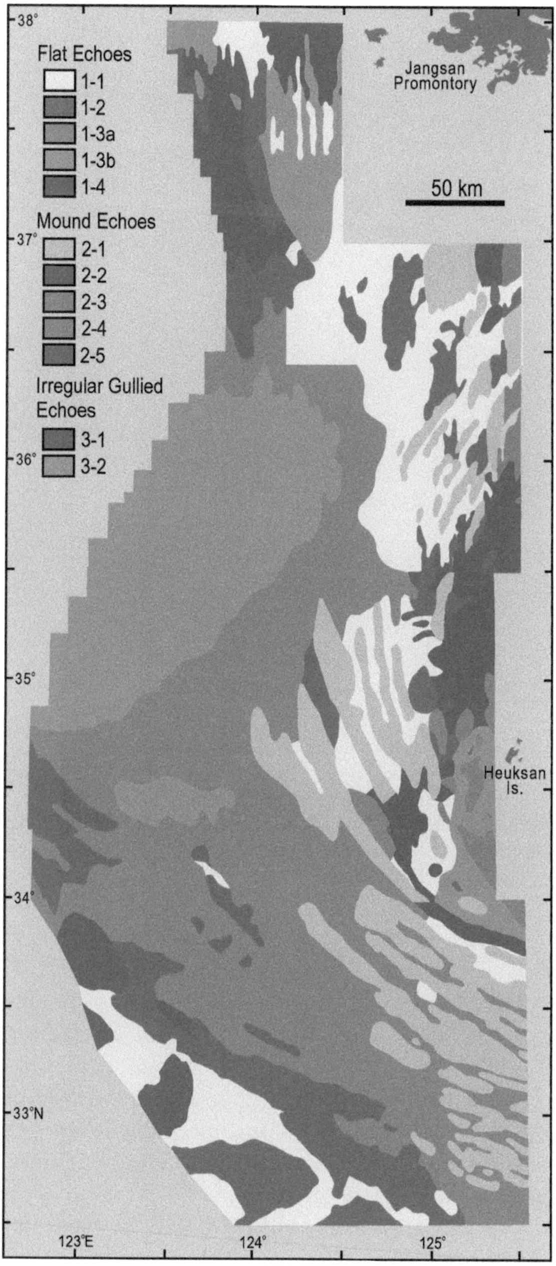

Figure 14.5 Distribution of echo types in the eastern part of the Yellow Sea.
Source: Chough et al. (2004) by permission of *Geosciences Journal*.

gravel, laminated muddy sand, and laminated mud, which are indicative of deposition in tide-dominated environments (Jin and Chough, 2002). The overlying unit comprises sandy mud with interlayered sand and silt, which were most likely deposited in estuarine environments (tidal flats and bars). The topmost unit (3 m

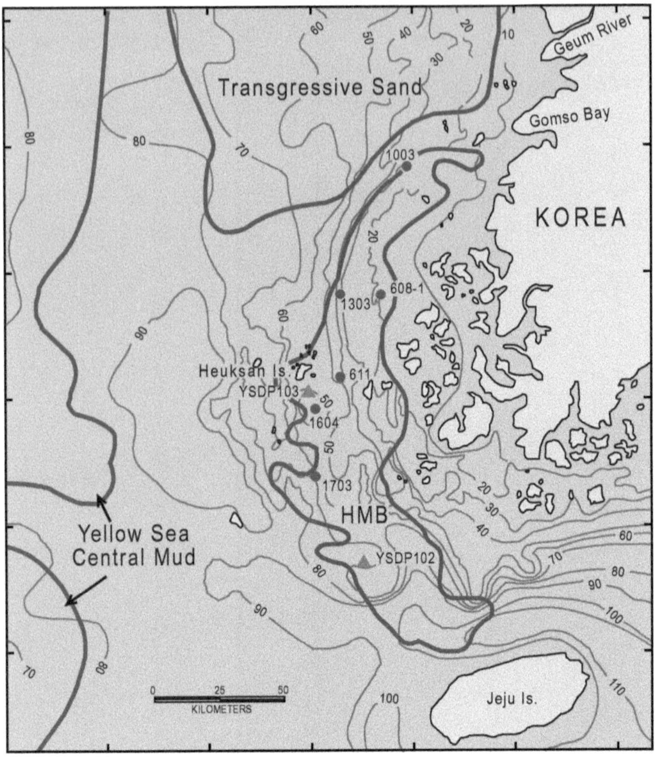

Figure 14.6 Map of the Heuksan mud belt (HMB) and its vicinities in the southeastern Yellow Sea. Drilled cores (filled triangles) and piston cores (dots) are described in Chang et al. (1996) and Lee et al. (1987), respectively.
Source: Lee and Chu (2001) by permission of the SEPM.

thick) consists of massive sands with scattered shell fragments, and overlies a stiff, yellow mud layer (about 28,000 years old). It suggests that the ridge sediments were reworked and subsequently covered with sands (Jin and Chough, 2002; Jung et al., 1998).

YSDP 105, retrieved from the eastern part of the sea (Figure 14.3), consists of six depositional units (Figure 14.8). Each unit was well constrained by seven age dates of peat, organic materials, and foraminifera (Figure 14.8C). The lowermost depositional unit (DU 6) consists of a mixture of gravel, sand, and mud with reddish mottles indicative of floodplain paleosols (Jin et al., 2002). The overlying unit (DU 5) comprises various sediment layers of rhythmic sand and mud as well as flaser and herringbone structures, which are indicative of deposition in tidal flat environments. The overlying unit (DU 4) consists of gravel and bioturbated mud and bedded peat, which are indicative of tidal inlet/channel and paralic swamp. The boundary between DU 5 and DU 4 most likely represents a period of lowered sea level. Peat at the top of unit 4 was dated at 20 ka, indicating formation

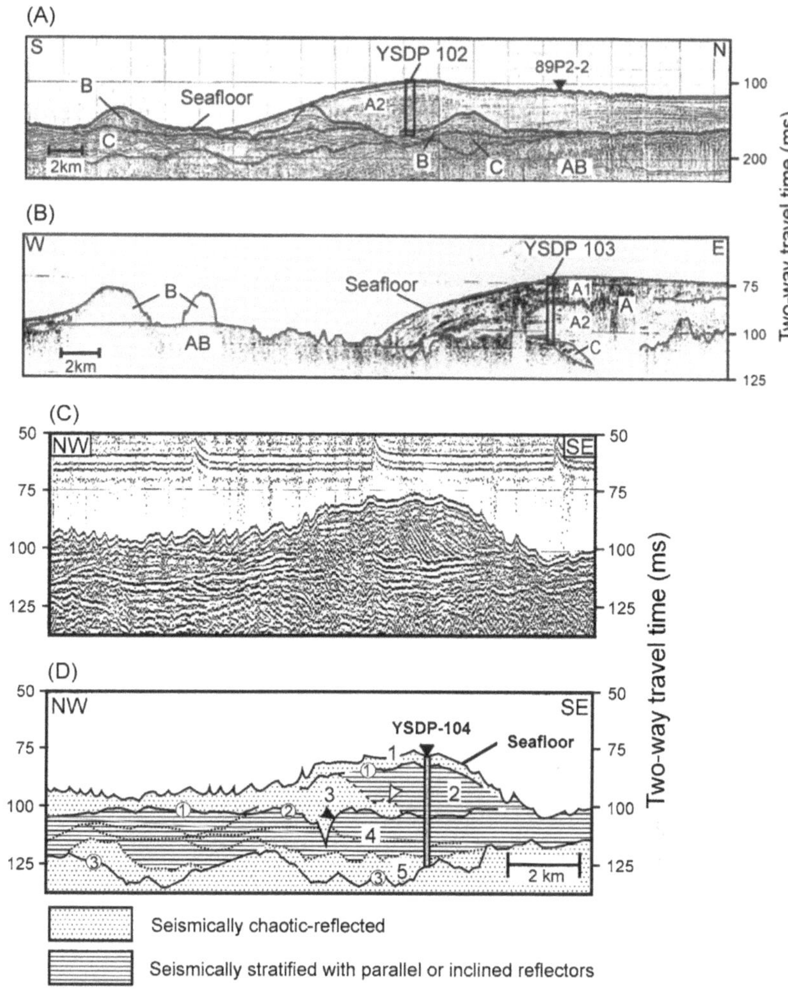

Figure 14.7 (A) and (B) Seismic profiles and interpretation of YSDP 102 and 103, respectively. Units C and B represent incised channel fills and ridge-and-swale topography, respectively. Note that Unit A can be divided into subunits A1 and A2 by an erosional boundary. (C) and (D) Seismic profile and interpretation of YSDP 104. Five distinct seismic units are shown. (E) and (F) Air-gun profile and interpretation of YSDP 105. Six seismo-stratigraphic units (SU 1–6) are recognized. (G) Seismic profile and interpretation of YSDP 106. Lobe II consists of SUs IIA and IIB. Note that SU IIA comprises two subunits IIA-1 and IIA-2. (H) YSDP 107 drilled on lobe I, which consists of unit IA. Unit IA consists of eight subunits bounded by local discontinuities.
Source: Jin (2001).

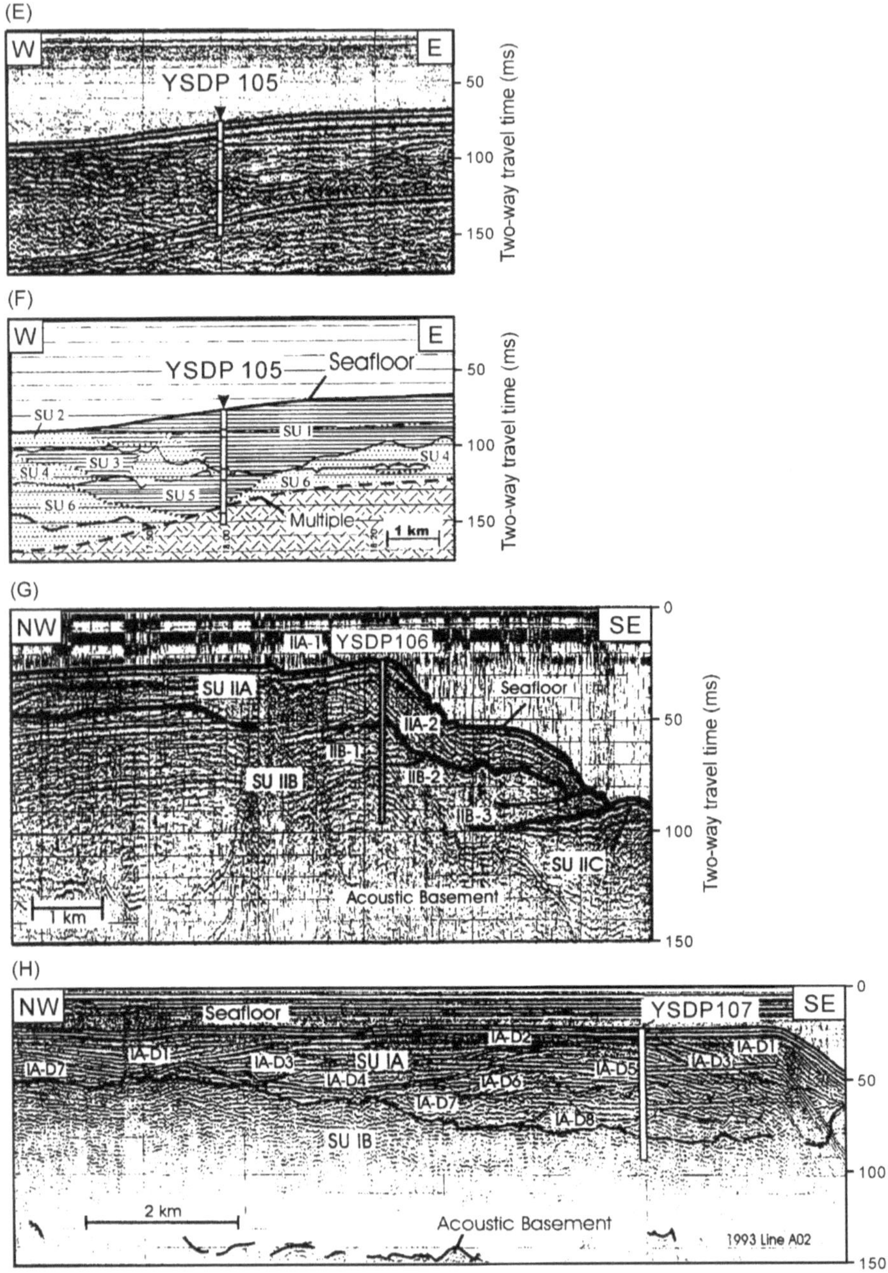

Figure 14.7 (Continued)

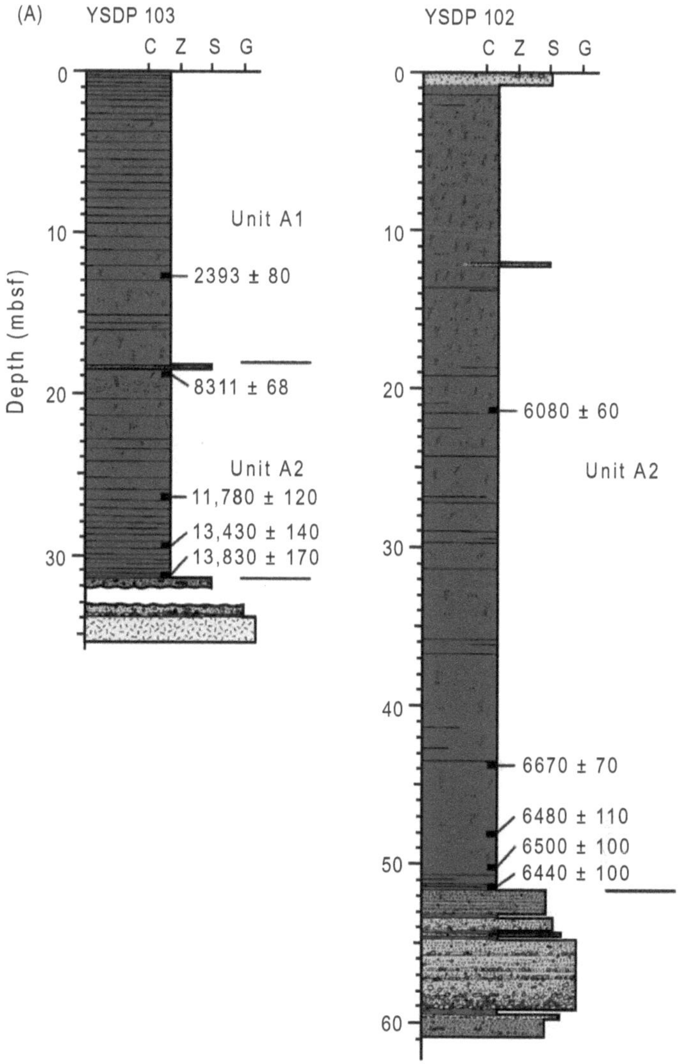

Figure 14.8 (A) Radiocarbon dates (yrs BP) of drill cores YSDP 102 and 103 with dominant lithofacies consisting of laminated muds in YSDP 103 and bioturbated mud in YSDP 102. C, clay; Z, silt; S, sand; G, gravel. (B) and (C) Descriptions of YSDP 104 and 105, respectively, including texture, sedimentary structures, radiocarbon ages and interpretation of depositional processes, environments and sequences. Seismically chaotic unit consists largely of clast-rich nonmarine to paralic sediments, whereas seismically stratified unit mostly comprises tidal fine-grained sediments. LST, lowstand systems tract; TST, transgressive systems tract; HST, highstand systems tract; SB, sequence boundary; TSE, transgressive surface of erosion; TR, tidal ravinement; WR, wave ravinement. (D) and (E) Facies descriptions, radiocarbon ages and interpretation of depositional environments of YSDP 106 and 107, respectively.
Source: Jin (2001).

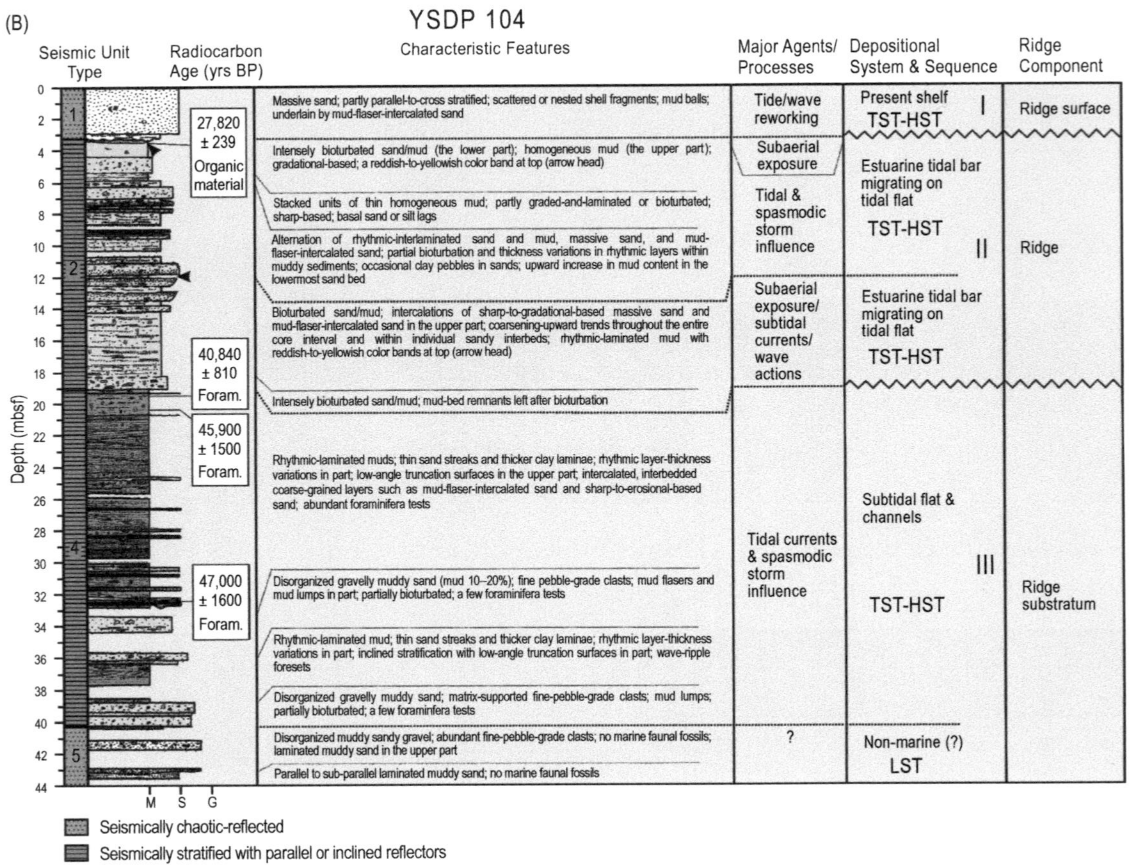

Figure 14.8 (Continued)

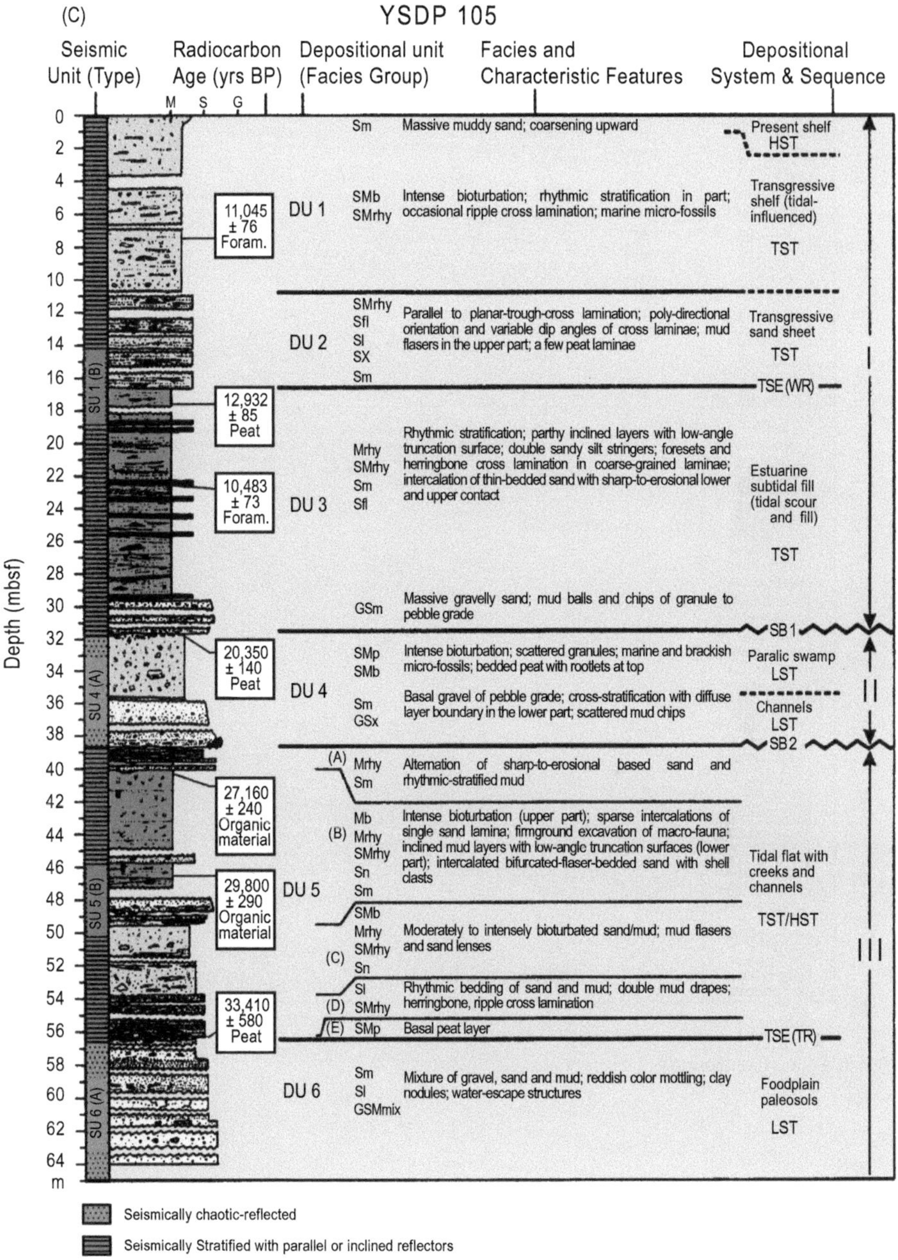

Figure 14.8 (Continued)

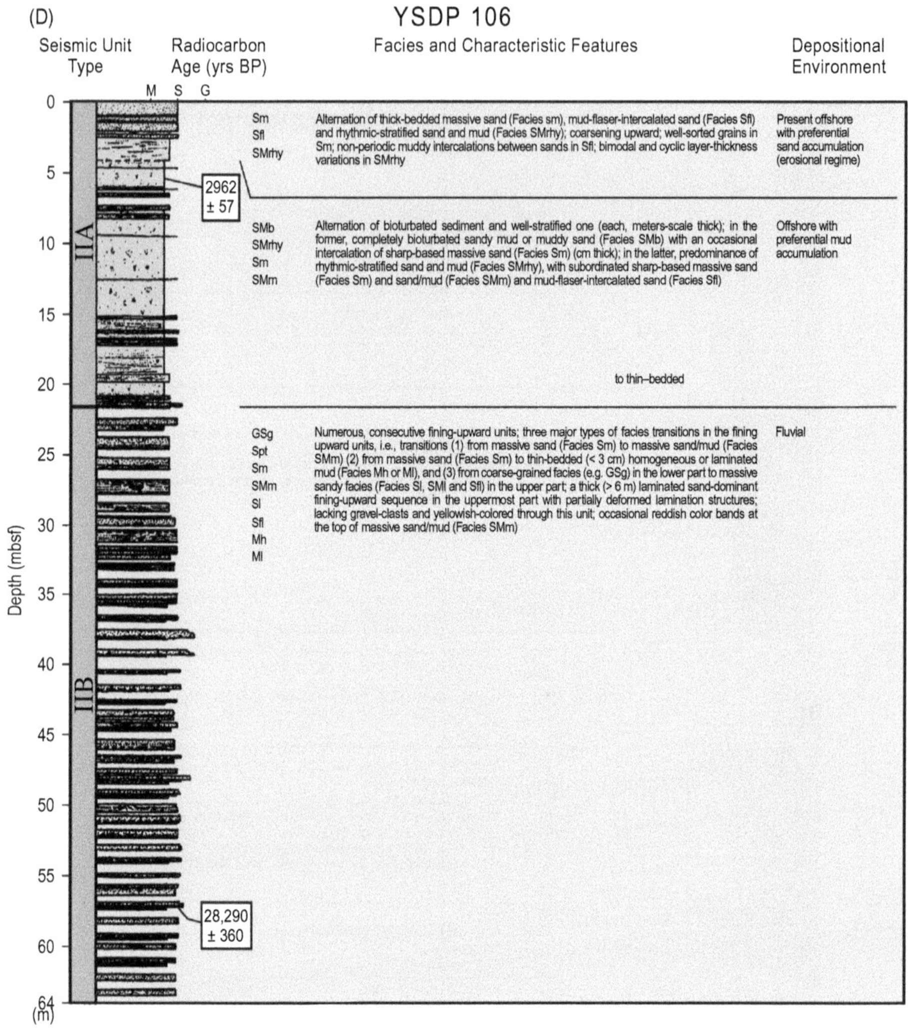

Figure 14.8 (Continued)

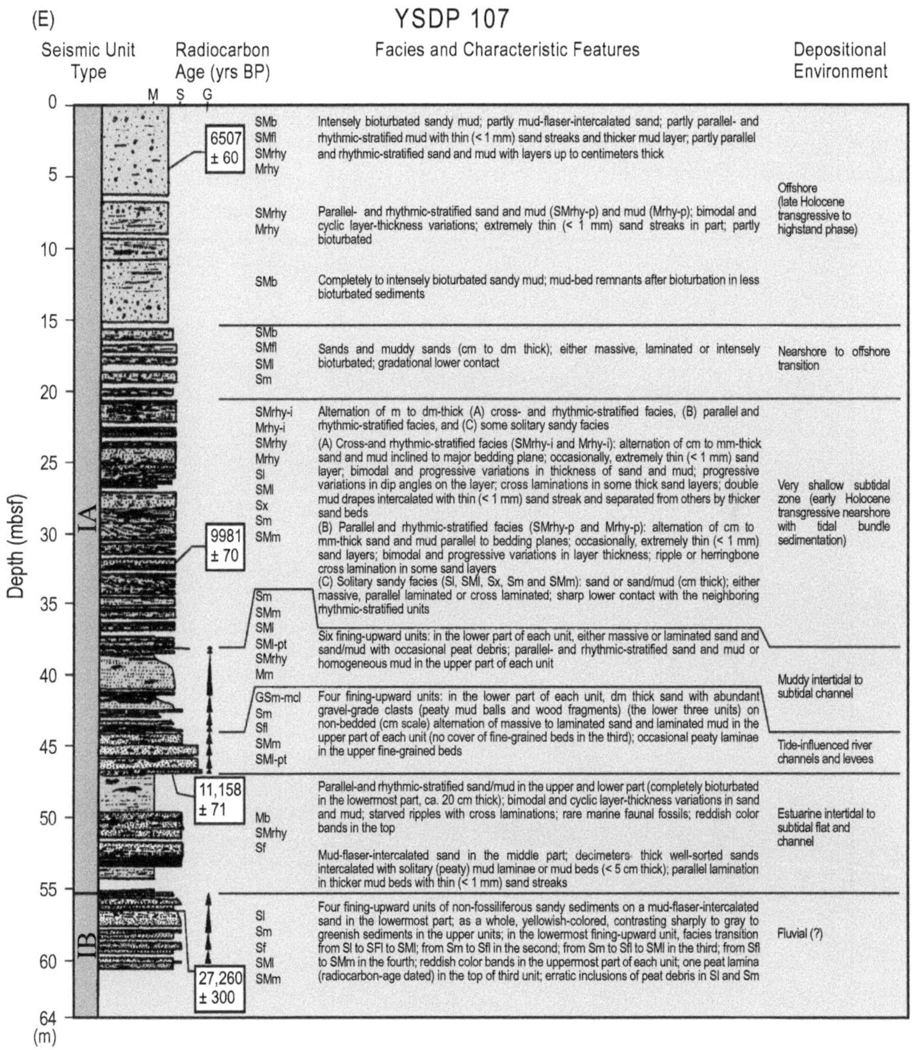

Figure 14.8 (Continued)

during the last glacial maximum. The overlying unit (DU 2) consists of laminated and massive sand, most likely deposited by a retreating shoreface during ensuing transgression. The lower boundary thus represents a ravinement surface (13 ka). The topmost unit (DU 1) consists of bioturbated mud (11 ka), which represents deposition during the Holocene transgression.

YSDP 106 and 107 were drilled on a sediment lobe in the Gyeonggi Bay (Figure 14.3) (Jin, 2001). In YSDP 106, the lower part (22–63 m) consists of reddish gravelly sands and peat sands (about 28 ka at 57 m depth). The overlying unit (4–22 m) comprises bioturbated and stratified sediment layers with massive sands and flaser beds with abundant fossils (foraminifera test, 3 ka at 5 m depth) (Figure 14.8D). YSDP 107 also consists of various sand and mud (massive, laminated, homogeneous, cross-laminated, and bioturbated). The lowermost unit (55–60 m) comprises unfossiliferous yellowish–reddish sand layers, indicative of subaerial exposure. The overlying unit (47–55 m) consists of bioturbated mud, sand/mud, and flaser-bedded sand with rare marine fauna. The overlying units

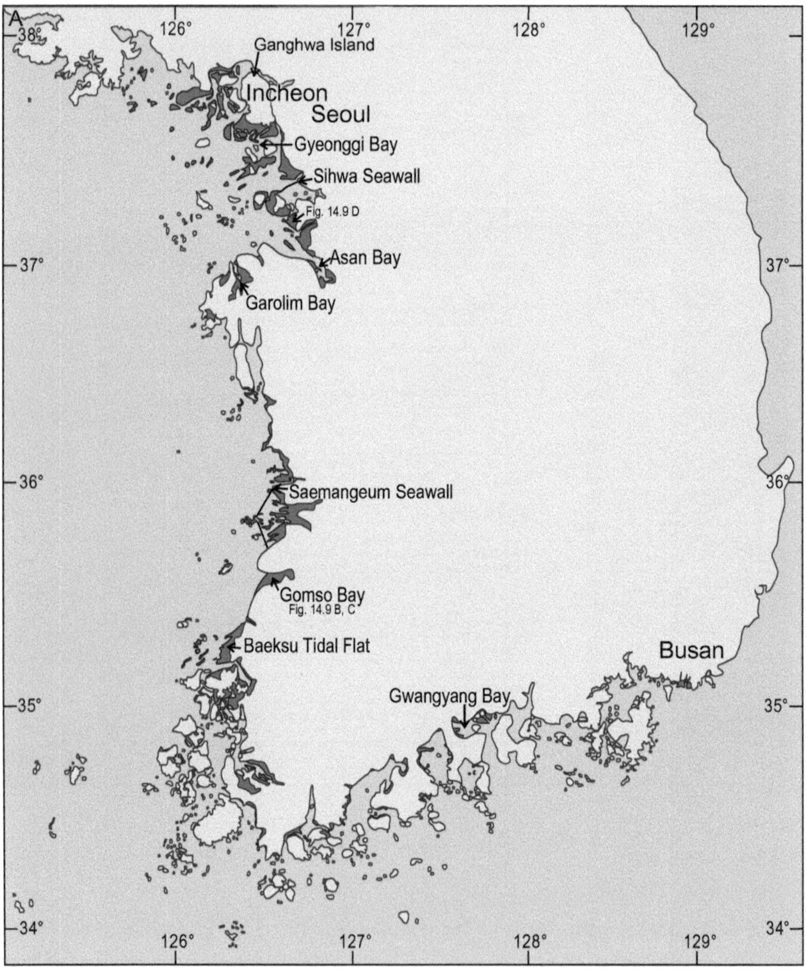

Figure 14.9 (A) Distribution of tidal flats along the west and south coasts of the Korean Peninsula. (B) Mud flat partly covered with incoming tides, showing abundant small-scale mounds, holes (burrows), and tracks. The main tidal channel runs east-west at the foot of the mountain in the background. (C) Meandering tidal creeks are connected to the main channel. (D) Mixed tidal flat, Tando Island. Mean tidal range is 4.3 m.

(44–47, 38–44, 21–38, and 16–21 m) are characterized by sand and mud of tidal channels and bundles as well as laminated mud of subtidal to offshore environments. The uppermost unit (0–16 m) consists of bioturbated sandy mud with minor flaser beds, again indicative of tidal influence. Tests of foraminifera at a depth of 4 m are dated at 6.5 ka.

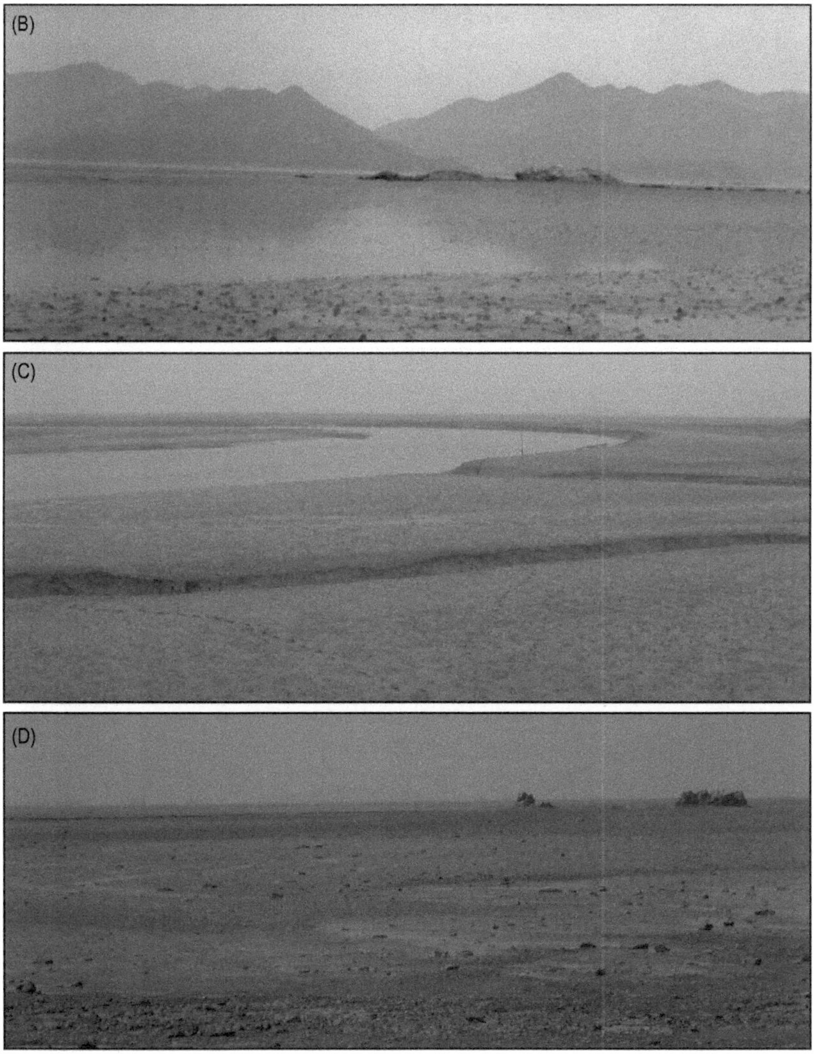

Figure 14.9 (Continued)

14.4 Tidal Flats

The west and south coasts of the Korean Peninsula are characterized by extensive development of tidal flats (Figure 14.9). A large number of tidal flats occur along the embayed shorelines or face to the open sea; others are bounded by artificial dams and dykes that were built to reclaim intertidal flats and tidal marshes for land use (Figure 14.9). Tidal flats are under macrotidal (or mesotidal) regime and semi-diurnal (diurnal inequality of approximately 1 m) with a mean spring range of 4.3–9.1 m (Chough et al., 2004). Tidal currents are up to 2 and 2.2 m/s during spring and neap tides, respectively, which are also influenced by the magnitude and direction of prevailing winds (Yang and Chun, 2001). In winter, northwest winds prevail with a mean speed of approximately 5–6 m/s; in summer, south winds are dominant (mean speed of 2–3 m/s). Storms ($>$13.9 m/s in wind speed) are common in winter ($>$9 days/month), but occur on less than about 2–3 days/month in summer. The tidal flats are also prevailed by typhoons ($>$17.2 m/s in wind speed), which occur in summer and autumn. They produce waves up to 5 m high (commonly 1.5–2 m in wave height) (Chun et al., 2000).

Tidal flats act as a principal repository for littoral sediments carried by tidal currents (Chough et al., 2004). Due to low sediment input from the hinterland, most tidal flats along the coast of the Korean Peninsula have formed a retrogradational, coarsening-upward succession during sea-level rise in the Holocene (Kim et al., 1999). Sediments are generally bioturbated, showing certain zonation with seasonal variations (Baek et al., 2009, 2010). Sedimentation in the open-coast macrotidal flats has been controlled mainly by the combined effect of waves and tides with distinctive seasonal cycles. Owing to the monsoonal climate, mud-dominated facies formed mainly by tidal processes in summer and sand-dominated facies mainly by waves in winter (Yang and Chun, 2001; Yang et al., 2005, 2006). Sedimentary facies analysis indicates that winter storms play a major role in sedimentation in intertidal flats. Quantitative measurements of sediment transport on tidal flats indicate that wind-generated waves during winter interrupt sedimentation (Chun and Yang, 2004). Ebb currents often generate more turbidity than flood currents when waves are high. They are a major source for the nearshore sediment plume during winter.

Tidal flats of the open coasts are largely composed of sands (Yang and Chun, 2001). These flats are exposed to significant wave action during the winter season (November–February), i.e., they are wave dominated, despite the existence of a large tidal range (Yang et al., 2005, 2008). On a steep-gradient shoreline, the tidal flat forms sandy beach, whereas on a low-gradient shoreline, it becomes progressively finer in a landward direction. Sediments commonly reveal storm-generated graded beds and contain hummocky cross-stratification and wave ripples (Yang et al., 2006, 2007). Wave-generated tidal bundles also occur in these wave-dominated tidal flats (Yang et al., 2008). Summer mud deposits are commonly extensively bioturbated (Baek et al., 2009).

During the later stages of Holocene transgression, the retrogradational coarsening-upward succession in the back-barrier tidal marsh region is overlain by

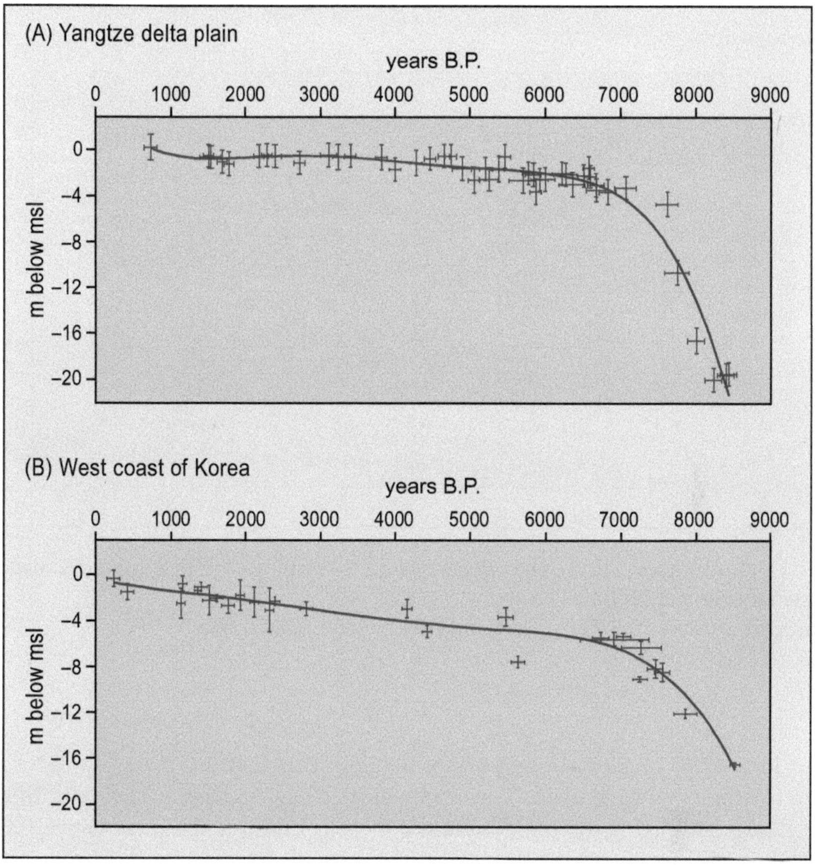

Figure 14.10 Holocene relative sea-level curves based on radiocarbon dates and estimated paleo-mean sea level in the Yangtze delta plain (A) and the western coast of Korea (B). *Source*: (A) Zong, 2004 and (B) Chough et al. (2004) by permission of *Geosciences Journal*.

a fining-upward progradational succession of highstand conditions due to landward retreat of the barriers, forming a wave ravinement surface (Yang et al., 2006).

14.5 Rising Sea Level

The sea-level curves of the Holocene epoch, derived from the Chinese coasts, generally show a rapid rise from about −40 m (about 10 ka) to −10 m (about 7 ka), followed by a gradual rise (Feng and Wang, 1986; Pirazzoli, 1991; Yang and Xie, 1984). A refined curve of the Yangtze delta plain, based on corrected

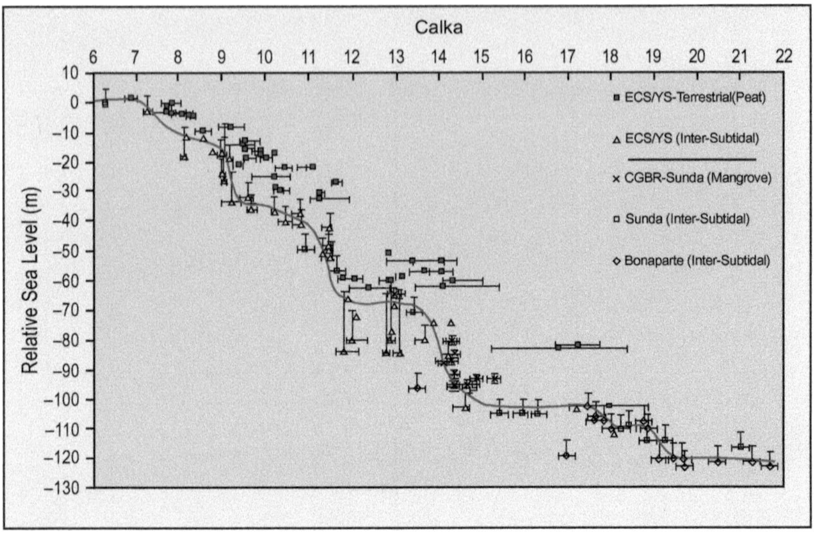

Figure 14.11 A sea-level curve showing the western Pacific post-glacial sea-level history suggested by Liu and Milliman (2004).
Source: Chough et al. (2004) by permission of *Geosciences Journal*.

data of long-term subsidence and tidal range, reveals that there was an early rapid rise up to about −4 m at about 7 ka and a slow rise since then (Chen and Stanley, 1998; Zong, 2004) (Figure 14.10). On the Korean coast, a sea-level curve (Bloom and Park, 1985) shows a gradual rise of mean high-water level from −8 m at about 8.5 ka to −2 m (about 5 ka) and a gradual rise thereafter.

An integrated curve of the Korean coast shows a relatively rapid rise, up to −5 m about 7 ka, followed by a gradual rise to the present level without discernable fluctuations (Figure 14.10). The curve is based on additional radiocarbon dates of plant remains, peat, and shells of the Gomso Bay and other coastal areas. The paleo-mean sea level was determined under the assumption that the tidal range (and frequency) and the bathymetric slopes were constant throughout the Holocene. The curve is similar to those given by Kim et al. (1999) and that of the Yangtze delta plain.

Based on radiocarbon dating on benthic foraminifera, Kim and Kennett (1998) have demonstrated that Holocene marine transgression started in the central Yellow Sea at about 11.3 ka and approached the present lower tidal flat level about 7.5 ka. Based on shallow-marine, intertidal, and submerged terrestrial data of the East China Sea, the Yellow Sea, and the Sunda shelf, Liu and Milliman (2004) have suggested a step-wise sea-level curve for the western Pacific coast (Figure 14.11). It shows a series of rapid flooding events (as fast as 80 mm/year) that were separated by long-term slow rises (2–10 mm/year).

Recent measurements of sea level using tide gauges and satellites suggest that the curves over the past decades have been on the rise, especially since the industrial revolution. With the explosive increase in human population and subsequent consumption of fossil fuels, greenhouse gases have begun to build up in the atmosphere, leading to global warming. Although the rate of sea level rise is different from place to place, it is about a few millimeters (3 mm on average) per year for the last decades around the globe (Pilkey and Young, 2009). They suggest that some unstable parts of the globe can be submerged at higher rate (1–2 m/100 years). Although the crust around the Korean Peninsula is tectonically stable, local erosion and deposition of sand and mud will occur in the coastal area, especially along the artificial seawalls and dams, depending on the dominant coastal currents and storm waves. Besides, a large part of the coastlines around the peninsula will be submerged under water as deep as 30 cm in 100 years. If this rate were constant in the next decades, property owners on the coastline and society in general would be able to manage possible defect of the rising sea. The best way to cope with such gradual environmental change is, however, to prepare for retreat toward higher ground and pursue whatever merits the submerged lands may offer. The submerged sea can be effectively used for fish and seaweed cultures and production of natural salt as well as for natural seashore parks.

14.6 Depositional Processes

The Yellow Sea floor was subaerially exposed during the last glacial maximum (20–18 ka) in which the shoreline was about 120 m below present sea level (Figure 14.12). Dry plains were most likely incised by channels, forming a wandering network of drainage systems. The basal surface of the incised channel-fills (echo type 1-2) thus represents a sequence boundary. The incised channels were filled with deltaic/estuarine deposits during the subsequent transgression (Figure 14.12A), which most likely commenced about 11.5 ka in the central part of the Yellow Sea. Unconsolidated red-stained pebbly sand lags at the base of these deposits are suggestive of reworking of sediments. The seafloor progressively eroded by strong waves and currents along the advancing shorefront. Sea level advanced landward at a rate of about 500 m/year. This transgressive surface of erosion commonly appears flat in seismic profiles, separating the transparent sheet above (echo type 1-3a) from the lowermost acoustic unit below (echo type 1-2) (Figure 14.12B).

In the eastern part of the sea, extensive coastal erosion occurred. Tidal currents were extensive and caused reworking of surface sediments, forming transgressive lags and ridges (echo types 1-1, 2-1, and 2-3) (Figure 14.5) (Shinn et al., 2004). The elongated ridges are oriented subparallel to the dominant tidal currents, showing seaward-inclined internal reflectors. Most ridges show an undulatory base ranging in age from about 47 to 28 ka, which originated from erosion of the substratum.

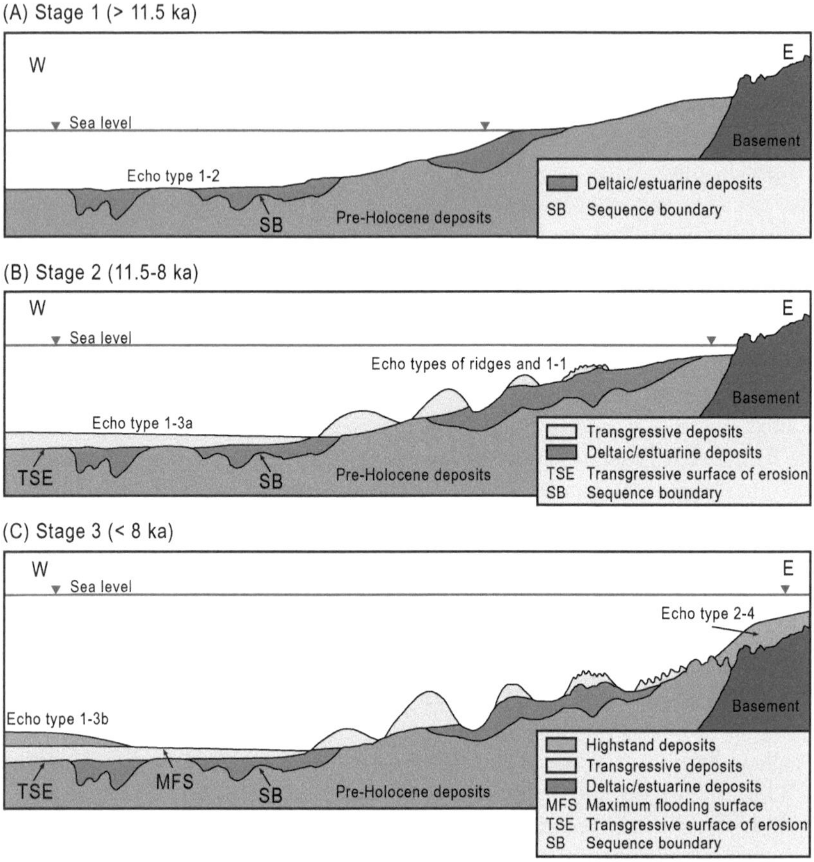

Figure 14.12 Schematic diagrams showing echo-type development in the central-eastern Yellow Sea. (A) During the last glacial maximum, the eastern Yellow Sea was entirely exposed and pre-Holocene deposits were incised. The basal surface of the incised channel fill (echo type 1-2) represents sequence boundary. (B) During the transgressive period, rapid retreat of shoreface and tidal processes formed tidal ridges (echo types of ridges and 1-1) and transgressive lag deposits (echo type 1-3a) in the inner-shelf area. A transgressive sediment sheet extensively covers the transgressive surface of erosion in the central Yellow Sea. (C) As the rate of sea-level rise decreased, the Huanghe-derived mud unit (echo type 1-3b) and the Heuksan mud belt formed in the central Yellow Sea and the southwestern tip of the Korean Peninsula, respectively.
Source: Shinn et al. (2004) by permission of the AGU.

In the western part of the sea off the Jiangsu coast, extensive lag deposits and isolated sediment bodies (echo types 1-1 and 2-5) formed on rather flat surfaces (40–50 m deep). The flat ravinement surface formed at water depths of 60–40 m by erosion along the shoreface during rapid rise of sea level (11–10 ka), which prevailed with strong waves and currents. The lag deposits resulted from landward retreat of the shoreface. On the other hand, the isolated sediment bodies were due to subaqueous progradation of a mud wedge of the paleo-Huanghe River during the period of 9.5–7.5 ka when sea-level rise was temporarily at a standstill (Milliman et al., 1989).

As the rate of sea-level rise decreased (about 8 ka), large amounts of mud from the Huanghe River were transported southward to the central part of the Yellow Sea, forming the mud wedge (echo type 1-3b) (Figure 14.12C). Relatively small amounts of mud were delivered to the eastern part of the sea along the shore; they mainly emanated from the Geum River, forming the Heuksan mud belt (part of echo type 2-4). The mud overlies large-scale transgressive dunes (echo type 1-4), formed about 6-7 ka (Chough et al., 2002; Jin and Chough, 1998). The elongated morphology was due to sediment dispersal controlled by the physical barriers of the tidal and thermohaline fronts (Lee and Chu, 2001).

15 Summary

The Korean Peninsula is part of the Eurasian Plate. The basement rocks consist of high-grade gneiss and schist as old as 2.51 Ga, including the Nangrim, Gyeonggi, and Yeongnam massifs. In the late Proterozoic, sedimentation was initiated in the Okcheon Basin of the Gyeonggi Massif, accompanied by deposition of siliciclastic and volcaniclastic sediments as well as carbonates. The Nangrim and Yeongnam massifs were submerged in the Early Paleozoic during the greenhouse period, forming a shallow marine platform and the associated environments in the Pyeongnam and Taebaeksan basins, respectively.

The Cambrian–Ordovician succession in the Taebaek area, Taebaek Group of the Taebaeksan Basin, unconformably overlies granite gneiss of the Yeongnam Massif. It consists of mixed carbonate-siliciclastic rocks of sandstone, shale, and shallow-marine carbonates. Sedimentation was initiated in the Early Cambrian with a global rise in sea level on the stable craton of the Sino-Korean Block. During the long-standing transgression for tens of million years, the shallow platform was under mild climatic conditions with an explosive production of calcium carbonates by organisms such as algae, crinoids, trilobites, and brachiopods as well as microbes. The entire succession is represented by four second-order sequences. Sequence I represents an initial inundation and subsequent drowning during the late Early to Middle Cambrian. Sequence II was formed by prolonged marine flooding during the Middle and Late Cambrian and was terminated by a large supply of siliciclastic sediments due to rejuvenation of the drainage area at the Cambrian–Ordovician boundary. Sequence III comprises the lowstand siliciclastics, transgressive shale by marine flooding, and platform carbonates by highstand progradation in the Early Ordovician. Following the large-scale eustatic fall in early Middle Ordovician, sequence IV was formed on the subaerial surface by resumed marine flooding in the Middle Ordovician.

The Cambrian succession of the North China Platform in Shandong Province comprises a thick mixed carbonate-siliciclastic sequence, which formed in an extensive epeiric sea. The lower part of the succession (Early Cambrian) represents initial flooding and stabilization of shallow marine environments. Siliciclastic sediments were derived from unsubmerged archipelago and deposited in coastal floodplains under strong tidal regime, whereas carbonates and microbialites formed in flooded peritidal environments. With the ensuing rise in sea level, carbonate production kept pace and formed numerous depositional cycles that comprised lime mudstone, dolomite, stromatolite, and mudstone, showing a general shallowing-upward trend in

Geology and Sedimentology of the Korean Peninsula. DOI: http://dx.doi.org/10.1016/B978-0-12-405518-6.00015-2

response to an increase in accommodation. In the Middle Cambrian, carbonates kept up with the rise in sea level and were dominated by oolite and microbialite facies, representing optimum conditions for carbonate production and microbial activity during highstand in sea level. The platform was then drowned during the subsequent rise in eustatic sea level. Bounded by drowning unconformity, the overlying succession consists of dominant shale and lime mudstone. Carbonate production then caught up with the sea level rise, accumulating more carbonates in the late Furongian. Microbialites resurged and flourished, and then demised as water depth increased beyond the critical level for photosynthesis.

The lithologic units of the Cambrian succession in Shandong Province is well correlated with that of the Taebaeksan Basin, more than 1000 km apart. The quartzite deposit occurs at the base of the succession in both areas. The overlying unit of siliciclastics and carbonates (Zhushadong and Mantou formations) is equivalent to that of shale-dominant Myobong Formation. The overlying carbonate-dominant succession (Zhangxia Formation) is similar to that of the Daegi Formation. The overlying shale and lime-mudstone units are generally well correlated in both the lithology and the biozone of trilobites. Both areas experienced marine flooding at about the same time, although transgression seems to have progressed northwestwards. During the ensuing transgression, water increased slowly in the Shandong region, but abruptly in the Taebaek area. The sedimentary facies are thus distinctive from one another. There are high variabilities of bounding surfaces within the comparable units as well as significant diachroneity, which are due to local changes in accommodation and sediment supply.

There was a major break in sedimentation during the Silurian and Devonian periods in the entire platform, and in the North China Platform and the Taebaeksan Basin. During the Carboniferous to early Triassic, sedimentation was resumed in coastal plain and swamp environments with progradation of deltas.

Major tectonic events were initiated in the Triassic when the South China Block collided with the Sino-Korean Block along the Imjingang Belt, an extension of the Qinling-Dabie-Sulu Belt in China. The eastern part of the Sino-Korean Block rotated clockwise and moved southward relative to the South China Block along the SKTL. Along the Imjingang Belt, blocks of the Paleozoic succession formed thrust and were preserved as tectonic windows in the periphery of the Pyeongnam Basin. High mountain chains formed along the suture belt. The Yeongwol Subblock was pushed north- and eastwards, and bounded by the Taebaek Group.

In the Middle–Late Jurassic, orthogonal subduction of the paleo-Pacific plate under the Asian continent caused compression and thrust deformation. A number of piggyback basins formed along the thrust faults in the east of the SKTL. At the same time, the entire peninsula was prevailed by granite batholiths, especially along the northeast–southwest-trending tectonic belt.

In the Cretaceous Period, the paleo-Pacific Plate subducted northward under the Asian continent, forming numerous extensional (left-lateral strike-slip) basins in the southern part of the peninsula and the Yellow Sea. A large backarc basin was initiated in the southeastern part, Gyeongsang Basin, in which sediments were largely deposited in fluvial and lacustrine environments with streamflow-dominated alluvial fans adjacent to the fault-bounded basin margin in the north. Sediments were also derived from the western highlands, draining Precambrian to Jurassic basement rocks. The arc platform also played an important role in supplying volcaniclastic materials and accreted sediments into the backarc basin via extensive fluvial networks. During basin expansion, basaltic volcanic rocks were intermittently extruded. In the Late Cretaceous, andesitic to rhyolitic volcanism was extensive in the arc platform, forming numerous calderas and volcaniclastic depressions.

In the Paleogene, both the volcanic arc and the backarc basin ceased to develop, as volcanic activities shifted eastward, accompanied by rollback of the subduction of the Pacific plate. In the Miocene, pull-apart (right-lateral) basins formed in the eastern continental margin, including the Pohang and Ulleung basins and the associated backarc basins (Youngduk and Hupo basins). The Korea Plateau experienced continental rifting accompanied by extensive volcanism during the extensional opening of the Ulleung Basin. It subsided more than 1000 m below sea level. The Ulleung Interplain Gap represents an extensional opening of the crust between the South Korea Plateau and the Oki Bank.

Jeju Island is an intraplate volcanic island. The island is a shield volcano of basaltic lava flows and pyroclastic rocks, underlain by a succession of shallow platforms (nearshore, inner shelf, and skeletal banks) (about 1.8 Ma). Over 300 scoria cones occur in the island, which resulted from single explosion of basaltic magma with abundant volatile component. Tuff rings and cones occur along the shore, which were produced by powerful explosion at deeper levels and emplaced by wet and dry pyroclastic surges.

In the Yellow Sea, analysis of unprecedented high-resolution subbottom profiles and drill cores has revealed the subsurface distribution of quaternary sediments and dynamic processes of the transgressive deposits during the Holocene sea-level rise. It was revealed that with continued rise in sea level, the transgression ravinement and remoulding processes propagated to the northern part of the sea. The sheetlike mud wedge in the central region represents deposition of suspended sediments derived from the Huanghe River. The Huksan mud belt originates from the Geum River and is controlled by tidal and thermohaline fronts, i.e., partitioning of sedimentary facies.

The west and south coasts of the peninsula are characterized by numerous embayments with extensive development of tidal flats under a macrotidal regime. During the sea-level rise in the Holocene, most tidal flats formed a retrogradational, coarsening-upward succession, controlled by the combined effect of waves and tides with seasonal cycles. Tidal flats are a major source for the nearshore sediment plume and act as a principal repository for littoral sediments around the Korean Peninsula.

References

Aigner, T., 1985. Storm Depositional Systems. Lecture Notes in Earth Sciences. Springer, Berlin, 174 pp.

Allen, J.R.L., 1964. Primary current lineation in the Lower Old Red Sandstone (Devonian), Anglo-Welsh Basin. Sedimentology. 3, 89–108.

Allen, J.R.L., 1968. Current Ripples. North-Holland, 433 pp.

Allen, J.R.L., 1973. Phase differences between bed configurations and flow in natural environments, and their geological relevance. Sedimentology. 20, 323–329.

Allen, J.R.L., 1982. Sedimentary Structures: Their Character and Physical Basis. Developments in Sedimentology. Elsevier, New York, 593 pp.

Allen, J.R.L., 1983. Studies in fluviatile sedimentation: bars, bar-complexes and sandstone sheets (low-sinuosity braided streams) in the Brownstones (L. Devonian), Welsh Borders. Sediment. Geol. 33, 237–293.

Allen, J.R.L., 1984. Parallel lamination developed from upper stage plane beds: a model based on the larger coherent structures of the turbulent boundary layer. Sediment. Geol. 39, 227–242.

Allen, P.A., 1981. Sediments and processes on a small stream-flow dominated, Devonian alluvial fan, Shetland Islands. Sediment. Geol. 29, 31–66.

Allen, M.B., Macdonald, D.I.M., Xun, Z., Vincent, S.J., Brouet-Menzies, C., 1997. Early Cenozoic two-phase extension and late Cenozoic thermal subsidence and inversion of the Bohai basin, northern China. Mar. Petrol. Geol. 14, 951–972.

Aoti, K., 1942. Geology of Bunkei District in Tyosen with special references to the stratigraphy of the Tyosen Group. J. Geol. Soc. Jpn. 49, 279–281, in Japanese.

Arnott, R.W.C., Hand, B.M., 1989. Bedforms, primary structures and grain fabric in the presence of suspended sediment rain. J. Sediment. Petrol. 59, 1062–1069.

Baek, Y.S., Chun, S.S., Shin, S.J., 2009. Seasonal variation of sedimentary structures and ichnofacies in the Doouri open-coast tidal flat, southwestern Korea. J. Geol. Soc. Kor. 45, 187–208.

Baek, Y.S., Chun, S.S., Kim, J.K., 2010. Seasonal sedimentation of intertidal flat revealed by distribution of ichonofacies: southern tidal flat (Yeochari) of Ganghwa Island. J. Geol. Soc. Kor. 46, 661–674.

Bahk, J.J., Chough, S.K., 1996. An interplay of syn- and intereruption depositional processes: the lower part of the Jangki Group (Miocene), SE Korea. Sedimentology. 43, 421–438.

Bahk, J.J., Lee, S.H., Yoo, H.S., Back, G.G., Chough, S.K., 2005. Late quaternary sedimentary processes and variations in bottom-current activity in the Ulleung Interplain Gap, East Sea (Korea). Mar. Geol. 217, 119–142.

Baker, V.R., 1973. Paleohydrology and sedimentology of Lake Missoula flooding in eastern Washington. Geol. Soc. Am. 79 (Special Paper 14).

Bentham, P.A., Talling, P.J., Burbank, D.W., 1993. Braided stream and floodplain deposition in a rapidly aggrading basin: the Escanilla Formation, Spanish Pyrenees. In: Best, J.L., Bristow, C.S. (Eds.), Braided Rivers. Geological Society of London, London, pp. 177–194.

Besly, B.M., Turner, P., 1983. Origin of red beds in a moist tropical climate (Eturia Formation, Upper Carboniferous, UK). In: Wilson, R.C.L. (Ed.), Residual Deposits: Surfaces Related Weathering Processes and Material. Geological Society of London, London, pp. 131–147.

Blackwelder, E., Willis, B., 1907. Stratigraphy of Shantung. In: Willis, B., Blackwelder, E., Sargent, R.H. (Eds.), Descriptive Topography and Geology. Research in China, vol. 1. Pt. 1. Carnegie Institution of Washington, Washington D.C., pp. 19–58.

Blair, T.C., 1987. Sedimentary processes, vertical stratification sequences, and geomorphology of the Roaring River alluvial fan, Rocky Mountain National Park, Colorado. J. Sediment. Petrol. 57, 1–18.

Blair, T.C., 1999. Sedimentary processes and facies of the waterlaid Anvil Spring Canyon alluvial fan, Death Valley, California. Sedimentology. 46, 941–965.

Blair, T.C., 2000. Sedimentology and progressive tectonic unconformities of the sheetflood-dominated Hell's Gate alluvial fan, Death Valley, California. Sediment. Geol. 132, 233–262.

Blair, T.C., Bilodeau, W.L., 1988. Development of tectonic cyclothems in rift, pull-apart, and foreland basins: sedimentary response to episodic tectonism. Geology. 16, 517–520.

Blair, T.C., McPherson, J.G., 1994. Alluvial fan processes and forms. In: Abrahams, A.D., Parsons, A.J. (Eds.), Geomorphology of Desert Environments. Chapman & Hall, London, pp. 354–402.

Bloom, A.L., Park, Y.A., 1985. Holocene sea-level history and tectonic movements, Republic of Korea. Quat. Res. Jpn. 24, 77–84.

Bluck, B.J., 1980. Structure, generation and preservation of upward fining braided stream cycles in the Old Red Sandstone of Scotland. Trans. R. Soc. Edinburgh. 71, 29–46.

Bond, G.C., Nickeson, P.A., Kominz, M.A., 1984. Breakup of a supercontinent between 625 Ma and 555 Ma: new evidence and implications for continental histories. Earth Planet. Sci. Lett. 70, 325–345.

Boothroyd, J.C., Ashley, G.M., 1975. Process, bar morphology and sedimentary structures on braided outwash fans, North-eastern Gulf of Alaska. In: Jopling, A.V., McDonald, B.C. (Eds.), Glaciofluvial and Glaciolacustrine Sedimentation. SEPM, Tulsa, pp. 193–222.

Bouma, A.H., 1962. Sedimentology of Some Flysch Deposits. Elsevier, Amsterdam, 168 pp.

Brenna, M., Cronin, S.J., Smith, I.E.M., Maas, R., Sohn, Y.K., 2012. How small-volume basaltic magmatic systems develop: a case study from the Jeju Island volcanic field, Korea. J. Petrol. 53, 985–1018.

Bridge, J.S., 1978. Origin of horizontal lamination under turbulent boundary layers. Sediment. Geol. 20, 1–18.

Bridge, J.S., 1981. Bed shear stress over subaqueous dunes, and the transition to upper-stage plane beds. Sedimentology. 28, 33–36.

Bridge, J.S., 1993. Description and interpretation of fluvial deposits: a crtical perspective. Sedimentology. 40, 801–810.

Bridge, J.S., Best, J.L., 1988. Flow, sediment transport and bedform dynamics over the transition from dunes to upper-stage plane beds: implications for the formation of planar laminae. Sedimentology. 35, 753–763.

Bristow, C.S., Skelly, R.L., Ethridge, F.G., 1999. Crevasse splays from the rapidly aggrading, sand-bed, braided Niobrara River, Nebraska: effect of base-level rise. Sedimentology. 46, 1029–1047.

Bull, W.B., 1977. The alluvial-fan environment. Prog. Phys. Geog. 1, 222–270.

Bureau of Geology and Mineral Resources of Shandong Province, 1996. Stratigraphy (Lithostratigraphic) of Shandong Province. China University of Geoscience Press, Beijing, 328 pp. (in Chinese).

Busby, C., Azor, A., 2011. Recent Advances in the Tectonics of Sedimentary Basins. Wiley-Blackwell, 656 pp.
Cabrera, L., Saez, A., 1987. Coal deposition in carbonate-rich shallow lacustrine system: the Calaf and Mequinenza sequences (Oligocene, eastern Ebro Basin, NE Spain). J. Geol. Soc. London. 144, 451–461.
Cant, D.J., Walker, R.G., 1976. Development of a braided-fluvial facies model for the Devonian Battery Point Sandstone, Quebec. Can. J. Earth Sci. 13, 102–119.
Catuneanu, O., 2006. Principles of Sequence Stratigraphy. Elsevier, Amsterdam, 375 pp.
Catuneanu, O., Abreu, V., Bhattacharya, J.P., Blum, M.D., Dalrymple, R.W., Eriksson, P.G., et al., 2009. Towards the standardization of sequence stratigraphy. Earth Sci. Rev. 92, 1–33.
Catuneanu, O., Bhattacharya, J.P., Blum, M.D., Dalrymple, R.W., Eriksson, P.G., Fielding, C.R., et al., 2010. Sequence stratigraphy: common ground after three decades of development. First Break. 28, 21–34.
Cha, M.S., Yun, S.H., 1988. Cretaceous volcanic cauldrons and ring complexes in Korea. J. Geol. Soc. Korea. 24, 67–86 (in Korean with English abstract).
Chang, K.H., 1975. Cretaceous stratigraphy of southeast Korea. J. Geol. Soc. Korea. 11, 1–23.
Chang, K.H., 1977. Late Mesozoic stratigraphy, sedimentation and tectonics of southeastern Korea. J. Geol. Soc. Kor. 13, 76–90 (in Korean with English abstract).
Chang, K.H., 1988. Cretaceous stratigraphy and paleocurrent analysis of Kyongsang Basin, Korea. J. Geol. Soc. Kor. 24, 194–205.
Chang, K.H., Woo, B.G., Lee, J.H., Park, S.O., Yao, A., 1990. Cretaceous and Early Cenozoic stratigraphy and history of eastern Kyongsang Basin, S. Korea. J. Geol. Soc. Kor. 26, 471–487.
Chang, J.H., Lee, C.W., Jin, J.H., Kim, S.P., Park, Y.S., Shin, W.C., et al., 1996. Yellow Sea Drilling Program for Studies on Quaternary Geology (Analyses of YSDP-102, YSDP-103, YSDP-104, YSDP-105 Cores). Report KR-96(T)-18, Korea Institute of Geoscience and Mineral Resources, Daejeon, 595 pp.
Chang, K.H., Suzuki, K., Park, S.-O., Ishida, K., Uno, K., 2003. Recent advances in the Cretaceous stratigraphy of Korea. J. Asian Earth Sci. 21, 937–948.
Charvet, J., Lapierre, H., Yu, Y., 1994. Geodynamic significance of the Mesozoic volcanism of southeastern China. J. Southeast Asian Earth Sci. 9, 387–396.
Cheel, R.J., Middleton, G.V., 1986. Horizontal laminae formed under flow regime plane bed conditions. J. Geol. 94, 489–504.
Chen, H., Qin, D., 1989. Unstable cratonic and Paleozoic basins of China. In: Zhu, X. (Ed.), Chinese Sedimentary Basins. Elsevier, Amsterdam, pp. 7–16.
Chen, Z., Stanley, D.J., 1998. Sea-level rise on eastern China's Yangtze delta. J. Coastal Res. 14, 360–366.
Chen, J., Chough, S.K., Han, Z., Lee, J.-H., 2011. An extensive erosion surface of a strongly deformed limestone bed in the Gushan and Chaomidian formations (late Middle Cambrian to Furongian), Shandong Province, China: sequence–stratigraphic implications. Sediment. Geol. 233, 129–149.
Chen, J., Chough, S.K., Lee, J.-H., Han, Z., 2012. Sequence-stratigraphic comparison of the upper Cambrian Series 3 to Furongian succession between the Shandong region, China and the Taebaek area, Korea: high variability of bounding surface in an epeiric platform. Geosci. J. 16, 357–379.
Cheong, C.H., 1969. Stratigraphy and paleontology of the Samcheog Coalfield, Gangweondo, Korea (I). J. Geol. Soc. Kor. 5, 13–56.
Cheong, C.H., 1973. A paleontological study of the fusulinids from the Samcheog Coalfield, Korea. J. Geol. Soc. Kor. 9, 47–48.

Cheong, S.W., 1987. The Structural Analysis of the Northern Part of the Cretaceous Sedimentary Basin in Jeungpyeong-Eumseong Area, Korea. M.S. Thesis, Seoul National University, Seoul, p. 68 (in Korean with English abstract).

Cheong, C.H., Park, Y.A., 1977. Fusulinids and depositional conditions of the marine sediment (Carboniferous) of the Yongweol Coalfield, Korea. J. Geol. Soc. Korea. 13, 151–178 (in Korean with English abstract).

Cheong, C.S., Kim, N., 2012. Review of radiometric ages for Phanerozoic granitoids in southern Korean Peninsula. J. Petrol. Soc. Korea. 21, 173–192.

Cheong, C.S., Kim, Y.I., 1996. Sedimentological study of the Nakdong Formation to analyse the forming and evolving tectonics of the Cretaceous Gyeongsang Basin, I: depositional setting, source, and paleocurrent analyses of the Nakdong Formation in the southwestern Gyeongsang Basin. Econ. Environ. Geol. 29, 639–660 (in Korean with English abstract).

Cheong, C.H., Lee, D.Y., Yu, Y.S., Kang, K.W., 1979a. Explanatory text of the geological map of Pyongchang–Yongwol sheets (1:50,000). Korea Institute of Geosciences and Mineral Resources, Seoul, 19 pp. (in Korean with English summary).

Cheong, C.H., Lee, H.Y., Ko, I.S., Lee, J.D., 1979b. A study on stratigraphy and sedimentological environments of the lower Paleozoic sequences in South Korea (chiefly in Jeongseon area). J. Natl. Acad. Sci., Republic Kor., Natl. Sci. Ser. 18, 123–159.

Cho, D.R., 1992. Mineralogical and Petrological Study of the Phanerozoic Granitoids in South Korea. Ph.D. Thesis, Yonsei University, Seoul, p. 189 (in Korean with English abstract).

Cho, D.R., Kwon, S.T., 1994. Hornblende geobarometry of the Mesozoic granitoids in South Korea and the evolution of the crustal thickness. J. Geol. Soc. Korea. 30, 41–61 (in Korean with English abstract).

Cho, M., Kim, H., 2005. Metamorphic evolution of the Ogcheon belt, Korea: a review and new age constraints. Int. Geol. Rev. 47, 41–57.

Cho, M., Kwon, S.T., Ree, J.H., Nakamura, E., 1995. High-pressure amphibolite of the Imjingang belt in the Yeoncheon-Cheongok area. J. Petrol. Soc. Korea. 4, 1–19.

Cho, D.L., Kim, Y.-J., Armstrong, R., 2006a. SHRIMP U–Pb geochronology of detrital zircons from iron-bearing quartzite of the Seosan Group: constraints on age and stratigraphy. J. Petrol. Soc. Korea. 15, 119–127 (in Korean with English abstract).

Cho, H.-M., Baag, C.-E., Lee, J.M., Moon, W.I., Jung, H., Kim, K.Y., et al., 2006b. Crustal velocity structure across the southern Korean Peninsula from seismic refraction survey. Geophys. Res. Lett. 33, L06307.

Cho, M., Kim, Y., Ahn, J., 2007. Metamorphic evolution of the Imjingang belt, Korea: implications for Permo–Triassic collisional orogeny. Int. Geol. Rev. 49, 30–51.

Cho, M., Kim, H., Lee, Y., Horie, K., Hidaka, H., 2008. The oldest (ca. 2.51 Ga) rock in South Korea: U–Pb zircon age of a tonalitic migmatite, Daeijak Island, western Gyeonggi massif. Geosci. J. 12, 1–6.

Cho, M., Na, J., Yi, K., 2010. SHRIMP U–Pb ages of detrital zircons in metasandstones of the Taean Formation, western Gyeonggi massif, Korea: tectonic implications. Geosci. J. 14, 99–109.

Cho, H., Kim, J.-S., Son, M., Sohn, Y.K., Kim, I.-S., 2011. Petrology and $^{40}Ar/^{39}Ar$ ages of volcanic rocks in the Cretaceous Dadaepo Basin, Busan: accumulation time and correlation of the Dadaepo Formation. J. Geol. Soc. Korea. 47, 1–18 (in Korean with English abstract).

Choe, M.Y., Chough, S.K., 1988. The Hunghae Formation, SE Korea: Miocene debris aprons in a back-arc intraslope basin. Sedimentology. 35, 239–255.

Choi, D.K., 1985. Spores and pollen from the Gyeongsang Supergroup, southeastern Korea and their chronologic and paleoecologic implications. J. Paleontol. Soc. Korea. 1, 33–50.

Choi, D.K., 1998. The Yongwol Group (Cambrian–Ordovician) redefined: a proposal for the stratigraphic nomenclature of the Choson Supergroup. Geosci. J. 2, 220–234.

Choi, H.I., 1986. Sedimentation and evolution of the Cretaceous Gyeongsang Basin, southeastern Korea. J. Geol. Soc. London. 143, 29–40.

Choi, S.-J., 1987. Study on the Lower Cretaceous charophytes from the upper Gyeongsang Supergroup. J. Paleontol. Soc. Korea. 3, 79–92.

Choi, S.J., 1990. Some nonmarine fossils from the central part of Gyeongsang Basin, S. Korea. J. Paleontol. Soc. Korea. 6, 165–179.

Choi, Y.S., 1996. Structural Evolution of the Cretaceous Eumseong Basin, Korea. Ph.D. Thesis, Seoul National University, Seoul, p. 159 (in Korean with English abstract).

Choi, D.K., Chough, S.K., 2005. The Cambrian–Ordovician stratigraphy of the Taebaeksan Basin, Korea: a review. Geosci. J. 9, 187–214.

Choi, D.K., Jeong, K.W., 1990. Discovery of conulariid from the Yeongheung Formation (Ordovician), Korea. J. Geol. Soc. Korea. 26, 494–499.

Choi, D.K., Kim, E.Y., 2006. Occurrence of *Changshania* (Trilobita, Cambrian) in the Taebaeksan Basin, Korea and its stratigraphic and paleogeographic significance. Palaeogeogr. Palaeoclimatol. Palaeoecol. 242, 343–354.

Choi, D.K., Park, J.B., 1987. Palynology of the Jinju Formation (Lower Cretaceous), Waegwan-Daegu and Jinju areas, Korea. J. Paleontol. Soc. Korea. 3, 28–43.

Choi, S.J., Woo, K.S., 1993. Depositional environment of the Ordovician Yeongheung Formation near Machari area, Yeongweol, Kangweondo, Korea. J. Geol. Soc. Korea. 29, 375–386.

Choi, H.I., Kim, D.S., Seo, H.G., 1987. Stratigraphy, depositional environment and basin evolution of the Daedong strata in the Chungnam coalfield. Report KR-87-(B)-3, Korea Institute of Energy and Resources, Taejon, p. 97 (in Korean with English abstract).

Choi, S., Kim, B.C., Chun, H.Y., Kim, Y.B., 1995. Charophytes from the Chopyeong Formation (Cretaceous) of the Eumsung Basin, Korea. J. Geol. Soc. Korea. 31, 523–528 (in Korean with English abstract).

Choi, D.K., Lee, J.G., Choi, S.Y., 1999. Middle Cambrian trilobites from the Sambangsan Formation in Yongwol area, Korea. J. Paleontol. Soc. Korea. 15, 134–144.

Choi, D.K., Kim, D.H., Sohn, J.W., 2001. Ordovician trilobite faunas and depositional history of the Taebaeksan Basin, Korea: implications for palaeogeography. Alcheringa: Australas. J. Palaeontol. 25, 53–68.

Choi, D.K., Kim, D.H., Sohn, J.W., Lee, S.B., 2003. Trilobite faunal successions across the Cambrian–Ordovician boundary intervals in Korea and their correlation with China and Australia. J. Asian Earth Sci. 21, 781–793.

Choi, D.K., Chough, S.K., Kwon, Y.K., Lee, S.B., Woo, J., Kang, I., et al., 2004. Taebaek group (Cambrian–Ordovician) in the Seokgaejae section, Taebaeksan basin: a refined Lower Paleozoic stratigraphy in Korea. Geosci. J. 8, 125–151.

Choi, S.H., Mukasa, S.B., Kwon, S.T., Andronikov, A.V., 2006. Sr, Nd, Pb and Hf isotopic compositions of late Cenozoic alkali basalts in South Korea: evidence for mixing between the two dominant asthenospheric mantle domains beneath East Asia. Chem. Geol. 232, 134–151.

Choi, P.-Y., Rhee, C.W., Lim, S.-B., So, Y., 2008. Subdivision of the Upper Paleozoic Taean Formation in the Anmyeondo-Boryeong area, west Korea: a preliminary approach to the sedimentary organization and structural features. Geosci. J. 12, 373–384.

Choi, D.K., Woo, J., Park, T.-Y., 2012. The Okcheon Supergroup in Lake Chungju area, Korea: Neoproterozoic volcanic and glaciogenic sedimentary successions in a rift basin. Geosci. J. 12, 207–227.

Chough, S.K., 1981. Submarine debris flow deposits in the Ogcheon Basin, Korean Peninsula. UN ESCAP, CCOP Tech. Bull. 14, 17–29.

Chough, S.K., Bahk, K.S., 1984–1985. Deposition of muds in the Ulleung marginal basin. Geo-Mar. Lett. 4, 235–241.

Chough, S.K., Bahk, K.S., 1992. The Hwangkangri Formation in the Okchon Basin. In: Chough, S.K. (Ed.), Sedimentary Basins in the Korean Peninsula and Adjacent Seas, Korean Sedimentology Research Group Special Publication. Hanlimwon Publishers, Seoul, pp. 77–101.

Chough, S.K., Barg, E., 1987. Tectonic history of Ulleung Basin margin, East Sea (Sea of Japan). Geology. 15, 45–48.

Chough, S.K., Chun, S.S., 1988. Intrastratal rip-down clasts, Late Cretaceous Uhangri Formation, southwest Korea. J. Sediment. Petrol. 58, 530–533.

Chough, S.K., Hwang, I.G., 1997. The Duksung fan delta, SE Korea: growth of delta lobes on a Gilbert-type topset in response to relative sea-level rise. J. Sediment. Res. 67, 725–739.

Chough, S.K., Sohn, Y.K., 1990. Depositional mechanics and sequences of base surges, Songaksan tuff ring, Cheju Island, Korea. Sedimentology. 37, 1115–1135.

Chough, S.K., Sohn, Y.K., 2010. Tectonic and sedimentary evolution of a Cretaceous continental arc–backarc system in the Korean peninsula: new view. Earth Sci. Rev. 101, 225–249.

Chough, S.K., Mosher, D.C., Srivastava, S.P., 1985. Ocean Drilling Program (ODP) site survey (Hudson 84–30) in the Labrador Sea: 3.5 kHz profiles. Geol. Surv Can. 33–41 (Paper 85-1B).

Chough, S.K., Hwang, J.G., Choe, M.Y., 1990. The Miocene Doumsan fan-delta, southeast Korea: a composite fan-delta system in back-arc margin. J. Sediment. Petrol. 60, 445–455.

Chough, S.K., Kim, S.B., Chun, S.S., 1996. Sandstone/chert and laminated chert/black shale couplets, Cretaceous Uhangri Formation (southwest Korea): depositional events in alkaline lake environments. Sediment. Geol. 104, 227–242.

Chough, S.K., Lee, S.H., Kim, J.W., Park, S.C., Yoo, D.G., Han, H.S., et al., 1997. Chirp (2–7 kHz) echo characters in the Ulleung Basin. Geosci. J. 1, 143–153.

Chough, S.K., Kwon, S.T., Ree, J.H., Choi, D.K., 2000. Tectonic and sedimentary evolution of the Korean peninsula: a review and new view. Earth-Sci. Rev. 52, 175–235.

Chough, S.K., Kwon, Y.K., Choi, D.K., Lee, D.J., 2001. Autoconglomeration of limestone. Geosci. J. 5, 159–164.

Chough, S.K., Kim, J.W., Lee, S.H., Shinn, Y.J., Jin, J.H., Suh, M.C., et al., 2002. High-resolution acoustic characteristics of epicontinental sea deposits, central-eastern Yellow Sea. Mar. Geol. 188, 317–331.

Chough, S.K., Lee, H.J., Chun, S.S., Shinn, Y.J., 2004. Depositional processes of late Quaternary sediments in the Yellow Sea: a review. Geosci. J. 8, 211–264.

Chough, S.K., Kim, H., Woo, J., Lee, H.S., 2006. Tectonic implications of quartzite-shale and phyllite beds in the Seochangri Formation (Okcheon Group), Bonghwajae section, mid-Korea. Geosci. J. 10, 403–421.

Chough, S.K., Lee, H.S., Woo, J., Chen, J., Choi, D.K., Lee, S.-b., et al., 2010. Cambrian stratigraphy of the North China Platform: revisiting principal sections in Shandong Province, China. Geosci. J. 14, 235–268.

Chun, S.S., 1990. Sedimentary Processes, Depositional Environments and Tectonic Settings of the Cretaceous Uhangri Formation, Southwest Korea. Ph.D. Thesis, Seoul National University, Seoul, p. 328.

Chun, S.S., Chough, S.K., 1992. Depositional sequences from high-concentration turbidity currents, Cretaceous Uhangri Formation (SW Korea). Sediment. Geol. 77, 225–233.
Chun, S.S., Chough, S.K., 1995. The Cretaceous Uhangri Formation, SW Korea—lacustrine margin facies. Sedimentology. 42, 293–322.
Chun, S.S., Yang, B.C., 2004. Sedimentation and Holocene evolution of wave-dominated, macrotidal-flat depositional system in the southern open coasts of Korean Peninsula. J. Sedimentol. Soc. Jpn. 58, 65–83.
Chun, H.Y., Choi, Y.S., Choi, S., Kim, Y.B., Kim, B.C., Lee, B.S., et al., 1994. Fossil floral and faunal assemblages and paleoenvironmental modelling study (II) on the Cretaceous sedimentary basins scattered in/near the Ogcheon Belt (Report KR-94(T)-11). Korea Institute of Geology, Mining and Materials, Seoul, p. 101 (in Korean with English abstract).
Chun, S.S., Yang, B.C., Lee, I.T., Lee, H.J., 2000. Non-barred, open macrotidal flat strongly influenced by wave action, Gomso Bay and Baeksu Coast, Southwest Korea: depositional processes, seasonal evolution and transgressive stratigraphy. Tidalite-2000. 5th International Conference on Tidal Environments Field Guide Book (B2), p. 101.
Clark, J.D., Pickering, K.T., 1996. Submarine Channels: Processes and Architecture. Vallis Press, London, p. 231.
Cluzel, D., 1992. Ordovician bimodal magmatism in the Ogcheon belt (South Korea): intracontinental rift-related volcanic activity. J. Southeast Asian Earth Sci. 7, 195–209.
Cluzel, D., Cadet, J.P., Lapierre, H., 1990. Geodynamics of the Ogcheon belt (South Korea). Tectonophysics. 183, 41–56.
Cluzel, D., Lee, B.J., Cadet, J.P., 1991. Indosinian dextral ductile fault system and synkinematic plutonism in the southwest of the Ogcheon belt (South Korea). Tectonophysics. 194, 131–151.
Collinson, J.D., Thompson, D.B., 1989. Sedimentary Structures. second ed. Unwin Hyman, London, 207 pp.
Costa, J.E., 1988. Rheologic, geomorphic, and sedimentologic differentiation of water floods, hyperconcentrated flows, and debris flows. In: Baker, V.R., Kochel, R.C., Patton, P.C. (Eds.), Flood Geomorphology. John Wiley & Sons, New York, pp. 113–122.
Costello, W.R., Southard, J.B., 1981. Flume experiments on lower-flow-regime bed forms in coarse sand. J. Sediment. Petrol. 51, 849–864.
Dalrymple, R., 2010. Interpreting sedimentary successions: facies, facies analysis and facies models. In: James, N.P., Dalrymple, R. (Eds.), Facies Models, vol. 4. Geological Association of Canada, St. John's, pp. 3–18.
Damuth, J.E., 1980. Use of high-frequency (3.5–12 kHz) echograms in the study of near-bottom sedimentation processes in the deep-sea: a review. Mar. Geol. 38, 51–75.
Damuth, J.E., Embley, R.W., 1981. Mass-transport processes on Amazon Cone: western equatorial Atlantic. Am. Assoc. Pet. Geol. Bull. 65, 629–643.
Damuth, J.E., Hayes, D.E., 1977. Echo character of the east Brazilian continental margin and its relationship to sedimentary processes. Mar. Geol. 24, 73–95.
Doh, S.-J., Piper, J.D.A., 1994. Paleomagnetism of the (Upper Paleozoic–Lower Mesozoic) Pyeongan Supergroup, Korea: a Phanerozoic link with the North China Block. Geophys. J. Int. 117, 850–863.
Egawa, K., Lee, Y.I., 2009. Jurassic synorogenic basin filling in western Korea: sedimentary response to inception of the western Circum-Pacific orogeny. Basin Res. 21, 407–431.
Embley, R.W., 1976. New evidence for ocurrence of debris flow deposits in the deep sea. Geology. 4, 371–374.
Embley, R.W., Jacobi, R.D., 1977. Distribution and morphology of large submarine sediment slides and slumps on Atlantic continental margins. Mar. Geotechnol. 2, 205–228.

Emery, D., Myers, K.J., 1996. Sequence Stratigraphy. Blackwell, Oxford, 297 pp.

Engebretson, D.C., Cox, A., Gordon, R.G., 1985. Relative motions between oceanic and continental plates in the Pacific Basin. Geological Society of America Special Papers, Geological Society of America, Boulder, 59 pp.

Eyles, C.H., Eyles, N., 2010. Glacial deposits. In: James, N.P., Dalrymple, R. (Eds.), Facies Models, vol. 4. Geological Association of Canada, St. John's, pp. 73–104.

Falk, P.D., Dorsey, R.J., 1998. Rapid development of gravelly high-density turbidity currents in marine Gilbert-type fan deltas, Loreto Basin, Baja California Sur, Mexico. Sedimentology. 45, 331–350.

Faugères, J.-C., Stow, D.A.V., Imbert, P., Viana, A., 1999. Seismic features diagnostic of contourite drifts. Mar. Geol. 162, 1–38.

Feary, D.A., 1984. The Boambolo Formation—A Silurian prograding fan-delta sequence in southeastern New South Wales, Australia. Sediment. Geol. 39, 169–195.

Feng, H., Wang, Z., 1986. Holocene Sea-Level Changes and Coastline Shifts in Zhejiang Province. Department of Geography, Hangzhou University, Hangzhou, 23 pp.

Fisher, R.V., 1971. Features of coarse-grained, high-concentration fluids and their deposits. J. Sediment. Petrol. 41, 916–927.

Flood, R.D., Damuth, J.E., 1987. Quantitative characteristics of sinuous distributary channels on the Amazon Deep-Sea Fan. Geol. Soc. Am. Bull. 98, 728–738.

Friend, P.F., 1983. Towards the field classification of alluvial architecture or sequence. In: Collinson, J.D., Lewin, J. (Eds.), Modern and Ancient Fluvial Systems. International Association of Sedimentologists, pp. 345–349.

Gardner, J.V., Prior, D.B., Field, M.E., 1999. Humbolt Slide—a large shear-dominated retrogressive slope failure. Mar. Geol. 154, 323–338.

Gawthorpe, R.L., Colella, A., 1990. Tectonic controls on coarse-grained delta depositional systems in rift basins. In: Colella, A., Prior, D.B. (Eds.), Coarse-Grained Deltas. International Association of Sedimentologists, Cambridge, pp. 113–128.

GICTR (Geological Investigation Corps of the Taebaegsan Region), 1962. Report on the Geology and Mineral Resources of the Taebaegsan Region, Geological Society of Korea, Seoul, p. 89.

Gloppen, T.G., Steel, R.J., 1981. The deposits, internal structure and geometry in six alluvial fan–fan delta bodies (Devonian–Norway)—a study in the significance of bedding sequence. In: Ethridge, F.G., Flores, R.M. (Eds.), Recent and Ancient Non-Marine Depositional Environments: Models for Exploration. SEPM, Tulsa, pp. 49–69.

Glumac, B., Walker, K.R., 2000. Carbonate deposition and sequence stratigraphy of the terminal cambrian grand cycle in the Southern Appalachians, U.S.A.. J. Sediment. Res. 70, 952–963.

Gottsche, C., 1884. Auffindung Cambrischer Schichten in Korea. Z. der. Dtsch. Geologischen Ges. 34, 875–876.

Grotzinger, J., Jordan, T.H., Press, F., Siever, R., 2007. Understanding Earth. fifth ed. W.H. Freeman, New York, 672 pp.

Hampton, M.A., 1970. Subaqueous Debris Flow and Generation of Turbidity Currents. Ph.D. Thesis, Stanford University, Stanford, p. 180.

Hampton, M.A., 1972. The role of subaqueous debris flow in generating turbidity currents. J. Sediment. Petrol. 42, 775–793.

Hampton, M.A., 1975. Competence of fine-grained debris flows. J. Sediment. Petrol. 45, 834–844.

Hampton, M.A., 1979. Buoyancy in debris flows. J. Sediment. Petrol. 49, 753–758.

Han, R., Ree, J.H., Cho, D.L., Kwon, S.T., Armstrong, R., 2006. SHRIMP U–Pb zircon ages of pyroclastic rocks in the Bansong Group, Taebaeksan Basin, South Korea and their implication for the Mesozoic tectonics. Gondwana Res. 9, 106–117.

Hardie, L.A., 1977. Introduction. In: Hardie, L.A. (Ed.), Sedimentation on the Modern Carbonate Tidal Flats of Northwest Andros Island, Bahamas. Johns Hopkins Studies in Geology 22. Johns Hopkins University Press, Baltimore, MD, pp. 1–3.

Harms, J.C., Fahnestock, R.K., 1965. Stratification, bed forms, and flow phenomena with an example from the Rio Grande. In: Middleton, G.V. (Ed.), Primary Sedimentary Structures and Their Hydrodynamic Interpretation. SEPM, Tulsa, pp. 84–115.

Harms, J.C., Southard, J.B., Spearing, D.R., Walker, R.G., 1975. Depositional environments as interpreted from primary sedimentary structures and stratification sequences. SEPM Short Course Notes No. 2, p. 161.

Harms, J.C., Southard, J.B., Walker, R.G., 1982. Structures and Sequences in Clastic Rocks. SEPM Short Course Notes No. 9, p. 249.

Heckel, P.H., 1972. Possible inorganic origin for stromatactis in calcilutite mounds in the Tully Limestone, Devonian of New York. J. Sediment. Petrol. 42, 7–18.

Hein, F.J., 1982. Depositional mechanisms of deep-sea coarse clastic sediments, Cap Enrage Formation, Quebec. Can. J. Earth Sci. 19, 267–287.

Heinz, J., Aigner, T., 2003. Hierarchical dynamic stratigraphy in various Quaternary gravel deposits, Rhine glacier area (SW Germany): implications for hydrostratigraphy. Int. J. Earth Sci. 92, 923–938.

Hesse, R., Chough, S.K., 1980. The Northwest Atlantic Mid-Ocean channel of the labrador Sea: II. Deposition of parallel laminated levee-muds from the viscous sublayer of low density turbidity currents. Sedimentology. 27, 697–711.

Hisakoshi, S., 1943. Geology of Seizen District, Kogendo, Tyosen. J. Geol. Soc. Jpn. 50, 269–277 (in Japanese).

Hiscott, R.N., Middleton, G.V., 1979. Depositional mechanics of thick-bedded sandstones at the bases of a submarine slope, Tourelle Formation (Lower Ordovician), Quebec, Canada. In: Doyle, L.J., Pilkey, O.H. (Eds.), Geology of Continental Slopes. SEPM, Tulsa, pp. 75–82.

Hjellbakk, A., 1997. Facies and fluvial architecture of a high-energy braided river: the Upper Proterozoic Seglodden Member, Baranger Peninsula, northern Norway. Sediment. Geol. 114, 131–161.

Hoffman, P.F., 2008. Snowball Earth: status and new developments. GEO. 11, 44–46 (IGC Special Climate Issue).

Hong, Y.K., 1987. Geochemical characteristics of Precambrian, Jurassic and Cretaceous granites in Korea. J. Korea. Inst. Min. Geol. 20, 35–60.

Hong, J., Cho, S.-H., Choh, S.-J., Woo, J., Lee, D.-J., 2012. Middle Cambrian siliceous sponge-calcimicrobe buildups (Daegi Formation, Korea): metazoan buildup constituents in the aftermath of the Early Cambrian extinction event. Sediment. Geol. 253–254, 47–57.

Houck, K.J., Lockley, M.G., 2006. Life in an active volcanic arc: petrology and sedimentology of dinosaur track beds in the Jindong Formation (Cretaceous), Gyeongsang Basin, South Korea. Cretaceous Res. 27, 102–122.

Howe, J.A., 1996. Turbidite and contourite sediment waves in the northern Rockall Trough, North Atlantic Ocean. Sedimentology. 43, 219–234.

Hubert, J.F., Filipov, A.J., 1989. Debris-flow deposits in alluvial fans on the west flank of the White Mountains, Owens-Valley, California, USA. Sediment. Geol. 61, 177–205.

Hukasawa, T., 1943. Geology of Heisyo District, Kogendo, Tyosen. J. Geol. Soc. Jpn. 50, 29–43 (in Japanese).

Hwang, I.G., Chough, S.K., 1990. The Miocene Chunbuk Formation, southeastern Korea: marine Gilbert-type fan-delta system. In: Colella, A., Prior, D.B. (Eds.), Coarse-Grained Deltas. International Association of Sedimentologists, Cambridge. pp. 29–73.

Hwang, I.G., Chough, S.K., 2000. The Maesan fan delta, Miocene Pohang Basin, SE Korea: architecture and depositional processes of a high-gradient fan-delta-fed slope system. Sedimentology. 47, 995–1010.

Hwang, S.K., Cheong, C.S., 1998. Magmatic evolution of igneous rocks related with the Samrangjin caldera, southeastern Korea. J. Petrol. Soc. Korea. 7, 161–176 (in Korean with English abstract).

Hwang, S.K., Kim, S.H., 2006. Magmatic processes of the Muposan Tuff, southern and eastern Cheongsong, Korea. J. Geol. Soc. Korea. 42, 253–271 (in Korean with English abstract).

Hwang, S.K., Kim, S.W., 1990. Silicic ash-flow tuffs with calderas in the terrane between Miryang and Yangsan, Korea. J. Geol. Soc. Korea. 26, 329–340.

Hwang, S.K., Kim, S.W., 1992. Silicic volcanism of Yangsan Caldera, Korea. J. Geol. Soc. Korea. 28, 491–503.

Hwang, I.G., Chough, S.K., Hong, S.W., Choe, M.Y., 1995. Controls and evolution of fan delta systems in the Miocene Pohang Basin, SE Korea. Sediment. Geol. 98, 147–179.

Hwang, J.H., Kim, D.H., Cho, D.R., Song, K.Y., 1996. Geological Map of Andong sheet (1:250,000). Korea Institute of Geoscience and Mineral Resources, Daejeon.

Hwang, S.K., Son, J.D., Lee, B.J., Reedman, A.J., 2002. Eruptive phases and volcanic processes of the Guamsan caldera, southeastern Cheongsong, Korea. J. Petrol. Soc. Korea. 11, 74–89 (in Korean with English abstract).

Hwasung City, 2005. Report on the Preservation and Characteristics of the Dinosaur Egg Fossil Localities in Hwasung City, Gyeonggi-do. Hwasung City, p. 172 (in Korean).

Ingle, J.C., 1992. Subsidence of the Japan Sea: stratigraphic evidence from ODP Sites and onshore sections. In: K. Tamaki, K. Suyehiro, J. Allen, M. McWilliams (Eds.), Proceedings of the Ocean Drilling Program, Scientific Results 127/128 (part 2), Ocean Drilling Program, College Station, TX, pp. 1197–1218.

Jeon, Y.M., Sohn, Y.K., 2003. Sedimentary characteristics and stratigraphic implications of the Kusandong Tuff, Cretaceous Gyeongsang Basin, Korea. Geosci. J. 7, 53–64.

Jeon, Y.M., Sohn, Y.K., 2008. Characteristics, emplacement processes, and stratigraphic implications of the basalts intercalated in the Hayang Group, Cretaceous Gyeongsang Basin, SE Korea. J. Geol. Soc. Korea. 44, 707–727 (in Korean with English abstract).

Jeong, J.O., Jwa, Y.J., 2000. Volcanic stratigraphy and petrology of Cretaceous volcanic rocks in the eastern part of the Euiseong Basin. J. Petrol. Soc. Korea. 9, 238–253 (in Korean with English abstract).

Jeon, H., Cho, M., Kim, H., Horie, K., Hidaka, H., 2007. Early Archean to Middle Jurassic evolution of the Korean Peninsula and its correlation with Chinese cratons: SHRIMP U–Pb zircon age constraints. J. Geol. 115, 525–539.

Jeong, J.O., Kwon, C.W., Sohn, Y.K., 2008. Lithofacies and architecture of a basinwide tuff unit in the Miocene Eoil Basin, SE Korea: modes of pyroclastic sedimentation, changes in eruption style, and implications for basin configuration. Geol. Soc. Am. Bull. 120, 1263–1279.

Jin, J.H., 2001. A Sedimentological Study of Long Sediment Cores in the Eastern Yellow Sea. Ph.D. Thesis, Seoul National University, Seoul, p. 176.

Jin, M.S., 1980. Geological and isotopic contrasts of Jurassic and Cretaceous granites in south Korea. J. Geol. Soc. Korea. 16, 205–215.

Jin, J.H., Chough, S.K., 1998. Partitioning of transgressive deposits in the southeastern Yellow Sea: a sequence stratigraphic interpretation. Mar. Geol. 149, 79–92.

Jin, J.H., Chough, S.K., 2002. Erosional shelf ridges in the mid-eastern Yellow Sea. Geo-Mar. Lett. 21, 219–225.

Jin, M.S., Kim, S.J., Shin, S.C., 1988. K/Ar and fission-track datings for volcanic rocks in the Pohang-Kampo area. Report KD-87-27, Korea Institute of Energy and Resources, Taejon (in Korean).

Jin, M.S., Kim, S.J., Shin, S.C., Lee, J.Y., 1989. K/Ar and fission-track datings for granites and volcanic rocks in the southeastern part of Korean Peninsula. Report KD-88-6D, Korea Institute of Energy and Resources, Taejon (in Korean).

Jin, J.H., Chough, S.K., Ryang, W.H., 2002. Sequence aggradation and systems tracts partitioning in the mid-eastern Yellow Sea: roles of glacio-eustasy, subsidence and tidal dynamics. Mar. Geol. 184, 249–271.

Jo, H.R., 2003a. Depositional environments, architecture, and controls of Early Cretaceous non-marine successions in the northwestern part of Kyongsang Basin, Korea. Sediment. Geol. 161, 269–294.

Jo, H.R., 2003b. Non-marine successions in the northwestern part of Kyongsang Basin (Early Cretaceous): fluvial styles and stratigraphic architecture. Geosci. J. 7, 89–106.

Jo, H.R., Chough, S.K., 2001. Architectural analysis of fluvial sequences in the northwestern part of Kyongsang Basin (Early Cretaceous), SE Korea. Sediment. Geol. 144, 307–334.

Jo, H.R., Rhee, C.W., Chough, S.K., 1997. Distinctive characteristics of a streamflow-dominated alluvial fan deposit: Sanghori area, Kyongsang Basin (Early Cretaceous), southeastern Korea. Sediment. Geol. 110, 51–59.

Johnson, A.M., 1970. Physical Processes in Geology. Freeman, Cooper & Company, San Francisco, CA, 571 pp.

Johnson, D.A., 1984. The Vema Channel: physiography, structure, and sediment–current interactions. Mar. Geol. 58, 1–34.

Johnson, D.A., Damuth, J.E., 1979. Deep thermohaline flow and current-controlled sedimentation in the Amirante Passage, western Indian Ocean. Mar. Geol. 33, 1–44.

Johnson, H.D., Baldwin, C.T., 1996. Shallow clastic seas. In: Reading, H.G. (Ed.), Sedimentary Environments: Processes, Facies and Stratigraphy. Blackwell, London, pp. 232–280.

Jolivet, L., Tamaki, K., 1992. Neogene kinematics in the Japan Sea region and volcanic activity of the northeast Japan arc. In: K. Tamaki, K. Suyehiro, J. Allen, M. McWilliams (Eds.), Proceedings of the Ocean Drilling Program, Scientific Results 127/128 (part 2), Ocean Drilling Program, College Station, TX, pp. 1311–1331.

Jones, B., Desrochers, A., 1992. Shallow platform carbonates. In: Walker, R.G., James, N.P. (Eds.), Facies Models: Response to Sea Level Change. Geological Association of Canada, St. John's, pp. 277–301.

Jopling, A.V., 1967. Origin of laminae deposited by the movement of ripples along a streambed: a laboratory study. J. Geol. 75, 287–305.

Jung, W.Y., Suk, B.C., Min, G.H., Lee, Y.K., 1998. Sedimentary structure and origin of a mud-cored pseudo-tidal sand ridge, eastern Yellow Sea, Korea. Mar. Geol. 151, 73–88.

Jwa, Y.J., Nakajima, T., Uchiumi, S., Shibata, K., 1990. Geochronology and cooling history of Mesozoic granitic rocks in the Inje-Hongcheon district, South Korea. Geochem. J. 24, 93–103.

Kamata, Y., Hisada, K., Lee, Y.I., 2000. Late Jurassic radiolarians from pebbles of Lower Cretaceous conglomerates of the Hayang Group, southeastern Korea. Geosci. J. 4, 165–174.

Kaneoka, I., 1990. Radiometric age and Sr isotope characteristics of volcanic from the Japan Sea floor. Geochem. J. 24, 7–19.

Kaneoka, I., Notsu, K., Takigami, Y., Fujioka, K., Sakai, H., 1990. Constraints on the evolution of the Japan Sea based on $^{40}Ar-^{39}Ar$ ages and Sr isotopic ratios for volcanic rocks of the Yamato Seamount chain in the Japan Sea. Earth Planet. Sci. Lett.97.

Kang, S., 2003. Benthic foraminiferal biostratigraphy and paleoenvironments of the Seogwipo Formation, Jeju Island, Korea. J. Paleontol. Soc. Korea. 19, 63–153.

Kang, I., Choi, D.K., 2007. Middle cambrian trilobites and biostratigraphy of the daegi formation (Taebaek Group) in the Seokgaejae section, Taebaeksan Basin, Korea. Geosci. J. 11, 279–296.

Kang, S., Jung, K.K., Yoon, S., 1999. Benthic foraminiferal fauna from the Seoguipo Formation of Cheju Island, Korea. J. Paleontol. Soc. Korea. 15, 95–108.

Kang, S., Jung, K.K., Yoon, S., 2002. Benthic foraminiferal fauna from drilled cores of Jeju Island, Korea. J. Paleontol. Soc. Korea. 18, 1–10.

Kang, J.-H., Hayasaka, Y., Ryoo, C.-R., 2012. Tectonic evolution of the central Ogcheon Belt, Korea. J. Petrol. Soc. Korea. 21, 129–150.

Kano, A., Lee, D.J., Choi, D.K., Yoo, C.M., 1994. Ordovician (Llanvirnian) stromatoporoids from the Yeongweol area, southern Korea. Trans. Proc. Palaeontol. Soc. Jpn, New Ser., 174, 449–457.

Karpeta, W.P., 1993. Sedimentology and gravel bar morphology in an Archean braided river sequence: the Witpan Conglomerate Member (Witwatersrand Supergroup) in the Welkom Goldfield, South Africa. In: Best, J.L., Bristow, C.S. (Eds.), Braided Rivers. Geological Society of London, London, pp. 369–388.

Kawasaki, S., 1927. The flora of the Heian System. Part 1. Bull. Geol. Surv. Chosen 6.

Kee, W.-S., Kim, S.W., Jeong, Y.-J., Kwon, S., 2010. Characteristics of Jurassic continental arc magmatism in South Korea: tectonic implications. J. Geol. 118, 305–323.

Khadkikar, A.S., 1999. Trough cross-bedded conglomerate facies. Sediment. Geol. 128, 39–49.

Khim, B.K., Woo, K.S., Sohn, Y.K., 2001. Sr isotopes of the Seoguipo Formation (Korea) and their application to geologic age. J. Asian Earth Sci. 19, 701–711.

KIGAM (Korea Institute of Geoscience and Mineral Resources), 1979. Geological Atlas of the Samcheog Coalfield (1:25,000). Daejeon.

Kihm, Y.H., Kim, J.H., 2003. Structural characteristics of the central Ogcheon Belt, South Korea: orogen-parallel tectonic transport model. J. Asian Earth Sci. 22, 41–57.

Kim, H.M., 1994. Paleocurrent analysis and depositional history of the Cretaceous Jinju Formation in Jinju-Sacheon area, Korea. J. Geol. Soc. Korea. 30, 506–519.

Kim, J.H., 1996. Mesozoic tectonics in Korea. J. Southeast Asian Earth Sci. 13, 251–265.

Kim, O.J., 1970. Reply to the article "On the geologic age of the Ogcheon Group" by C.M. Son. J. Korea. Inst. Min. Geol. 3, 187–191.

Kim, Y.H., 2012. Depositional Environments of the Lower Ordovician Mungok Formation in Yeongwol, Korea. M.S. Thesis, Chungbuk National University, p. 135.

Kim, D.H., Choi, D.K., 2000. Lithostratigraphy and biostratigraphy of the Mungok Formation (Lower Ordovician), Yongwol, Korea. Geosci. J. 4, 301–311.

Kim, I.S., Kang, H.C., 1989. Palaeomagnetism of Tertiary rocks in the Oil (Eoil) Basin and its vicinities, southeast Korea. J. Geol. Soc. Korea. 25, 273–293 (in Korean with English abstract).

Kim, I.S., Kang, H.C., 1996. Palaeomagnetism of Tertiary basins in southern Korea: 1. Changgi Basin. Econ. Environ. Geol. 29, 357–367 (in Korean with English abstract).

Kim, J.C., Lee, Y.I., 1998. Cyclostratigraphy of the Lower Ordovician Dumugol Formation, Korea: meter-scale cyclicity and sequence stratigraphic interpretation. Geosci. J. 2, 134–147.

Kim, J.-M., Kennett, J.P., 1998. Paleoenvironmental changes associated with the Holocene marine transgression, Yellow Sea (Hwanghae). Mar. Micropaleontol. 34, 71–89.

Kim, J.W., Chough, S.K., 2000. A gravel lobe deposit in the prodelta of the Doumsan fan delta (Miocene), SE Korea. Sediment. Geol. 130, 183–203.

Kim, J.Y., Paik, I.S., 1997. Origin of graded cherty beds containing Skolithos ichnosp. from the Cretaceous Yucheon Group in Mt. Hwangryeong area, Pusan, Korea. J. Korea. Earth Sci. Soc. 18, 51–59.

Kim, N.J., Lee, H.K., 1970. Geological Map of Jungpyeongdong Sheet (1:50,000). Geological Survey of Korea, Seoul.

Kim, S.W., Park, B.S., 1970. Geological Map of Yongyang Sheet (1:50,000). Geological Survey of Korea, Seoul.

Kim, Y., Lee, Y.I., 2004. Diagenesis of shallow marine sandstones, the Lower Ordovician Dongjeom Formation, Korea: response to relative sea-level changes. J. Asian Earth Sci. 23, 235–245.

Kim, O.J., Hong, M.S., Park, H.I., Won, J.K., Park, Y.D., Kim, K.T., 1963a. Geological Map of Dogyedong Sheet (1:50,000). Geological Survey of Korea, Seoul.

Kim, O.J., Hong, M.S., Park, H.I., Won, J.K., Park, Y.D., Kim, K.T., 1963b. Geological Map of Pyonghae Sheet (1:50,000). Geological Survey of Korea, Seoul.

Kim, K.W., Park, B.S., Lee, H.K., 1967. Geological Map of Jecheon Sheet (1:50,000). Geological Survey of Korea, Seoul.

Kim, O.J., Lee, H.Y., Lee, D.S., Yun, S., 1973. The stratigraphy and geologic structure of the Great Limestone Series in South Korea. J. Korea. Inst. Min. Geol. 6, 81–114 (in Korean).

Kim, I.S., Cheong, C.H., Lee, H.Y., 1985. Trilobites from the Sambangsan Formation in the eastern side of Pyongchang area, Kangweon-do, Korea. J. Geol. Soc. Korea. 21, 45–49.

Kim, K.H., Choi, D.K., Lee, C.Z., 1991a. Trilobite biostratigraphy of the Dumugol Formation (Lower Ordovician) of Dongjeom area, Korea. J. Paleontol. Soc. Korea. 7, 106–116.

Kim, S.W., Lee, J.Y., Kim, Y.K., Koh, I.S., 1991b. Petrology of the Cretaceous volcanic rocks in southern Yuchon minor basin. J. Geol. Soc. Korea. 27, 24–39 (in Korean with English abstract).

Kim, G.S., Kim, J.Y., Jung, K.K., Hwang, J.Y., Lee, J.D., 1995a. Rb–Sr whole rock geochronology of the granitic rocks in the Kyeongju-Gampo area, Kyeongsangbugdo, Korea. J. Korea. Earth Sci. Soc. 16, 272–280 (in Korean with English abstract).

Kim, H., Cho, M., Koh, H.J., 1995b. Tectonometamorphic evolution of the central Ogcheon belt in the Jeungpyeong–Deokpyeong area. J. Geol. Soc. Korea. 31, 299–314 (in Korean with English abstract).

Kim, S.B., Chough, S.K., Chun, S.S., 1995c. Bouldery deposits in the lowermost part of the Cretaceous Kyokpori Formation, SW Korea: cohesionless debris flows and debris falls on a steep-gradient delta slope. Sediment. Geol. 98, 97–119.

Kim, S.B., Chun, S.S., Chough, S.K., 1997. Discussion on structural development and stratigraphy of the Kyokpo Pull-Apart Basin, South Korea and tectonic implications for inverted extensional basins. J. Geol. Soc. London. 154, 369–372.

Kim, D.H., Hwang, J.H., Park, K.H., Song, K.Y., 1998. Geological Map of Pusan sheet (1:250,000). Korea Institute of Geoscience and Mineral Resources, Daejeon.

Kim, Y.H., Lee, H.J., Chun, S.S., Han, S.J., Chough, S.K., 1999. Holocene transgressive stratigraphy of a macrotidal flat in the southeastern Yellow Sea: Gomso bay, Korea. J. Sediment. Res. 69, 328–337.

Kim, S.W., Hwang, S.K., Lee, Y.J., Koh, I.S., 2000. Diversity of the Cretaceous basaltic volcanics in Gyeongsang Basin, Korea. J. Petrol. Soc. Korea. 9, 1–12 (in Korean with English abstract).

Kim, S.B., Chough, S.K., Chun, S.S., 2003. Tectonic controls on spatio-temporal development of depositional systems and generation of fining-upward basin fills in a strike-slip setting: Kyokpori Formation (Cretaceous), south-west Korea. Sedimentology. 50, 639–665.

Kim, H.J., Lee, G.H., Jou, H.T., Cho, H.M., Yoo, H.S., Park, G.T., et al., 2007. Evolution of the eastern margin of Korea: constraints on the opening of the East Sea (Japan Sea). Tectonophysics. 436, 37–55.

Kim, J.Y., Kim, K.S., Lockley, M.G., Yang, S.Y., Seo, S.J., Choi, H.I., et al., 2008. New didactyl dinosaur footprints (*Dromaeosauripus hamanensis* ichnogen. et ichnosp. nov.) from the Early Cretaceous Haman Formation, south coast of Korea. Palaeogeogr. Palaeoclimatol. Palaeoecol. 262, 72–78.

Kim, S.B., Kim, Y.-G., Jo, H.R., Jeong, K.S., Chough, S.K., 2009. Depositional facies, architecture and environments of the Sihwa Formation (Lower Cretaceous), mid-west Korea with special reference to dinosaur eggs. Cretaceous. Res. 30, 100–126.

Kim, G.B., Yoon, S.H., Chough, S.K., Kwon, Y.K., Ryu, B.J., 2011. Seismic reflection study of acoustic basement in the South Korea Plateau, the Ulleung Interplain Gap, and the northern Ulleung Basin: volcano-tectonic implications for Tertiary back-arc evolution in the southern East Sea. Tectonophysics. 504, 43–56.

Kleinspehn, K.L., Steel, R.J., Johannessen, E., Netland, A., 1984. Conglomeratic fan-delta sequences, Late Carboniferous–Early Permian, western Spitsbergen. In: Koster, E.H., Steel, R.J. (Eds.), Sedimentology of Gravels and Conglomerates, Canadian Society of Petroleum Geologists. pp. 279–294.

Klimetz, M., 1983. Speculations on the Mesozoic plate tectonic evolution of Eastern China. Tectonics. 2, 139–166.

Kobayashi, T., 1927. Geological history of Gwangwon Province, Korea. Geol. Mag. Jpn. 34, 239 (in Japanese).

Kobayashi, T., 1930. Significance of unconformity at the base of the lower Daedong Formation. J. Geol. Soc. Jpn. 37, 593–607.

Kobayashi, T., 1933. A sketch of Korean Geology. Am. J. Sci. 26, 585–606.

Kobayashi, T., 1961. The Cambro-Ordovician formations and faunas of South Korea, Part VIII, Paleontology VII, Cambrian faunas of the Mungyeong (Bunkei) District and the Samposan Formation of the Yeongweol (Neietsu) District. J. Fac. Sci., Univ. Tokyo, Sect. II. 13, 181–241.

Kobayashi, T., 1962. The Cambro-Ordovician formations and faunas of South Korea, Part IX, Paleontology VIII, The Machari fauna. J. Fac. Sci., Univ. Tokyo, Sect. II. 14, 1–152.

Kobayashi, T., 1966. The Cambro-Ordovician formations and faunas of South Korea, Part X, Stratigraphy of the Chosen Group in Korea and South Manchuria and its relation to the Cambro-Ordovician formations of other areas, Section A, The Chosen Group of South Korea. J. Fac. Sci., Univ. Tokyo, Sect. II. 16, 1–84.

Kobayashi, T., 1967. Geology of South Korea. In: T. Ogura (Ed.), Geology and Mineral Resources of the Far East, University of Tokyo Press, Tokyo, pp. 25–134.

Kobayashi, T., Yosimura, I., Iwaya, Y., Hukasawa, T., 1942. The Yokusen geosyncline in the Chosen period—Brief notes on the geologic history of the Yokusen orogenic zone, 1. Proc. Imperial Acad. Tokyo. 18, 579–584.

Kodaira, R., 1924. Note on a new species of *Schizoneura* from Chosen (Korea). Jpn. J. Geol. Geogr. 3, 3–4.

Koh, G.W., 1997. Characteristics of the groundwater and hydrogeologic implications of the Seoguipo Formation in Cheju Island. Ph.D. Thesis, Pusan National University, Pusan, p. 326 (in Korean with English abstract).

Korea Institute of Energy and Resources, 1981. Geological Map of the Korea (121,000,000). Korea Institute of Geoscience and Minerals Resources, Daejeon.

Kuenen, P.H., Menard, H.W., 1952. Turbidity currents, graded and non-graded deposits. J. Sediment. Petrol. 22, 83–96.

Kwon, Y.K., 2005. Sequence Stratigraphy of the Taebaek Group (Cambrian–Ordovician), Mideast Korea and Seismic Stratigraphy of the Western South Korea Plateau, East Sea. Ph.D. Thesis, Seoul National University, Seoul, p. 205.

Kwon, C.W., Sohn, Y.K., 2008. Tephra-filled volcanic neck (diatreme) of a mafic tuff ring at Maegok, Miocene Eoil Basin, SE Korea. Geosci. J. 12, 317–329.

Kwon, S.T., Lan, C.Y., 1991. Sm-Nd isotopic study of the Ogcheon amphibolite, Korea: preliminary report. J. Korea Inst. Min. Geol. 24, 277–285 (in Korean with English abstract).

Kwon, S.T., Lee, D.H., 1992. Petrology and geochemistry of the Ogcheon metabasites in Poun, Korea. J. Petrol. Soc. Korea. 1, 104–123 (in Korean with English abstract).

Kwon, S.T., Ree, J.H., 1997. A note on the age of the Honam Shear Zone. J. Geol. Soc. Korea. 33, 183–188.

Kwon, S.T., Sagong, H., 1998. Geochemical and Sr–Nd Isotopic Study of Phanerozoic Granites in South Korea: Temporal and Spatial differences. Invited speaker SRG/SEG Joint Symposium "Granite Types and Mineralization" (17–19 June), Tyoko, Japan. The Society of Resource Geology (Abstracts with Program) S-05.

Kwon, Y.K., Chough, S.K., 2005. Sequence stratigraphy of the cyclic successions in the Dumugol Formation (Lower Ordovician), mideast Korea. Geosci. J. 9, 305–324.

Kwon, S.T., Ree, J.H., Park, K.H., Jeon, E.Y., 1995a. Nature of contact between the Ogcheon belt and Yeongnam massif and the Pb–Pb age of granitic gneiss in Cheondong-ri, Danyang. J. Petrol. Soc. Korea. 4, 144–152.

Kwon, S.T., Shin, K.B., Park, H.K., Mertzman, S.A., 1995b. Geochemistry of the Kwanaksan alkali fledspar granite: a-typegranite? J. Petrol. Soc. Korea. 4, 31–48.

Kwon, Y.K., Chough, S.K., Choi, D.K., Lee, D.J., 2002. Origin of limestone conglomerates in the Choson Supergroup (Cambro-Ordovician), mid-east Korea. Sediment. Geol. 146, 265–283.

Kwon, Y.K., Lee, D.J., Choi, D.K., Chough, S.K., 2003. Lower Ordovician sponge bioherms in the Makkol Formation, Taebaeksan Basin, mideast Korea. Facies. 48, 79–90.

Kwon, Y.K., Chough, S.K., Choi, D.K., Lee, D.J., 2006. Sequence stratigraphy of the Taebaek Group (Cambrian–Ordovician), mideast Korea. Sediment. Geol. 192, 19–55.

Kwon, Y.K., Yoon, S.H., Chough, S.K., 2009. Seismic stratigraphy of the western South Korea Plateau, East Sea: implications for tectonic history and sequence development during back-arc evolution. Geo-Mar. Lett. 29, 181–189.

Kwon, C.W., Jeong, J.O., Sohn, Y.K., 2011. Sedimentary records of rift to pull-apart tectonics in the Miocene Eoil Basin, SE Korea. Sediment. Geol. 236, 256–271.

Laine, E.P., Damuth, J.E., Jacobi, R., 1986. Surficial sedimentary processes revealed by echo-characters mapping in the western North Atlantic Ocean. In: P.R. Vogt, B.E. Tucholke (Eds.), The Western North Atlantic Region, Geology of North America, Geological Society of America, Boulder, pp. 427–436.

Lambert, A.M., Kelts, K.R., Marshall, N.F., 1976. Measurements of density underflows from Walensee, Switzerland. Sedimentology. 23, 87–105.

Lambiase, J.J., Bosworth, W.P., 1995. Structural development and stratigraphy of the Kyokpo Pull-Apart Basin, South Korea and tectonic implications for inverted extensional basins. In: Buchanan, J.G., Buchanan, P.G. (Eds.), Basin Inversion. Geological Society of London, London, pp. 457–471.

Lapierre, H., Jahn, B.M., Charvet, J., Yu, Y.W., 1997. Mesozoic felsic arc magmatism and continental olivine tholeiites in Zhejiang Province and their relationship with the tectonic activity in southeastern China. Tectonophysics. 274, 321–338.

Larsen, V., Steel, R.J., 1978. The sedimentary history of a debris-flow dominated, Devonian alluvial fan—a study of textural inversion. Sedimentology. 25, 37–59.

Lavoie, D., Asselin, E., 1998. Upper Ordovician facies in the Lac Saint-Jean outlier, Quebec (eastern Canada): palaeoenvironmental significance for Late Ordovician oceanography. Sedimentology. 45, 817–832.

Lee, C.Z., 1984. Bashkirian Fusulinids from the Lower Part of the Yobong Formation, Yeongweol Coalfield, Korea. J. Geol. Soc. Korea. 20, 345–359.

Lee, D.S., 1987a. Geology of Korea. Geological Society of Korea, Kyohaksa, Seoul, 514 pp.

Lee, H.Y., 1980. Discovery of Silurian conodont fauna from South Korea. J. Geol. Soc. Korea. 16, 114–123.

Lee, H.Y., 1982a. Conodonts from the Hoedongri Formation (Silurian), western Chongson area, Kangwondo, South Korea. J. Natl. Acad. Sci., Republic Korea, Natl. Sci. Ser. 21, 43–131.

Lee, H.Y., 1987b. Paleozoic Erathem, Choson Supergroup. In: Lee, D.S. (Ed.), Geology of Korea. The Geological Society of Korea, Seoul, pp. 49–82.

Lee, J.G., 1995. Late Cambrian Trilobites from the Machari Formation, Yeongweol, Korea. Ph.D. Thesis, Seoul National University, Seoul, p. 418.

Lee, J.I., 1992a. Mineralogy and petrology of the shallow-depth emplaced granitic rocks distributed in the southern part of the Kyeongsang basin, Korea. J. Korea. Earth Sci. Soc. 13, 176–199.

Lee, K.E., 1992b. Geological Structure of Ulleung Back-Arc Basin, East Sea. M.S. Thesis, Seoul National University, Seoul, p. 121.

Lee, M.W., 1982b. Petrology and geochemistry of Jeju volcanic Island, Korea. Science Reports of the Tohoku University 15, 177–256.

Lee, S.J., 1989. Conodont Biostratigraphy of the Lower Paleozoic Yeongheung Formation in the Yeongweol Area, Kangweondo, Korea. M.S. Thesis, Yonsei University, Seoul, p. 107 (in Korean).

Lee, Y.I., 1990. Petrofacies of the Late Carboniferous Manhang Sandstones, Samcheog Coalfield and Yobong Sandstones, Yeongweol Coalfield. J. Geol. Soc. Korea. 26, 235–239.

Lee, B.S., Seo, K.S., 2004. Lower Paleozoic conodonts of Korea. Paleontol. Soc. Korea, Spec. Publ. 7, 189–215.

Lee, D.J., Yoo, C.M., 1993. Middle Ordovician stromatoporoids from the Yeongheung Formation and its biostratigraphic implication. J. Paleontol. Soc. Korea. 9, 131–142.

Lee, D.S., Lee, H.Y., 1963. Geological Map of Yean Sheet (1:50,000). Geological Survey of Korea, Seoul.

Lee, G.H., Suk, B.C., 1998. Latest Neogene-Quaternary seismic stratigraphy of the Ulleung Basin, East Sea (Sea of Japan). Mar. Geol. 146, 205–224.

Lee, H.J., Chough, S.K., 1989. Sediment distribution, dispersal and budget in the Yellow Sea. Mar. Geol. 87, 195–205.

Lee, H.J., Chu, Y.S., 2001. Origin of inner-shelf mud deposit in the southeastern Yellow Sea: Huksan mud belt. J. Sediment. Res. 71, 144–154.

Lee, H.S., Chough, S.K., 2006a. Lithostratigraphy and depositional environments of the Pyeongan Supergroup (Carboniferous-Permian) in the Taebaek area, mid-east Korea. J. Asian Earth Sci. 26, 339–352.

Lee, H.S., Chough, S.K., 2006b. Sequence stratigraphy of Pyeongan Supergroup (Carboniferous-Permian), Taebaek area, mideast Korea. Geosci. J. 10, 369–389.

Lee, H.S., Chough, S.K., 2011. Depositional processes of the Zhushadong and Mantou formations (Early to Middle Cambrian), Shandong Province, China: roles of archipelago and mixed carbonate-siliciclastic sedimentation on cycle genesis during initial flooding of the North China Platform. Sedimentology. 58, 1530–1572.

Lee, J.G., Choi, D.K., 1994. *Glyptagnostus* and associated trilobites from the Machari Formation, Yeongweol, Korea. J. Paleontol. Soc. Korea. 10, 117–136.

Lee, J.G., Choi, D.K., 1995. Late Cambrian trilobites from the Machari Formation, Yeongweol-Machari area, Korea. J. Paleontol. Soc. Korea. 11, 1–46.

Lee, J.G., Choi, D.K., 1996. Trilobites from the *Pseudoyuepingia asaphoides* Zone (Late Cambrian) of the Machari Formation near Gamaesil village, Yeongweol, Korea. J. Paleontol. Soc. Korea. 12, 145–167.

Lee, M.S., Park, B.S., 1965. Geological Map of Hwanggangri Sheet (1:50,000). Geological Survey of Korea, Seoul.

Lee, S.-B., Choi, D.K., 2011. Dikelocephalid Trilobites from the *Eosaukia* Fauna (Upper Furongian) of the Taebaek Group, Korea. J. Paleontol. 85, 279–297.

Lee, S.H., Chough, S.K., 1999. Progressive changes in sedimentary facies and stratal patterns along the strike–slip margin, northeastern Jinan Basin (Cretaceous), southwest Korea: implications for differential subsidence. Sediment. Geol. 123, 81–102.

Lee, S.H., Chough, S.K., 2001. High-resolution (2–7 kHz) acoustic and geometric characters of submarine creep deposits in the South Korea Plateau, East Sea. Sedimentology. 48, 629–644.

Lee, S.R., Cho, K., 2012. Precambrian crustal evolution of the Korean Peninsula. J. Petrol. Soc. Korea. 21, 89–112.

Lee, D.S., Chang, K.H., Lee, H.Y., 1972. Discovery of Archaeocyatha from Hyangsanri Dolomite Formation of the Ogcheon System and its significance. J. Geol. Soc. Korea. 8, 191–197.

Lee, C.H., Lee, M.S., Park, B.S., 1980. Geological Map of Miwon Sheet (1:50,000). Korea Institute of Geosciences and Mineral Resources, Seoul.

Lee, H.J., Chough, S.K., Jeong, K.S., Han, S.J., 1987. Geotechnical properties of sediment cores from the southeastern Yellow Sea: effects of depositional processes. Mar. Geotechnol. 7, 37–52.

Lee, B.J., Kim, Y.B., Lee, S.R., Kim, J.C., Kang, P.C., Choi, H.I., et al., 1999a. Geological Map of Seoul-Namchonjeom Sheet (1:250,000). Korea Institute of Geoscience and Mineral Resources, Daejeon.

Lee, S.H., Chough, S.K., Back, G.G., Kim, Y.B., Sung, B.S., 1999b. Gradual downslope change in high-resolution acoustic characters and geometry of large-scale submarine debris lobes in Ulleung Basin, East Sea (Sea of Japan), Korea. Geo-Mar. Lett. 19, 254–261.

Lee, D.J., Kwon, Y.K., Choi, D.K., 2001. Fossil sponges from the Makkol Formation (Lower Ordovician), Taebaek Group, Korea. J. Paleontol. Soc. Korea. 17, 77–85.

Lee, S.H., Chough, S.K., Back, G.G., Kim, Y.B., 2002. Chirp (2–7-kHz) echo characters of the South Korea Plateau, East Sea: styles of mass movement and sediment gravity flow. Mar. Geol. 184, 227–247.

Lee, S.R., Cho, M., Yi, K., Stern, R.A., 2000a. Early Proterozoic granulites in central Korea: tectonic correlation with Chinese cratons. J. Geol. 108, 729–738.

Lee, Y.-N., Jeong, K.S., Chan, S.K., Choi, M.Y., Choi, J.I., 2000b. The preliminary research on the dinosaur eggs and nests found in the reclaimed area south to the Siwha Lake, Gyeonggi Province, Korea. J. Paleontol. Soc. Korea. 16, 27–36.

Lee, S.H., Bahk, J.J., Chough, S.K., 2004. Late Quaternary sedimentation in the eastern continental margin of the Korean Peninsula. In: Peter, C., Kuhnt, W., Wang, P., Hayes, D. (Eds.), Continent-Ocean Interactions within East Asian Marginal Seas. Geophysics Monograph. 149, American Geophysical Union, Washington D.C., pp. 205–233.

Lee, J.-H., Chen, J., Chough, S.K., 2010a. Paleoenvironmental implications of an extensive maceriate microbialite bed in the Furongian Chaomidian Formation, Shandong Province, China. Palaeogeogr. Palaeoclimatol. Palaeoecol. 297, 621–632.

Lee, T.-H., Park, K.-H., Chun, J.-H., Yi, K., 2010b. SHRIMP U–Pb zircon ages of the Jinju Formation and Silla Conglomerate, Gyeongsang Basin. J. Petrol. Soc. Korea. 19, 89–101 (in Korean with English abstract).

Lee, J.-H., Chen, J., Chough, S.K., 2012a. Demise of an extensive biostromal microbialite in the Furongian (late Cambrian) Chaomidian Formation, Shandong Province, China. Geosci. J. 275–287.

Lee, Y.I., Choi, T., Orihashi, Y., 2012b. Depositional ages of upper Pyeongan Supergroup strata in the Samcheok coalfield, eastern central Korea. J. Geol. Soc. Korea. 48, 93–99 (in Korean with English abstract).

Li, X., 2000. Cretaceous magmatism and lithospheric extension in Southeast China. J. Asian Earth Sci. 18, 293–305.

Li, Z.X., Powell, C.M., 2001. An outline of the palaeogeographic evolution of the Australasian region since the beginning of the Neoproterozoic. Earth Sci. Rev. 53, 237–277.

Li, B., Park, B.K., Kim, D.S., Woo, H.J., 1999. The geologic age and paleoenvironment of the lower Seoguipo Formation, Cheju Island, Korea. Geosci. J. 3, 181–190.

Li, S.Z., Zhao, G.C., Santoshi, M., Liu, X., Dai, L.M., 2011. Palaeoproterozoic tectonothermal evolution and deep crustal processes in the Jiao-Liao-Ji Belt, North China Craton: a review. Geol. J. 46, 525–543.

Lim, C., Cho, M., 2012. Two-phase contractional deformation of the Jurassic Daebo Orogeny, Chungnam Basin, Korea, and its correlation with the early Yanshanian movement of China. Tectonics. 31, TC1004.

Lim, S.-B., Chun, H.Y., Kim, Y.B., Kim, B.C., Cho, D.-L., 2005. Geologic ages, stratigraphy and geological structures of the metasedimentary strata in Bibong–Yeonmu area, NW Okcheon belt, Korea. J. Geol. Soc. Korea. 41, 335–368 (in Korean with English abstract).

Lim, S.-B., Chun, H.Y., Kim, Y.B., Kim, B.C., Song, K.-Y., 2006. Stratigraphy and geological ages of the metasedimentary strata in Jinsan–Boksu area, Chungcheongnam-do, NW Okcheon belt. J. Geol. Soc. Korea. 42, 149–174 (in Korean with English abstract).

Lim, S.-B., Chun, H.Y., Kim, Y.B., Lee, S.R., Kee, W.-S., 2007. Geological ages and stratigraphy of the metasedimentary strata in Hoenam–Miwon area, NW Okcheon belt. J. Geol. Soc. Korea. 43, 125–150 (in Korean with English abstract).

Liu, J.P., Milliman, J.D., 2004. Western Pacific post-glacial rapid and stepwise sea-level rise, 5th International Conference on Asian Marine Geology, Bangkok, p. 99.

Lombard, A., 1963. Laminites: a structure of flysch-type sediments. J. Sediment. Petrol. 33, 14–22.

Lowe, D.R., 1976. Grain flow and grain flow deposits. J. Sediment. Petrol. 46, 188–199.

Lowe, D.R., 1979. Sediment gravity flows: their classification and some problems of application to natural flows and deposits. In: Doyle, L.J., Pilkey, O.H. (Eds.), Geology of Continental Slopes. SEPM, Tulsa, pp. 75–82.

Lowe, D.R., 1982. Sediment gravity flows. II. Depositional models with special reference to the deposits of high-density turbidity currents. J. Sediment. Petrol. 52, 279–297.

Lowe, D.R., 1988. Suspended-load fallout rate as an independent variable in the analysis of current structures. Sedimentology. 35, 765–776.

Lu, Y., Dong, N., 1953. Revision of the Cambrian type sections of Shantung. Acta. Geol. Sin. 32, 1–23.

Lyang, T.J., Liu, Y., Yang, J.H., Kim, H., Han, R.Y., Kim, J.N., 2009. Precambrian crustal evolution in Rangrim Massif, Korean Peninsula. Global Geology. 12, 57–63.

Maizels, J., 1989. Sedimentology and paleohydrology of Holocene flood deposits in front of a Jokuhlaup Glacier, South Iceland. In: Beven, K., Carling, P. (Eds.), Floods: Hydrological, Sedimentological and Geomorphological Implications. John Wiley & Sons, Chichester, pp. 239–252.

Maizels, J., 1993. Lithofacies variations within sandur deposits: the role of runoff regime, flow dynamics and sediment supply characteristics. Sediment. Geol. 85, 299–325.

Markello, J.R., Read, J.F., 1981. Carbonate ramp-to-deeper shale shelf transitions of an Upper Cambrian intrashelf basin, Nolichucky Formation, Southwest Virginia Appalachians. Sedimentology. 28, 573–597.

Marzo, M., Anadon, P., 1988. Anatomy of a conglomeratic fan-delta complex: the Eocene Montserrat Conglomerate, Ebro Basin, northeastern Spain. In: Nemec, W., Steel, R.J. (Eds.), Fan Deltas: Sedimentology and Tectonic Settings. Blackie and Son, Glasgow, pp. 318–340.

Massari, F., 1984. Resedimented conglomerates of a Miocene fan-delta complex, southern Alps, Italy. In: Koster, E.H., Steel, R.J. (Eds.), Sedimentology of Gravels and Conglomerates. Canadian Society of Petroleum Geologists, Calgary, pp. 259–278.

McBride, E.F., Shepherd, R.G., Crawley, R.A., 1975. Origin of parallel laminae by migration of bed forms in a small flume. J. Sediment. Petrol. 45, 132–139.

McCabe, P.J., 1984. Depositional environments of coal and coal bearing strata. In: Rahmani, R.A., Flores, R.M. (Eds.), Sedimentology of Coal and Coal Bearing Sequences. International Association of Sedimentologists, Cambridge, pp. 13–42.

McClay, K.R., 1995. Recent advances in analogue modeling: uses in section interpretation and validation. In: Buchanan, P.G., Nieuwland, D.A. (Eds.), Modern Developments in Structural Interpretation, Validation, and Modeling. Geological Society of London, London, pp. 201–225.

McClay, K.R., White, M.J., 1995. Analogue modeling of orthogonal and oblique rifting. Mar. Petrol. Geol. 12, 137–151.

McKenzie, N.R., Hughes, N.C., Myrow, P.M., Choi, D.K., Park, T.y., 2011. Trilobites and zircons link north China with the eastern Himalaya during the Cambrian, Geology, 39, 591–594.

Meng, X., Ge, M., Tucker, M.E., 1997. Sequence stratigraphy, sea-level changes and depositional systems in the Cambro-Ordovician of the North China carbonate platform. Sediment. Geol. 114, 189–222.

Meyerhoff, A.A., Kamen-Kaye, M., Chen, C., Taner, I., 1991. China—Stratigraphy, Paleogeography, and Tectonics. Kluwer Academic Publishers, Dordrecht, Netherlands, 188 pp.

Miall, A.D., 1977. A review of the braided-river depositional environment. Earth Sci. Rev. 13, 1–62.

Miall, A.D., 1978. Lithofacies types and vertical profile models in braided river deposits: a summary. In: Miall, A.D. (Ed.), Fluvial Sedimentology. Canadian Society of Petroleum Geologists, Calgary, pp. 596–604.

Miall, A.D., 1985. Architectural-element analysis: a new method of facies analysis applied to fluvial deposits. Earth Sci. Rev. 22, 261–308.

Miall, A.D., 1996. The Geology of Fluvial Deposits. Springer, Berlin, 582 pp.

Miall, A.D., 2000. Principles of Sedimentary Basin Analysis. third edition, Springer, Heidelberg, 616 pp.
Miall, A.D., 2010. The Geology of Stratigraphic Sequences. second edition, Springer, Heidelberg, 531 pp.
Middleton, G.V., 1966. Experiments on density and turbidity currents. II. Uniform flow of density currents. Can. J. Earth Sci. 3, 627–637.
Middleton, G.V., 1967. Experiments on density and turbidity currents. III. Deposition of sediment. Can. J. Earth Sci. 4, 474–505.
Middleton, G.V., Hampton, M.A., 1973. Sediment gravity flows: mechanics of flow and deposition. In: Middleton, G.V., Bouma, A.H. (Eds.), Turbidites and Deep-Water Sedimentation. SEPM, Tulsa, pp. 1–38.
Middleton, G.V., Hampton, M.A., 1976. Subaqueous sediment transport and deposition by sediment gravity flow. In: Stanley, D.J., Swift, D.J.P. (Eds.), Marine Sediment Transport and Environment Management. John Wiley & Sons, New York, pp. 197–218.
Middleton, L.T., Trujillo, A.P., 1984. Sedimentology and depositional setting of the Upper Proterozoic Scanlan Conglomerate, Central Arizona. In: Koster, E.H., Steel, R.J. (Eds.), Sedimentology of Gravels and Conglomerates. Canadian Society of Petroleum Geologists, Calgary, pp. 189–201.
Milliman, J.D., Qin, Y.S., Park, Y.A., 1989. Sediment and sedimentary processes in the Yellow and East China Seas. In: Taira, A., Masuda, F. (Eds.), Sedimentary Facies in the Active Plate Margin. Terra Scientific Publishing Company, Tokyo, pp. 233–249.
Mitsugi, T., Ishida, K., Woo, B.-G., Chang, K.-H., Park, S.-O., Hirano, H., 2001. Radiolarian-bearing conglomerate from the Hayang Group, the Kyongsang Supergroup, southeastern Korea. J. Asian Earth Sci. 19, 751–763.
Morison, S.R., Hein, F.J., 1987. Sedimentology of the White Channel Gravels, Klondike Area, Yukon territory: fluvial deposits of a confined valley. In: Ethridge, F.G., Flores, R.M., Harvey, M.D. (Eds.), Recent Developments in Fluvial Sedimentology. SEPM, Calgary, pp. 205–216.
Myrow, P.M., Southard, J.B., 1996. Tempestite deposition. J. Sediment. Res. 66, 875–887.
Na, K.C., 1987. Precambrian Eonothem. In: Lee, D.S. (Ed.), Geology of Korea. Geological Society of Korea, Kyohaksa, Seoul, pp. 17–47.
Na, K.C., Kim, H.S., Lee, S.H., 1982. Stratigraphy and metamorphism of Seosan Group. J. Econ. Environ. Geol. 15, 33–39 (in Korean with English abstract).
Nardin, T.R., Hein, F.J., Gorsline, D.S., Dewards, B.D., 1979. A review of mass movement processes, sediment and acoustic characteristics in slope and base-of-slope systems versus canyon-fan-basin floor systems. In: Doyle, L.J., Pilkey, O.H. (Eds.), Geology of Continental Slopes. SEPM, Tulsa, pp. 61–73.
Naylor, M.A., 1980. The origin of inverse grading in muddy debris flow deposits—a review. J. Sediment. Petrol. 50, 1111–1116.
Nemec, W., 1990. Aspects of sediment movement on steep delta slopes. In: Colella, A., Prior, D.B. (Eds.), Coarse-Grained Deltas. International Association of Sedimentologists, Cambridge, pp. 29–73.
Nemec, W., Postma, G., 1993. Quaternary alluvial fans in southwestern Crete: sedimentation processes and geomorphic evolution. In: Marzo, M., Puigdefabregas, C. (Eds.), Alluvial Sedimentation. International Association of Sedimentologists, Cambridge, pp. 235–276.
Nemec, W., Steel, R.J., 1984. Alluvial and coastal conglomerates: their significant features and some comments on gravelly mass-flow deposits. In: Koster, E.H., Steel, R.J. (Eds.),

Sedimentology of Gravels and Conglomerates. Canadian Society of Petroleum Geologists, Calgary, pp. 1–31.
Noh, J.H., Park, H.S., 1990. Mineral diagenesis of sandstones from the Kyongsang Supergroup in Goryeong area. J. Geol. Soc. Korea. 26, 371–392 (in Korean with English abstract).
Normark, W.R., Dickson, F.H., 1976. Man-made turbidity currents in Lake Superior. Sedimentology. 23, 815–832.
Normark, W.R., Hess, G.R., Stow, D.A.V., Bowen, A.J., 1980. Sediment waves on the Monterey Fan levee: a preliminary physical interpretation. Mar. Geol. 37, 1–18.
Normark, W.R., Posamentier, H.W., Mutti, E., 1993. Turbidite systems: state of the art and future directions. Rev. Geophys. 31, 91–116.
Northrup, C.J., Royden, L.H., Burchfiel, B.C., 1995. Motion of the Pacific plate relative to Eurasia and its potential relation to Cenozoic extension along the eastern margin of Eurasia. Geology. 23, 719–722.
O'Brien, N.R., 1996. Shale lamination and sedimentary processes. In: Kemp, A.E.S. (Ed.), Paleoclimatology and Paleoceanography from Laminated Sediments. Geological Society, pp. 23–36.
O'Leary, D.W., Laine, E., 1996. Proposed criteria for recognition of intrastratal deformation features in marine high-resolution seismic reflection profiles. Geo. Mar. Lett. 16, 305–312.
Otoh, S., Yanai, S., 1996. Mesozoic inversive wrench tectonics in far east Asia: examples from Korea and Japan. In: Yin, A., Harrison, M. (Eds.), The Tectonic Evolution of Asia. Cambridge University Press, Cambridge, pp. 401–419.
Overstreet, R.B., Oboh-Ikuenobe, F.E., Gregg, J.M., 2003. Sequence stratigraphy and depositional facies of Lower Ordovician cyclic carbonate rocks, southern Missouri, U.S.A. J. Sediment. Res. 73, 421–433.
Paek, R.J., Kang, H.G., Jon, G.P., 1996. The Geology of Korea. Foreign Languages Books Publishing House, Democratic Peoples Republic of Korea, Pyeongyang, 631 pp.
Paik, I.S., 1987. Depositional environments of the Middle Ordovician Maggol Formation, southern part of the Baegunsan Syncline area. J. Geol. Soc. Korea. 23, 360–373.
Paik, I.S., Chun, J.H., 1993. Laminar calcretes, calcrete pisoids and ooids, and rhizoliths from the Kyeongsang Supergroup, Korea. J. Geol. Soc. Korea. 29, 108–117.
Paik, I.S., Kim, H.J., 2003. Palustrine calcretes of the Cretaceous Gyeongsang Supergroup, Korea: variation and palaeoenvironmental implications. The Island Arc. 12, 110–124.
Paik, I.S., Kim, H.J., 2006. Playa lake and sheetflood deposits of the Upper Cretaceous Jindong Formation, Korea: occurrences and palaeoenvironments. Sediment. Geol. 187, 83–103.
Paik, I.S., Lee, Y.I., 1989. Storm deposits of the Lower Ordovician Mungog Formation in the Machari, Yeongweol, Kangweondo, Korea. J. Geol. Soc. Korea. 25, 337–346.
Paik, I.S., Lee, Y.I., 1994. Paleoclimatic records in floodplain lake deposits of the Cretaceous Hasandong Formation in Jinju area, Korea. J. Geol. Soc. Korea. 30, 410–424.
Paik, I.S., Lee, Y.I., 1998. Desiccation cracks in vertic palaeosols of the Cretaceous Hasandong Formation, Korea: genesis and palaeoenvironmental implications. Sediment. Geol. 119, 161–179.
Paik, K.-H., Kim, Y.-K., Huh, M., 1988. Ostracodes from the Chinju Formation (Cretaceous, non-marine) in Kyongsang Basin, Korea. J. Paleontol. Soc. Korea. 4, 94–109.
Paik, I.S., Woo, K.S., Chung, G.S., 1991. Stratigraphic, sedimentologic and paleontologic investigation of the Paleozoic sedimentary rocks in Yeongweol and Gabsan areas:

depositional environments of the Lower Ordovician Mungok Formation in the vicinity of Yeongweol. J. Geol. Soc. Korea. 27, 357–370.

Paik, I.S., Kim, H.J., Kim, J.Y., Jin, S.I., 1996. Paleoenvironments of the Cretaceous lacustrine deposits (Yucheon Group), Mt. Hwangryeong, Pusan, Korea. J. Paleontol. Soc. Korea. 12, 181–202.

Paik, I.S., Lee, J.D., Kim, J.J., Kim, I.S., Park, C.W., 1997. Compound calcrete deposits from the Dadaepo Formation in Dusong Peninsula, Pusan: origin, paleoenvironments, and stratigraphic implication. Econ. Environ. Geol. 30, 263–275 (in Korean with English abstract).

Paik, I.S., Kim, H.J., Lee, Y.I., 2001. Dinosaur track-bearing deposits in the Cretaceous Jindong Formation, Korea: occurrence, palaeoenvironments and preservation. Cretaceous Res. 22, 79–92.

Paik, I.S., Kang, H.C., Huh, M., Yang, S.-Y., 2006. Goseong Formation (Yucheon Group) in the southern part of the Gyeongsang Basin, Korea: occurrences and stratigraphy. J. Geol. Soc. Korea. 42, 483–505 (in Korean with English abstract).

Paik, I.S., Huh, M., So, Y.H., Lee, J.E., Kim, H.J., 2007. Traces of evaporites in Upper Cretaceous lacustrine deposits of Korea: Origin and paleoenvironmental implications. J. Asian Earth Sci. 30, 93–107.

Paola, C., Wiele, S.M., Reinhart, M.A., 1989. Upper-regime parallel lamination as the result of turbulent sediment transport and low-amplitude bed forms. Sedimentology. 36, 47–59.

Park, J.B., 1986. Fossil Spores and Pollen from the Jinju Formation in the Waegwan-Daegu and Jinju Area, Korea. M.S. Thesis, Seoul National University, Seoul, p. 77.

Park, S., 1998. Stratal Patterns in the Southwestern Margin of Ulleung Back-Arc Basin: A Sequence Stratigraphic Analysis. Ph.D. Thesis, Seoul National University, Seoul, p. 176.

Park, S.D., 2000. Analysis of Sedimentary Basin in the Sihwa Area, Hwasunggun and Tando area, Ansanshi, Kungki Province, Korea. M.S. Thesis, Chungnam National University, Cheongju, p. 119 (in Korean with English abstract).

Park, N.Y., Kim, J.H., 1972. Geological Map and Explanatory Text of Namyang Area (1:50,000). Korea Institute of Geology Mining and Materials, Teajon, 13 pp. (in Korean with English abstract).

Park, T.-Y., Choi, D.K., 2010. Two middle Cambrian diceratocephalid trilobites, Cyclolorenzella convexa and Diceratocephalus cornutus, from Korea: development and functional morphology. Lethaia. 43, 73–87.

Park, T.-Y., Choi, D.K., 2011. Trilobite faunal successions across the base of the Furongian Series in the Taebaek Group, Taebaeksan Basin, Korea. Geobios. 44, 481–498.

Park, K.H., Ahn, J.S., Kee, W.-S., Park, W.-B., 2006a. Guidebook for a Geological Tour of Jeju Island. Korea Institute of Geoscience and Mineral Resources, Daejeon, 183 pp. (in Korean).

Park, S.-O., Jang, Y.-D., Hwang, S.-K., Kim, J.-J., 2006b. Petrology of the Cretaceous volcanic rocks in eastern part of the Kyeongsan Caldera. J. Petrol. Soc. Korea. 15, 90–105 (in Korean with English abstract).

Park, T.-Y., Han, Z., Bai, Z., Choi, D., 2008. Two middle Cambrian trilobite genera, Cyclolorenzella Kobayashi, 1960 and Jiulongshania gen. nov., from Korea and China. Alcheringa. 32, 247–269.

Park, K.H., Lee, T.-H., Yi, K., 2011. SHRIMP U–Pb ages of detrital zircons in the Daehyangsan Quartzite of the Okcheon Metamorphic Belt, Korea. J. Geol. Soc. Korea. 47, 423–431 (in Korean with English abstract).

Picard, M.D., High, L.R., 1973. Sedimentary Structures of Ephemeral Streams. Developments in Sedimentology. Elsevier, Amsterdam, p. 223.

Pickering, K.T., Stow, D.A.V., Watson, M.P., Hiscott, R.N., 1986. Deep-water facies, processes and models: a review and classification scheme for modern and ancient sediments. Earth-Sci. Rev. 23, 75–174.

Pierson, T.C., 1981. Dominant particles supporting mechanisms in debris flows at Mt. Thomas, New Zealand, and implications for flow mobility. Sedimentology. 28, 49–60.

Pilkey, O.H., Young, R., 2009. The Rising Sea. Island Press, Washington D.C., 203 pp.

Pirazzoli, P.A., 1991. World Atlas of Holocene Sea-Level Changes. Elsevier, Amsterdam, 300 pp.

Posamentier, H.W., Jervey, M.T., Vail, P.R., 1988. Eustatic controls on clastic deposition, I. Conceptual framework. In: Wilgus, C.K., Hastings, B.S., Kendall, C.G.S.C., Posamentier, H.W., Ross, C.A., van Wagoner, J.C. (Eds.), Sea Level Change—An Integrated Approach. SEPM Special Publication. SEPM, Tulsa, OK, pp. 110–124.

Postma, G., 1986. Classification for sediment gravity-flow deposits based on flow conditions during sedimentation. Geology. 14, 291–294.

Postma, G., Roep, T.B., 1985. Bottomset-modified Gilbert-type deltas (Espiritu Santo Formation, Pliocene, Vera Basin, SE Spain). J. Sediment. Petrol. 55, 874–885.

Postma, G., Nemec, W., Kleinspehn, K.L., 1988. Large floating clasts in turbidites: a mechanism for their emplacement. Sediment. Geol. 58, 47–61.

Pratson, L.F., Laine, E.P., 1989. The relative importance of gravity-induced versus current-controlled sedimentation during the Quaternary along the mideast United-States outer continental-margin revealed by 3.5 kHz echo character. Mar. Geol. 89, 87–126.

Read, J.F., 1989. Controls on evolution of Cambrian–Ordovician passive margin, U.S. Appalachians. In: Crevello, P.D., Wilson, J.L., Sarg, J.F., Read, J.F. (Eds.), Controls on Carbonate Platform and Basin development. SEPM, Tulsa, pp. 147–165.

Ree, J.H., Cho, M., Kwon, S.T., Nakamura, E., 1996. Possible eastward extension of Chinese collision belt in South Korea: the Imjingang belt. Geology. 24, 1071–1074.

Ree, J.-H., Kwon, S.-H., Park, Y., Kwon, S.-T., Park, S.-H., 1996. Pre- and post-tectonic emplacements of the granitoids in the central-southern Okchon belt, South Korea: implications for the timing of the strike-slip shearing and thrusting. Tectonics. 20, 850–867.

Reedman, A.J., Um, S.H., 1975. The Geology of Korea. Geological and Mineral Institute of Korea, Seoul, 139 pp.

Reedman, A.J., Fletcher, C.J.N., Evans, R.B., Workman, D.R., Yoon, K.S., Rhy, H.S., et al., 1973. Geology of the Hwanggangri Mining District, Republic of Korea. Geological and Mineral Institute of Korea, Seoul, 118 pp.

Reedman, A.J., Park, K.H., Merriman, R.J., Kim, S.E., 1987. Welded tuff infilling a volcanic vent at Weolseong, Republic of Korea. Bull. Volcanol. 49, 541–546.

Reid, I., Frostick, L.E., 1984. Particle interaction and its effect on the thresholds of initial and final bedload motion in coarse alluvial channels. In: Koster, E.H., Steel, R.J. (Eds.), Sedimentology of Gravels and Conglomerates. Canadian Society of Petroleum Geologists, Calgary, pp. 61–68.

Reineck, H.E., Singh, I.B., 1980. Depositional Sedimentary Environments. Springer, New York/Heidelberg, 549 pp.

Ren, J., Tamaki, K., Li, S., Junxia, Z., 2002. Late Mesozoic and Cenozoic rifting and its dynamic setting in Eastern China and adjacent areas. Tectonophysics. 344, 175–205.

Rhee, C.W., Chough, S.K., 1993a. The Cretaceous Pyonghae Basin, southeast Korea: sequential development of crevasse splay and avulsion in a terminal alluvial fan. Sediment. Geol. 83, 37–52.

Rhee, C.W., Chough, S.K., 1993b. The Cretaceous Pyonghae sequence, southeast Korea: Terminal fan facies. Palaeogeogr. Palaeoclimatol. Palaeoecol. 105, 139–156.

Rhee, C.W., Ryang, W.H., Chough, S.K., 1993. Contrasting development patterns of crevasse channel deposits in Cretaceous alluvial successions, Korea. Sediment. Geol. 85, 401–410.

Rhee, C.W., Jo, H.R., Chough, S.K., 1998. An allostratigraphic approach to a non-marine basin: the north-western part of Cretaceous Kyongsang basin, SE Korea. Sedimentology. 45, 449–472.

Rhoads, D.C., 1967. Biogenic reworking of intertidal and subtidal sediments in Barnstable Harbor and Buzzards Bay, Massachusetts. J. Geol. 75, 461–476.

Ri, J.N., Ri, J.C., 1994. Tectonic Map of Korea (1:1,000,000 Scale) and Explanatory Text. Central Geological Survey of Mineral Resources, Pyongyang, 22 pp.

Rust, B.R., 1978. Depositional models for braided alluvium. In: Miall, A.D. (Ed.), Fluvial Sedimentology. Canadian Society of Petroleum Geologists, pp. 596–604.

Rust, B.R., 1979. Coarse alluvial deposits. In: R.G. Walker (Ed.), Facies Models, Geoscience Canada, pp. 9–21.

Ryang, W.H., Chough, S.K., 1997. Sequential development of alluvial/lacustrine system: Southeastern Eumsung Basin (Cretaceous), Korea. J. Sediment. Res. 67, 274–285.

Ryang, W.H., Chough, S.K., Kim, J.S., Shon, H., 1999. Three-dimensional configuration of a pull-apart basin from high-resolution magnetotelluric profiling: Eumsung Basin (Cretaceous), Korea. Sediment. Geol. 129, 101–109.

Ryu, I.C., Kim, B.Y., Kwak, W.J., Kim, G.H., Park, S.J., 2000. Stratigraphic response to tectonic evolution of sedimentary basins in the Yellow Sea and adjacent seas. Korea. J. Petroleum Geol. 8, 1–43 (in Korean with English abstract).

Sagong, H., Jwa, Y.J., 1997. Mineral chemistry and major elements geochemistry of the granitic rocks in the Cheongsan area. J. Petrol. Soc. Korea. 6, 185–209.

Sagong, H., Kwon, S.T., Jeon, E.Y., Mertzman, S.A., 1997. Geochemistry and petrology of the Hwacheon granite. J. Geol. Soc. Korea. 33, 99–110 (in Korean with English abstract).

Sanders, J.E., 1965. Primary sedimentary structures formed by turbidity currents and related re-sedimentation mechanisms. In: Middleton, G.V. (Ed.), Primary Sedimentary Structures and Their Hydrodynamic Interpretation. SEPM, Tulsa, pp. 192–219.

Schieber, J., 2003. Simple gifts and buried treasures—implication of finding bioturbation and erosional surfaces in black shales. Sediment. Rec. 2, 4–8.

Séguret, M., Moussine-Pouchkine, A., Gabaglia, G.R., Bouchette, F., 2003. Storm deposits and storm-generated coarse carbonate breccias on a pelagic outer shelf (South-East Basin, France). Sedimentology. 48, 231–254.

Seo, S.J., 1985. Lower Cretaceous Geology and Paleontology (Charophyta) of Central Kyongsang Basin, Korea. Ph.D. Thesis, Kyungpook National University, Daegu, p. 177 (in Korean with English abstract).

Shimizu, S., Ozaki, K., Obata, T., 1934. Gotlandian deposits of northwest Korea. Shanghai Science Institute Journal, Section. II, 59–85.

Shinn, E.A., 1983. Tidal flat environment. In: Scholle, P.A., Bebout, D.G., Moore, C.H. (Eds.), Carbonate Depositional Environments. American Association of Petroleum Geologists, Tulsa, pp. 173–210.

Shinn, Y.J., Chough, S.K., Kim, J.W., Lee, S.H., Woo, J., Jin, J.H., et al., 2004. High-resolution seismic reflection studies of late quaternary sediments in the Eastern Yellow Sea. In: Clift, P., Kuhnt, W., Wang, P., Hayes, D. (Eds.), Continent–Ocean Interactions within East Asian Marginal Seas. Geophysics Monograph. 149, American Geophysical Union, Washington, DC, pp. 175–192.

Shinn, Y.J., Chough, S.K., Hwang, I.G., 2010. Structural development and tectonic evolution of Gunsan Basin (Cretaceous–Tertiary) in the central Yellow Sea. Mar. Petrol. Geol. 27, 500–514.

Shiraki, T., 1930. Geological Map of the Samcheok Coalfield.

Shultz, A.W., 1984. Subaerial debris-flow deposition in the Upper Paleozoic Cutler Formation, western Colorado. J. Sediment. Petrol. 54, 759–772.

Siegenthaler, C., Huggenberger, P., 1993. Pleistocene Rhine gravel: deposits of a braided river system with dominant pool preservation. In: Best, J.L., Bristow, C.S. (Eds.), Braided Rivers. Geological Society of London, London, pp. 147–162.

Sim, M., Lee, Y., 2006. Sequence stratigraphy of the Middle Cambrian Daegi Formation (Korea), and its bearing on the regional stratigraphic correlation. Sediment. Geol. 191, 151–169.

Smith, G.A., 1986. Coarse-grained nonmarine volcaniclastic sediment: terminology and depositional process. Geol. Soc. Am. Bull. 97, 1–10.

Smith, N.D., 1971. Pseudo-planar stratification produced by very low amplitude sand waves. J. Sediment. Petrol. 41, 69–73.

Smith, G.A., Lowe, D.R., 1991. Lahars: volcano-hydrologic events and deposition in the debris flow-hyperconcentrated flow continuum. In: Fisher, R.V., Smith, G.A. (Eds.), Sedimentation in Volcanic Settings. SEPM, Tulsa, pp. 59–70.

Sohn, Y.K., 1996. Hydrovolcanic processes forming basaltic tuff rings and cones on Cheju Island, Korea. Geol. Soc. Am. Bull. 108, 1199–1211.

Sohn, J.W., Choi, D.K., 2007. Furongian trilobites from the *Asioptychaspis* and *Quadraticephalus* zones of the Hwajeol Formation, Taebaeksan Basin, Korea. Geosci. J. 11, 297–314.

Sohn, Y.K., Chough, S.K., 1989. Depositional processes of the Suwolbong tuff ring, Cheju Island (Korea). Sedimentology. 36, 837–855.

Sohn, Y.K., Chough, S.K., 1992. The Ilchulbong tuff cone, Cheju Island, South Korea: depositional processes and evolution of an emergent, Surtseyan-type tuff cone. Sedimentology. 39, 523–544.

Sohn, Y.K., Chough, S.K., 1993. The Udo tuff cone, Cheju Island, South Korea: transformation of pyroclastic fall into debris fall and grain flow on a steep volcanic cone slope. Sedimentology. 40, 769–786.

Sohn, Y.K., Park, K.H., 1994. Geology and evolution of Tok Island. J. Geol. Soc. Korea. 30, 242.

Sohn, Y.K., Park, K.H., 2004. Early-stage volcanism and sedimentation of Jeju Island revealed by the Sagye borehole, SW Jeju Island, Korea. Geosci. J. 8, 73–84.

Sohn, Y.K., Yoon, S.H., 2010. Shallow-marine records of pyroclastic surges and fallouts over water in Jeju Island, Korea, and their stratigraphic implications. Geology. 38, 763–766.

Sohn, Y.K., Kim, S.B., Hwang, I.G., Bahk, J.J., Choe, M.Y., Chough, S.K., 1999. Characteristics and depositional processes of large-scale gravelly Gilbert-type foresets in the Miocene Doumsan fan delta, Pohang Basin, SE Korea. J. Sediment. Res. 67, 130–141.

Sohn, Y.K., Rhee, C.W., Kim, B.C., 1999. Debris flows and hyperconcentrated floodflow deposits in an alluvial fan, northwestern part of the Cretaceous Yongdong Basin, central Korea. J. Geol. 107, 111–132.

Sohn, Y.K., Park, K.H., Yoon, S.H., 2008. Primary versus secondary and subaerial versus submarine hydrovolcanic deposits in the subsurface of Jeju Island, Korea. Sedimentology. 55, 899–924.

Sohn, Y.K., Son, M., Jeong, J.O., Jeon, Y.M., 2009. Eruption and emplacement of a laterally extensive, crystal-rich, and pumice-free ignimbrite (the Cretaceous Kusandong Tuff, Korea). Sediment. Geol. 220, 190–203.

Son, M., 1998. Formation and Evolution of the Tertiary Miocene Basins in Southeastern Korea: Structural and Paleomagnetic Approaches. Ph.D. Thesis, Pusan National University, Pusan, p. 233.

Son, M., Seo, H.-J., Kim, I.-S., 2000. Geological structures and evolution of the Miocene Eoil Basin, southeastern Korea. Geosci. J. 4, 73–88.

Son, M., Kim, I.S., Sohn, Y.K., 2005. Evolution of the Miocene Waup Basin, SE Korea, in response to dextral shear along the southwestern margin of the East Sea. J. Asian Earth Sci. 25, 529–544.

Son, M., Cho, H., Kim, I.S., 2009. Field Guide Book: Sedimentary and Volcanic Rocks in the Cretaceous Dadaepo Basin between Yangsan and Dongrae Faults, Pusan, Korea. Korean Society of Economic and Environmental Geology, Seoul, 42 pp. (in Korean).

Son, M., Kim, J.S., Jung, S., Ki, J.S., Kim, M.-C., Sohn, Y.K., 2012. Tectonically controlled vent migration during maar-diatreme formation: an example from a Miocene half-graben basin in SE Korea. J. Volcanol. Geotherm. Res. 223-224, 29–46.

Song, M.Y., Lee, C.Z., Lee, J.R., 1990. Integrated geotectonic study on the Mesozoic sedimentary basins in Umsong–Jeungpyeong and Kongju area (I. Umsong–Jeungpyeong). J. Korean Earth Sci. Soc. 11, 1–12.

Song, Y.S., Park, M.E., Park, K.H., 2006. Ages and evolutions of the volcanic rocks from Ulleung-do and Dok-do. J. Petrol. Soc. Korea. 15, 72–80.

Song, Y.S., Park, K.H., Lee, H.S., Cao, L., Orihashi, Y., 2009. LA-ICP-Ms U–Pb Zircon Age of the Granite Gneiss from Jeungsan-Pyeongwon Area of North Korea. J. Petrol. Soc. Korea. 18, 171–179 (in Korean with English abstract).

Southard, J.B., 1971. Representation of bed configurations in depth-velocity-size diagrams. J. Sediment. Petrol. 41, 903–915.

Stear, W.M., 1983. Morphological characteristics of ephemeral stream channel and overbank splay sandstone bodies in the Permian Lower Beaufort Group, Karoo Basin, South Africa. In: Collinson, J.D., Lewin, J. (Eds.), Modern and Ancient Fluvial Systems. International Association of Sedimentologists, Cambridge, pp. 405–420.

Steel, R.J., Thompson, D.B., 1983. Structures and textures in Triassic braided stream conglomerates ("Bunter" Pebble beds) in the Sherwood Sandstone Group, north Staffordshire, England. Sedimentology. 30, 341–367.

Sturm, M., Matter, A., 1978. Turbidites and varves in Lake Brienz (Switzerland): deposition of clastic detritus by density currents. In: Matter, A., Tucker, M.E. (Eds.), Modern and Ancient Lake Sediments. International Association of Sedimentologists, Cambridge, pp. 147–168.

Surlyk, F., 1984. Fan-delta to submarine fan conglomerates of the Volgian–Valanginian Wollaston Forland Group, East Greenland. In: Koster, E.H., Steel, R.J. (Eds.), Sedimentology of Gravels and Conglomerates. Canadian Society of Petroleum Geologists, pp. 359–382.

Syvitski, J.P.M., Burrell, D.C., Skei, J.M., 1987. Fjords: Processes and Products. Springer, New York, 379 pp.

Tamaki, K., 1988. Geological structure of the Japan Sea and its tectonic implications. Bull. Geol. Surv. Jpn. 39, 269–365.

Tamaki, K., Suyehiro, K., Allan, J., Ingle, J.C., Pisciotto, K.A., 1992. Tectonic synthesis and implications of Japan Sea ODP Drilling. In: K. Tamaki, K. Suyehiro, J. Allan (Eds.), Proceedings of the Ocean Drilling Program, Scientific Results, pp. 1333–1348.

Tamanyu, S., 1990. The K–Ar ages and their stratigraphic interpretation of the Cheju Island volcanics, Korea. Bull. Geol. Surv. Jpn. 41, 527–537.

Tapponnier, P., Molnar, P., 1977. Active faulting and tectonics in China. J. Geophys. Res. 82, 2905–2930.

Tateiwa, I., 1924. 1:50,000 Geological Atlas of Chosen, No 2, Ennichi, Kuryuho and Choyo sheets. Geol. Surv. Joseon. (in Japanese).

Tateiwa, I., 1931. Occurrence of ore deposit and stratigraphy of the Proterozoic Era to Jurassic Period in Sungcheon region, Pyangannam-do. J. Min. Chosen..

Tateiwa, I., 1976. The Korea–Japanese Geotectonic Zone, with Special Reference to the History of Geological Research in Korea. Tokyo University Press, Tokyo, 654 pp. (in Japanese).

Tatsumi, Y., Shukuno, H., Yoshikawa, M., Chang, Q., Sato, K., Lee, M.W., 2005. The petrology and geochemistry of volcanic rocks on Jeju Island: Plume magmatism along the Asian continental margin. J. Petrol. 46, 523–553.

Teraoka, Y., Okumura, K., 2003. Geological Map of East Asia. Geological Survey of Japan, AIST, Tsukuba, Japan.

Todd, S.P., 1989. Stream-driven, high-density gravelly traction carpets: possible deposits in the Trabeg Conglomerate Formation, SW Ireland and some theoretical considerations of their origin. Sedimentology. 36, 513–530.

Todd, S.P., Went, D.J., 1991. Lateral migration of sand-bed rivers: examples from the Devonian Glashabeg Formation, SW Ireland and the Cambrian Alderney Sandstone Formation, Channel Islands. Sedimentology. 38, 997–1020.

Tucker, M.E., Wright, V.P., 1990. Carbonate Sedimentology. Blackwell, Oxford, 482 pp.

Turner, P., 1980. Continental Red Beds. Elsevier, Amsterdam, 562 pp.

van Houten, F.B., 1968. Iron Oxides in Red Beds. Geol. Soc. Am. Bull. 79, 399–416.

Van Wagoner, J.C., Mitchum Jr., R.M., Campion, K.M., Rahmanian, V.D., 1990. Siliciclastic sequence stratigraphy in well logs, core, and outcrops: concepts for high-resolution correlation of time and facies. Am. Assoc. Petrol. Geol. Meth. Explorat. Ser. 7, 55 pp.

Walker, R.G., 1965. The origin and significance of the internal sedimentary structures of turbidites. Proc. Yorkshire Geol. Soc. 35, 1–32.

Walker, T.R., 1967. Formation of red beds in modern and ancient deserts. Geol. Soc. Am. Bull. 78, 353–368.

Watson, M.P., Hayward, A.B., Parkinson, D.N., Zhang, Z.M., 1987. Plate tectonic history, basin development and petroleum source rock deposition onshore China. Mar. Petrol. Geol. 4, 205–225.

Wells, S.G., Harvey, A.M., 1987. Sedimentologic and geomorphic variations in storm-generated alluvial fans, Howgill Fells, northwest England. Geol. Soc. Am. Bull. 98, 182–198.

Won, J.K., Lee, M.W., 1984. The volcanism and petrology of alkali volcanic rocks, Ulrung Island. J. Geol. Soc. Korea. 20, 296–305 (in Korean with English abstract).

Woo, J., 2009. The Middle Cambrian Zhangxia Formation, Shandong Province, China: Depositional Environments and Microbial Sedimentation. Ph.D. Thesis, Seoul National University, Seoul, p. 243.

Woo, K.S., 1999. Cyclic tidal successions of the Middle Ordovician Maggol Formation in the Taebaeg area, Kangwondo, Korea. Geosci. J. 3, 123–140.

Woo, J., Chough, S.K., 2007. Depositional processes and sequence stratigraphy of the Jigunsan Formation (Middle Ordovician), Taebaeksan Basin, mideast Korea: implications for basin geometry and sequence development. Geosci. J. 11, 331–355.

Woo, J., Chough, S.K., 2010. Growth patterns of the Cambrian microbialite: phototropism and speciation of *Epiphyton*. Sediment. Geol. 229, 1–8.

Woo, K.S., Park, B.K., 1989. Depositional environments and diagenesis of the sedimentary rocks, Choseon Supergroup, Korea: past, present, and future; the state of the art. J. Geol. Soc. Korea. 25, 347–363.

Woo, K.S., Lee, K.C., Paik, K.H., 1991. Cretaceous lacustrine radial ooids in the Kyongsang Basin, Korea: palaeoclimatic implications. J. Geol. Soc. Korea. 27, 171–176.

Woo, J., Shinn, Y.J., Kwon, Y.K., Chough, S.K., 2006. The Jangsan and Myeonsan Formations (Early Cambrian) of the Taebaek Group, mideast Korea: depositional processes and environments. Geosci. J. 10, 35–57.

Woo, J., Chough, S.K., Han, Z., 2008. Chambers of *Epiphyton* thalli in microbial buildups, Zhangxia Formation (Middle Cambrian), Shandong Province, China. Palaios. 23, 55–64.

Wu, F.Y., Lin, J.Q., Wilde, S.A., Zhang, X., Yang, J.H., 2005. Nature and significant of the early Cretaceous giant igneous event in eastern China. Earth Planet. Sci. Lett. 233, 103–119.

Wu, F.Y., Han, R.-H., Yang, J.-H., Wilde, S.A., Zhai, M.-G., Park, S.C., 2007. Initial constraints on the timing of granitic magmatism in North Korea using U–Pb zircon geochronology. Chem. Geol. 238, 232–248.

Wynn, R.B., Stow, D.A.V., 2002. Classification and characterisation of deep-water sediment waves. Mar. Geol. 192, 7–22.

Wynn, R.B., Masson, D.G., Stow, D.A.V., Weaver, P.P.E., 2000. Turbidity current sediment waves on the submarine slopes of the western Canary Islands. Mar. Geol. 163, 185–198.

Yanai, S., Park, B.S., Otoh, S., 1985. The Honam Shear Zone (South Korea): Deformation and Tectonic Implication in the Far East. Scientific Papers of the College of Arts and Sciences, The University of Tokyo, vol. 35, pp. 181–210.

Yanai, S.C., Jwa, Y.J., Otoh, S., Yamakita, S., Park, B.S., 1993. Honam intra-arc transcurrent fault and Jurassic Geodynamics in East Asia. In: Xu, J. (Ed.), The Tancheng-Lujiang Wrench Fault System. John Wiley & Sons, New York, pp. 213–224.

Yang, S.Y., 1976. On fossils from the Gyeongsang Group: especially on some of the molluscan fauna. J. Geol. Soc. Korea. 12, 23–30 (in Korean with English abstract).

Yang, S.Y., 1982. Geology around the type-locality of *Trigonioides* (s.s.) *kodairai* and age of the Nagdong Subgroup. J. Geol. Soc. Korea. 18, 67–72 (in Korean with English abstract).

Yang, B.C., Chun, S.S., 2001. A seasonal model of surface sedimentation on the Baeksu open-coast intertidal flat, southwestern coast of Korea. Geosci. J. 5, 251–262.

Yang, H., Xie, Z., 1984. Sea-level changes in East China over the past 20,000 years. In: Whyte, R.O. (Ed.), The Evolution of Eastern Asian Environment. University of Hong Kong, Hong Kong, pp. 288–308.

Yang, B.C., Dalrymple, R.W., Chun, S.S., 2005. Sedimentation on a wave-dominated, open-coast tidal flat, south-western Korea: summer tidal flat-winter shoreface. Sedimentology. 52, 235–252.

Yang, B.C., Dalrymple, R.W., Chun, S.S., 2006. The significance of hummocky cross-stratification (HCS) wavelengths: Evidence from an open-coast tidal flat, South Korea. J. Sediment. Res. 76, 2–8.

Yang, B.C., Dalrymple, R.W., Gingras, M.K., Chun, S.S., Lee, H.J., 2007. Up-Estuary variation of sedimentary facies and ichnocoenoses in an open-mouthed, macrotidal, mixed-energy estuary, Gomso Bay, Korea. J. Sediment. Res. 77, 757–771.

Yang, B.C., Gingras, M.K., Pemberton, S.G., Dalrymple, R.W., 2008. Wave-generated tidal bundles as an indicator of wave-dominated tidal flats. Geology. 36, 39–42.

Yi, S., Batten, D.J, 2002. Palynostratigraphy of upper Cretaceous (Maastrichtian) deposits in the Yellow Sea Basin, Korea. Cretaceous Res. 23, 687–706.

Yi, M.-S., Choi, S.-J., Yun, H., 1993. Cretaceous palynomorphs from the Iljik Formation in the Euiseong area, Korea. J. Paleontol. Soc. Korea. 9, 166–179.

Yi, S., Yun, H., Yoon, S., 1998. Calcareous nannoplankton from the Seoguipo Formation of Cheju Island, Korea and its paleoceanographic implications. Paleontol. Res. 2, 253–265.

Yi, S., Batten, D.J., Yun, H., Park, S.J., 2003. Cretaceous and Cenozoic non-marine deposits of the Northern South Yellow Sea Basin, offshore western Korea: palynostratigraphy and palaeoenvironments. Palaeogeogr. Palaeoclimatol. Palaeoecol. 191, 15–44.

Yin, A., Nie, S., 1993. An indentation model for the north and south China collision and development of the Tan-Lu and Honam fault systems, Eastern Asia. Tectonics. 12, 810–813.

Yoo, C.M., Lee, Y.I., Paik, I.S., 1994. Evidence for hypersaline conditions in the Middle Ordovician Yeongheung Formation, Korea. J. Geol. Soc. Korea. 30, 355–368.

Yoon, S.H., Chough, S.K., 1995. Regional strike-slip in the Eastern continental-margin of Korea and its tectonic implications for the evolution of Ulleung Basin, East Sea (Sea of Japan). Geol. Soc. Am. Bull. 107, 83–97.

Yoon, S.H., Chough, S.K., Thiede, J., Werner, F., 1991. Late Pleistocene sedimentation on the Norwegian continental slope between 67° and 71°N. Mar. Geol. 99, 187–207.

Yoon, S.H., Lee, H.J., Han, S.J., Kim, S.R., 1996. Quaternary sedimentary processes on the east Korean continental slope. J. Geol. Soc. Korea. 32, 250–266 (in Korean with English abstract).

Yosimura, I., 1940. Geology of the Neietsu District, Kogendo, Tyosen. J. Geol. Soc. Jpn. 40, 112–122 (in Japanese).

Yun, S.H., 1998. Volcanic activity and petrology of the Cretaceous volcanic complex in Yucheon volcanic field of southeastern Korea. J. Korea. Earth Sci. Soc. 19, 408–424 (in Korean with English abstract).

Yun, S.H., Lee, M.-W., Koh, J.-S., Kim, Y.-L., Han, M.-K., 2000. Petrology of the Bokyeongsa Volcanics in the northeast Gyeongsang Basin. J. Korea. Earth Sci. Soc. 21, 595–610 (in Korean with English abstract).

Zhai, M., Guo, J., Peng, P., Hu, B., 2007. U–Pb zircon age dating of a rapakivi granite batholith in Rangnim massif, North Korea. Geol. Mag. 144, 547–552.

Zhang, W.T., Jell, P.A., 1987. Cambrian trilobites of North China; Chinese Cambrian trilobites housed in the Smithsonian Institution. Science Press, Beijing, 459 pp.

Zhang, Z.M., Liu, J.G., Coleman, R.G., 1984. An outline of the plate tectonics of China. Geol. Soc. Am. Bull. 95, 295–321.

Zhao, G., Cao, L., Wilde, S.A., Sun, M., Choe, W.J., Li, S., 2006. Implications based on the first SHRIMP U–Pb zircon dating on Precambrian granitoid rocks in North Korea. Earth Planet. Sci. Lett. 251, 365–379.

Zhu, M.-Y., Zhang, J.-M., Li, G.-X., Yang, A.-H., 2004. Evolution of C isotopes in the Cambrian of China: implications for Cambrian subdivision and trilobite mass extinctions. Geobios. 37, 287–301.

Zong, Y., 2004. Mid-Holocene sea-level highstand along the southeast coast of China. Quatern. Int. 117, 55–67.

Kolibri 360 GmbH
Ferdinand-Jühlke-Straße 7
99095 Erfurt, Deutschland
produktsicherheit@kolibri360.de